SPRINGER HANDBOOK OF AUDITORY RESEARCH

Series Editors: Richard R. Fay and Arthur N. Popper

Springer

New York
Berlin
Heidelberg
Barcelona
Budapest
Hong Kong
London
Milan
Paris
Santa Clara
Singapore
Tokyo

SPRINGER HANDBOOK OF AUDITORY RESEARCH

Volume 1: The Mammalian Auditory Pathway: Neuroanatomy
Edited by Douglas B. Webster, Arthur N. Popper, and Richard R. Fay

Volume 2: The Mammalian Auditory Pathway: Neurophysiology
Edited by Arthur N. Popper and Richard R. Fay

Volume 3: Human Psychophysics
Edited by William Yost, Arthur N. Popper, and Richard R. Fay

Volume 4: Comparative Hearing: Mammals
Edited by Richard R. Fay and Arthur N. Popper

Volume 5: Hearing by Bats
Edited by Arthur N. Popper and Richard R. Fay

Volume 6: Auditory Computation
Edited by Harold L. Hawkins, Teresa A. McMullen, Arthur N. Popper, and Richard R. Fay

Volume 7: Clinical Aspects of Hearing
Edited by Thomas R. Van de Water, Arthur N. Popper, and Richard R. Fay

Volume 8: The Cochlea
Edited by Peter Dallos, Arthur N. Popper, and Richard R. Fay

Volume 9: Development of the Auditory System
Edited by Edwin W Rubel, Arthur N. Popper, and Richard R. Fay

---

Forthcoming volumes (partial list)

Plasticity in the Auditory System
Edited by Edwin W Rubel, Arthur N. Popper, and Richard R. Fay

Comparative Hearing: Amphibians and Fish
Edited by Arthur N. Popper and Richard R. Fay

Comparative Hearing: Insects
Edited by Ronald R. Hoy, Arthur N. Popper, and Richard R. Fay

Comparative Hearing: Birds and Reptiles
Edited by Robert Dooling, Arthur N. Popper, and Richard R. Fay

Edwin W Rubel
Arthur N. Popper
Richard R. Fay
Editors

Development of the Auditory System

With 75 Illustrations

 Springer

Edwin W Rubel
Department of Otolaryngology
University of Washington
Seattle, WA 98195, USA

Arthur N. Popper
Department of Zoology
University of Maryland
College Park, MD 20742-9566, USA

Richard R. Fay
Department of Psychology and
Parmly Hearing Institute
Loyola University of Chicago
Chicago, IL 60626, USA

The cover illustrations are redrawn from the classic lithographs of Gustaf Retzius (Das Gehhsrorgan der Wirbelthiere, Samson & Wallin, Stockholm, 1884) and show transverse sections of the organ of Corti from the basal (top), middle, and apical (bottom) turns of the cochlea from a 2-day-old (postnatal) rabbit. These sections demonstrate the general finding that the late differentiation of the cochlea generally follows a gradient from basal to apical. Note the more mature appearance of the most basal section (top) as compared to the apical section (bottom).

Library of Congress Cataloging in Publication Data
Development of the auditory system / Edwin W Rubel, Arthur N. Popper,
    Richard R. Fay, editors.
        p.   cm. — (Springer handbook of auditory research : v. 9)
    Includes index.
    ISBN 0-387-94984-4 (hc : alk. paper)
        1. Ear — Growth.   2. Auditory pathways.   3. Development biology.
    4. Developmental cytology.   I. Rubel, Edwin W.   II. Popper, Arthur
    N.   III. Fay, Richard R.   IV. Series.
    QP461.D487   1997
    573.8'938 — dc21   97-6235

Printed on acid-free paper.

QP
461
.D487
1998

Production managed by Terry Kornak; manufacturing supervised by Joe Quatela.
Typeset by TechType, Inc., Ramsey, NJ.
Printed and bound by Maple-Vail, York, PA.
Printed in the United States of America.

9 8 7 6 5 4 3 2 1

ISBN 0-387-94984-4 Springer-Verlag New York Berlin Heidelberg SPIN 10425561

R. Bruce Masterton 1932–1996

This volume is dedicated to the memory of our friend and colleague, R. Bruce Masterton, who passed away unexpectedly on December 7, 1996. Bruce was the consummate sensory neuroscientist, whose work ranged from the behavioral to the cellular. He trained all of his students, colleagues, and professional friends to think critically and creatively about how the phenotype emerges through phylogeny and ontogeny.

Bruce and his wife, Pauline, are shown here on a trip taken to Lake Ortta, Italy, in 1994.

Series Preface

The *Springer Handbook of Auditory Research* presents a series of comprehensive and synthetic reviews of the fundamental topics in modern auditory research. The volumes are aimed at all individuals with interests in hearing research including advanced graduate students, postdoctoral researchers, and clinical investigators. The volumes are intended to introduce new investigators to important aspects of hearing science and to help established investigators to better understand the fundamental theories and data in fields of hearing that they may not normally follow closely.

Each volume is intended to present a particular topic comprehensively, and each chapter will serve as a synthetic overview and guide to the literature. As such, the chapters present neither exhaustive data reviews nor original research that has not yet appeared in peer-reviewed journals. The volumes focus on topics that have developed a solid data and conceptual foundation rather than on those for which a literature is only beginning to develop. New research areas will be covered on a timely basis in the series as they begin to mature.

Each volume in the series consists of five to ten substantial chapters on a particular topic. In some cases, the topics will be ones of traditional interest for which there is a substantial body of data and theory, such as auditory neuroanatomy (Vol. 1) and neurophysiology (Vol. 2). Other volumes in the series will deal with topics that have begun to mature more recently, such as development, plasticity, and computational models of neural processing. In many cases, the series editors will be joined by a co-editor having special expertise in the topic of the volume.

Richard R. Fay
Arthur N. Popper

Preface

Development is one of the most exciting and productive areas of modern biology. This is particularly true for the auditory system, where great inroads have been made in recent years on all aspects of development, from the molecular genetics of the ear to ontogeny of auditory capabilities. The purpose of this volume is to provide some insight into our knowledge of the developing auditory system, by bringing together much of the literature on its normal structural and functional development. Together, the contributors to this volume provide a detailed and integrated introduction into the behavioral, anatomical, and physiological changes that occur in the auditory system during development. Each contribution also attempts to describe what are the greatest challenges for the future.

In the first chapter, Rubel provides a general overview of the rationale for investigating development of the auditory system and suggests questions that may be the most important in future research. In Chapter 2, Werner and Gray describe the behavioral changes that occur in hearing capabilities as the auditory system develops in the newborn. This is followed by a paper by Fritzsch, Barald, and Lomas (Chapter 3) who consider the early embryogenesis of the ear, as well as what is known about the genetic basis of ear development. This is followed in Chapter 4, by Pujol, Lavigne-Rebillard, and Lenoir, in a discussion of slightly later stages of development when the sensory epithelia and hair cells arise. The development of cochlear function is treated in Chapter 5 by Rübsamen and Lippe and this is complemented by a discussion of development of CNS function by Sanes and Walsh (Chapter 6). Finally, Cant, in Chapter 7, considers the anatomical and pharmacological development of the auditory portions of the CNS.

This volume is closely tied to other volumes in the *Springer Handbook of Auditory Research*. A discussion of the genetics of the auditory system, in Chapter 3 of this volume, is complemented by a chapter on the molecular genetics of hearing by Steel and Kimberling in Volume 7 of the series (*Clinical Aspects of Hearing*). The anatomy and physiology of the auditory CNS is extensively discussed in the first two volumes of the series (*The Mammalian Auditory Pathway: Neuroanatomy; The Mammalian Auditory*

Pathway: Neurophysiology) and these volumes lead into the discussion of developmental anatomy and physiology in this volume. Similarly, the anatomy of the adult cochlea is discussed in a chapter by Slepecky in Volume 8 of the Series (*The Cochlea*) and comparative anatomy is treated in a chapter by Echteler, Fay, and Popper in Volume 4 (*Comparative Hearing: Mammals*).

Contents

Contributors

Kate F. Barald
Department of Anatomy and Cell Biology, University of Michigan, Ann Arbor, MI 48109-0616, USA

Nell Beatty Cant
Department of Neurobiology, Duke University Medical Center, Durham, NC 27710, USA

Bernd Fritzsch
Department of Biomedical Sciences, Creighton University, Omaha, NE 68178, USA

Lincoln Gray
Department of Otolaryngology—Head and Neck Surgery, University of Texas Medical School, Houston, TX 77030, USA

M. Lavigne-Rebillard
Laboratoire de Neurobiologie de l'Audition, C.H.R. Hospital St. Charles, I.N.S.E.R.M.-U 254, 34059 Montpellier Cedex, France

M. Lenoir
Laboratoire de Neurobiologie de l'Audition, C.H.R. Hospital St. Charles, I.N.S.E.R.M.-U 254, 34059 Montpellier Cedex, France

William R. Lippe
Virginia Merrill Bloedel Hearing Research Center and Department of Otolaryngology, University of Washington, Seattle, WA 98195, USA

Margaret I. Lomax
Developmental Biology Program, Kresge Hearing Research Institute, University of Michigan, Ann Arbor, MI 48109-0616, USA

R. Pujol
Laboratoire de Neurobiologie de l'Audition, C.H.R. Hospital St. Charles,
I.N.S.E.R.M.–U 254, 34059 Montpellier Cedex, France

Rudolf Rübsamen
Neurobiology Laboratory and Department of Zoology, University of
Leipzig, Leipzig, D-04103 Germany

Edwin W Rubel
Department of Otolaryngology, University of Washington, Seattle, WA
98195, USA

Dan H. Sanes
Center for Neural Science, New York University, New York, NY 10003,
USA

Edward J. Walsh
Department of Physiology, Boys Town National Institute, Omaha, NE
68131, USA

Lynne A. Werner
Department of Speech and Hearing Sciences, University of Washington,
Seattle, WA 98105-6246, USA

1

Overview: Personal Views on the Study of Auditory System Development

EDWIN W RUBEL

Understanding the ontogeny of the vertebrate auditory system, although far from complete, has progressed and changed radically since 1978, when I attempted to write a comprehensive review on that subject (Rubel 1978). At that time, I could identify only a few hundred publications since 1851 when Alphonse Corti first described the peripheral organ that bears his name (Corti 1851). Now there are literally thousands of relevant papers published each decade, and the arsenal of techniques we can apply to this endeavor has exploded along with all of the biological sciences. Yet the overall goals of these studies have not changed, nor have the central problems we hope to solve. Before addressing some of the progress that has been made and the challenges that lie ahead, I will briefly summarize some of the reasons why we may want to investigate this subject and some general approaches.

1.1 Why Study Ontogeny of the Auditory System

Humans excel above other vertebrates in few physical attributes. Other animals can jump higher, run faster, walk farther, and swim better. The attribute at which we excel, and probably the central reason for our unchallenged control of this planet, is the ability to communicate. First language and then written communication allowed us to pass on complex concepts across individuals, families, cultures, and generations. Communication has allowed the exponential growth of knowledge; the database is not lost with an individual or a generation. Verbal communication is also, of course, the medium by which we usually share emotions and attitudes, as well as ideas, facts, and abstract concepts. The loss of hearing during development or at maturity must therefore have profound impact on the ability to communicate with most other members of the species. It can take away or, at best, severely limit the most human of all attributes.

The astounding realization that over 24 million people in the United States alone and that approximately 1 in 10 people in the world are hearing handicapped is one compelling reason to study the development of hearing. By studying its development, we hope to better understand, devise ways to

1

prevent, and invent ways to treat hearing loss. In this endeavor, careful investigations of development provide a means of dissecting the system into critical elements, be it at the cellular and molecular level of gene action, or at the behavioral level of examining the atrributes of sound, or language that an infant can process. Thus ontogenetic studies with a medical objective are not exclusively aimed at congenital hearing losses due to the expression of abnormal genes. Although it is certainly of great interest and importance to understand the genetic principles underlying syndromic hearing impairments, it is our hope that a detailed understanding of development will facilitate an appreciation of the biological and psychological elements of hearing and how they must interact to create the emergent properties of human auditory perception. As the developing animal forms these fundamental elements and as we manipulate each piece, we can begin to discover how they relate to a functioning system. Even more importantly, we can begin to understand the dynamic interactions required to form and maintain a perceptually competent system. This arsenal of knowledge is a necessary prerequisite to devising medical interventions taking advantage of modern biomedical science.

Two disparate examples, one at the cellular level and one at the behavioral level, may serve to exemplify these points. The discovery of hair cell regeneration in the inner ear of birds (Cotanche 1987; Cruz, Lambert, and Rubel 1987) has stimulated the hope that ultimately it may be possible to restore hearing and balance to impaired individuals by mimicking this process. To achieve such a goal, we will have to initiate mitotic activity in the mature human sensory epithelium and stimulate a portion of the new cells to differentiate into hair cells. At this time, several laboratories are successfully investigating genes and proteins involved in the commitment of immature cells to a hair cell phenotype and in their subsequent differentiation (Fritzsch, Barald, and Lomax, Chapter 3). Promising candidates include, but are not limited to, brain-derived neurotropic factor (BDNF), neurotrophin 3 (NT3), retinoic acid, transforming growth factors, POU-domain transcription factors, and brain morphogenic factors. Understanding the cascade of events involving these and other factors is essential for tricking the mature, injured cochlea to make a new functional transduction system, i.e., for achieving hair cell regeneration in humans.

At the behavioral level, work by Werner and Gray (Chapter 2) and others has been systematically dissecting the stimulus attributes to which human infants and young animals can selectively respond, and electrophysiological studies have examined the differential firing properties of central nervous system neurons to stimuli ranging along these same dimensions (Sanes and Walsh, Chapter 6). Both types of investigations have revealed substantial developmental delays in the precision of encoding temporal properties of the stimulus (e.g., phase locking and maintaining a response to rapid trains of stimuli in the central nervous system and gap detection or repetition rate detection in behavior). Recently, two groups of investigators (Kraus et al.

1996; Merzenich et al. 1996; Tallal et al. 1996) have provided evidence that some learning-impaired children have deficiencies in temporal processing. These studies suggest that interventions, particularly at the time deficiencies first arise, may be useful in preventing lasting handicaps.

A second and equally valid reason to study development of the auditory system stems from the intrinsic fascination of understanding the complex cascade of sequentially inductive events responsible for the formation of a neurobehavioral system. Although the research emanating from this fascination may not be obviously applicable to human disease, we now have many examples of how untargeted fundamental science has changed mankind's future. Furthermore, I am reminded of the quote from an unidentified journalist who stated, "Albert Einstein's theories have altered human existence not at all. But they have revolutionized human understanding of existence" (*Time* magazine 1929).

It is important to recognize that science shares much in common with art. Both build on the successes, failures, and innovations of past masters and, when successful, present new ways to extend our comprehension of the universe in which we reside. Understanding the developing nervous system in general, and the developing auditory system in particular, is in a renaissance. In the last 50 years, science has enjoyed the patronage of a huge number of benefactors (taxpayers of industrialized countries) that have provided a relative plethora of commissions (research grants and contracts from the National Institutes of Health, National Science Foundation, Institut National de la Santé et de la Recherche Médicale, and private corporations). At the same time, we've enjoyed the explosion of media available for expression (the modern tools of genetics, pharmacology, molecular biology, electronics, optics, and computer science). We are just beginning to reap the rewards. For example, 20 years ago, the developmental neurobiologist was content to describe sequential steps in the development of an organ or a system and perform experiments asking the extent to which such events intrinsic to the tissue under investigation or required extrinsic influences. Today, the description of developmental events serves as the canvas on which we draw. It is still a necessary and critical element, but we now have the tools to explore intracellular and intercellular cascades of events involving the genes and gene products interacting at specific membrane surfaces or specific locations within a cell, an organ, or an organism. This is not merely increased power through reductionism; it is the ability to go from abstract concepts, such as "critical periods" or "experience," to concrete biological steps, such as a developmental change in an intracellular signaling pathways or activation of a specific set of events through binding of a known membrane spanning receptor complex.

Perceptual systems in general, and the auditory system in particular, provide an outstanding substrate to take advantage of this renaissance. In the auditory system, the structural and functional organization of the receptor and peripheral elements have been extensively investigated in a variety of species. Compared with other sensory systems (e.g., the retina), they are relatively simple and well understood (see Dallos, Popper, and Fay 1996). We also benefit from years of painstaking, careful neuroanatomy and neurophysiology describing cellular connections, pharmacology, and basic functional properties of the cellular elements in the adult (see Popper and Fay 1992; Webster, Popper, and Fay 1992). In addition, some basic descriptions of developmental changes in function and structure are available for several vertebrate species. Finally, thanks to years of incredibly tedious acoustics and psychoacoustics, the physical nature of stimulus attributes and the perceptual abilities a few species (particularly humans) have provided a relatively thorough description of the phenotype we wish to understand (see Yost, Popper, and Fay 1993).

1.2 Investigative Approaches

In Chapters 2–7, the reader will find summaries of research involving a large variety of scientific methods, ranging from controlled behavioral studies to manipulations of individual proteins and genes. Understanding the biological and psychological aspects of hearing development requires of all of these approaches and others that are not found here. Whether the primary objective of the research is the hearing-impaired patient or the acquisition of fundamental knowledge, modern biomedical research usually involves the application of multiple methodologies. This can be accomplished by teams of investigators with diverse scientific backgrounds or through collaborations among specialized laboratories.

More important than the methodology is a clear understanding of the investigative framework of developmental research. What is the goal of developmental investigation? Is the goal to document the steps and sequence of changes in a behavior or cellular characteristic (e.g., neuronal firing property) as an animal matures from nonresponsive to maturity? Certainly that is a necessary *step* in any developmental research program. But I would contend that it is not the goal of developmental investigations; it is a first step, providing the critically important information on which to proceed. It provides the developmental "tick marks" that identify when interesting changes are occurring; the time points on which to concentrate future studies.

In one sense, devotion to documenting the stages of development of particular systems, sensory responses, or behaviors is a futile and never-ending process. Animals *will* develop; we know the end product. And it is not feasible to investigate all of the developmental changes that are occurring in all species, all vertebrates, or even all mammals. Thus certain species are chosen for concentration because of their unique behavioral or biological attributes, out of convenience, or through historical accident. Biological studies of auditory system development in vertebrates have generally concentrated on chicks and laboratory rodents (rat, mouse, and gerbil), although several laboratories have concentrated physiological investigations on domestic cats because of the large amount of such data in mature cats. Thus developmental investigation requires much more than doing the same thing on young animals than we do on adults.

Whether the ultimate goal is to cure disease or the acquisition of fundamental knowledge, there are essentially two frameworks of developmental investigation. One is to understand the component processes underlying the mature system by isolating developmental changes in structure and function. In this regard, the approach is usually to isolate structural and functional changes that temporally coincide. We then must attempt to understand their causal relationships by experimental manipulations that alter one event and measure the outcome of the other. It is important to stress the opinion that temporal coincidence may suggest causality but does not provide strong evidence. Only through experimentally isolating and manipulating one event and observing concomitant changes in the other is it possible to validate structure-function relationships.

The second and most common goal of developmental research is to understand the events that cause, guide, and regulate the ontogeny of a cell, tissue, organ system, or behavior, that is, to understand the developmental process itself. The developmental process can be studies at any level of analysis from behavior to molecules. What is important here is the difference between documentation of developmental changes, a necessary beginning, and designing experiments to understand the causal relationships between developmental events. By design, the present volume primarily provides information related to the preliminary work of documenting developmental changes. There are two reasons. First, this information is critically important for experiments attempting to define the causal relationships underlying development. Just as success in the identification of genes related to human disease is totally dependent on accurate identification of the phenotype, so is understanding developmental causation dependent on thorough and accurate descriptive analyses.

The second reason is necessity. In developmental neurobiology, there are many topics in which the cascades of developmental dependencies are being actively investigated at the cellular and molecular levels. Probably the two most advanced areas involve examination of molecular events underlying ontogeny of the fruit fly eye (e.g., Bonini and Choi 1995; MacDougall and Waterfield 1996) and the maturation of neural crest cell derivatives (e.g., Groves and Anderson 1996). However, there are relatively few examples where investigators have successfully isolated causally related events in the development of the auditory system. Recent work using genetic manipulations in mice (see Fritzsch, Barald, and Lomax, Chapter 3) provides evidence of genes that are important for the maturation or survival of hair cells and/or ganglion cells. However, like most gene "knock-out" manipulations, it is not yet clear where these gene products affect the developmental cascade of event. Classic work identifying a role for the neural tube in the induction of the labyrinth (see Rubel 1978) and recent work suggesting a role for retinoic acid receptors in hair cell phenotype commitment (Kelley et al. 1993) provide examples of experimental designs that address this issue.

It should be stressed that experimental investigations seeking to understand developmental regulation are not limited to cellular and molecular studies. In an elegant series of experiments spanning over 30 years, Gilbert Gottlieb (e.g., Gottlieb 1988) has identified specific environmental stimuli that regulate differential responsiveness to maternal vocalizations in embryonic and hatching ducklings during development (see Chapter 2 and references therein). This research program serves as an outstanding example of how behavioral science can use a developmental strategy to unravel a complex series of interactions between the organism and its environment.

1.3 The Present and the Future

In contrast to the relative dearth of information on causal events regulating auditory system ontogeny, there is a wealth of information documenting anatomic, physiological, and behavioral changes as the system matures in a variety of species. As I noted above, this material is critically important for progress. The chapters in this volume bring much of that information into one location.

Werner and Gray (Chapter 2) review and discuss the information available on behavior development in humans and other vertebrates. By organizing the chapter primarily along the dimension of stimulus complexity, they are able to compare developmental trends across species. It is particularly noteworthy that most of the data available is on humans and chicks, with relatively little psychophysical data available on other nonhuman mammals. It is also apparent that there is no nonhuman species that uses complex communication signals on which developmental psychophysical testing has been extensively performed using both simple, artificial stimuli such as pure tones and complex naturalistic stimuli. This is unfortunate because it is important to have more complete information directly comparing developmental capacities with these two types of stimuli. Studies in humans and a variety of other vertebrates that use complex communication signals (most notably songbirds) have shown that hearing components of these signals during a specific period of development is essential for later perceptual behavior. However, aside from the studies by Gottlieb (see Refs. in Chapter 2), there is little information available on the specific stimulus attributes or sensory abilities required at the time of the critical period in question.

Werner and Gray (Chapter 2) briefly discuss the state of knowledge regarding the development of "perceptual maps." This area of research holds perhaps the greatest promise for the future. Studies by Kuhl and colleagues

(1992) on human infants and Gray and colleagues on both humans and chicks, among others, have provided important, tantalizing new information suggesting that perceptual maps develop well after the ability to make precise discriminations between the elements. This result not only challenges classic notions of perceptual development, it also challenges investigators interested in behavior development and developmental neurobiology to unravel the mechanisms underlying perceptual map development.

Fritzsch, Barald and Lomax (Chapter 3) consider two major issues related to the early development of the vertebrate ear. The first is a review of genes known to play a role in early development of the inner ear and statoacoustic ganglion. They correctly point out that new information on this subject is being accumulated at a rapid rate. Yet classic transplantation and ablation experiments, coupled with new data on the expression of specific genes and data from genetic manipulations, are beginning to provide the tools to make specific, testable hypotheses about inductive interactions regulating inner ear formation. Progress in this area is likely to be rapid over the coming decade. At present, it is hampered by the lack of early expressed phenotypic markers for specific cell types, but several laboratories are currently addressing this problem; here too, progress is likely to be rapid. The ear is a complex structure, and it will require a large number of cooperating laboratories to unravel the developmental cascades of intercellular and intracellular events regulating its development. But some of the foundation has been laid by careful development anatomy and by identification of the time course of expression of genes known to be important for the formation of other tissues. This information, the identification of expression patterns of other genes, the development of cell-specific markers, and the prudent use of in vitro preparations is likely to yield remarkable progress over the next decade or two.

The second and more fascinating problem posed by Fritsch, Barald, and Lomax is the molecular and developmental bases underlying evolutionary differences in inner ear structure and function. Here we are just beginning to know how to pose the questions and have few, if any, clues toward solutions. Stated simply, how are evolutionary differences in inner ear biology manifest in the developmental program of gene expression and tissue interactions? Answers to this perplexing question certainly lie primarily in early ontogeny and are fertile ground for future investigation.

Professor Rémy Pujol has led a team of investigators that have spent over two decades investigating the development of hair cells in the mammalian cochlea. Of particular emphasis has been the development pharmacology and anatomy of connections between the hair cells and their neural connections. In Chapter 4 Pujol, Lavigne-Rebillard, and Lenoir summarize their contributions and those of other investigators on these topics. This body of work provides one of the most careful and thorough descriptions of cellular and pharmacological development found in any vertebrate sensory organ, thereby providing a firm basis for two areas of future research:

What molecular and cellular events are responsible? How are these late changes in the morphology and pharmacology of the synapse reflected in the developmental information transmission?

Rübsamen and Lippe (Chapter 5) examine the major developmental trends in cochlear function. Comprehensive tables summarizing information on common laboratory animals provide a time line for developmental changes in cochlear function in these species. Of particular interest are two trends. First, as pointed out earlier (Rubel 1978), the final stages of cochlear development are marked by remarkable synchrony in the final maturation of a variety of cell types. This synchrony involves diverse changes in a variety of cell types and tissues including synaptic maturation, maturation of endocochlear potential, steriocilia development, regression of marginal pillars of the tympanic membrane, growth and death of different classes of supporting cells, etc. This synchrony obviates attributing the onset and maturation of function to a single cell type or cellular attribute. It also suggests that cellular interactions must regulate such changes. To date, these interactions have escaped investigation, which is unfortunate.

A second area that should receive increased attention during the next decade is the cellular physiology of developing cochlear cells. Increased availability of *in vitro* preparations, important technical advances in cellular physiology, and the ability to regenerate hair cells and supporting cells in some species make this an area that will receive increased attention. Some progress along these lines has already been seen by recent studies on the development of ion-channel properties in chick hair cells and the development of electromotility in mammalian outer hair cells. Finally, these authors point out that rhythmic spontaneous electrical activity in the auditory nerve has been observed in several mammalian as well as avian species before acoustically driven responses or at early stages of their development. As yet, we have little information regarding the role of this activity in regulating inner ear or central nervous system development.

The final two chapters of this volume explore our understanding of the ontogeny of mammalian central nervous system pathways involved in the processing of auditory information. In each of these chapters, it is apparent that we are beginning to have a foundation of descriptive data on the development of response properties and morphology of cells in central auditory structures of several laboratory animals. In addition, several investigators have begun to correlate these attributes with the expression patterns and/or functional attributes of molecules known to be expressed. Sanes and Walsh (Chapter 6) fortell the future by stressing recent studies investigating relationships between development of cellular physiological processes and ontogenetic changes in the response properties of auditory neurons to acoustic signals. Cant (Chapter 7) considers both classic and recent developmental anatomic information in light of where there are obvious holes in descriptions of central nervous system development.

Numerous areas where modern research techniques can be applied to important developmental problems are pointed out.

Two general conclusions from these chapters may be surprising to some readers. First, despite many statements to the contrary, it is clear from both chapters that the fundamental organization of central pathways is laid down before development of the ability of the inner ear to transduce and transmit information to the brain. This includes both the establishment of connections and the transmission of information through these circuits. Thus immaturities of central nervous system auditory processing in neonatal animals usually represent a *cumulative effect* or peripheral and central developmental conditions rather than a peripheral-to-central sequence of development. Second, the overwhelming weight of evidence suggests that both the connections between the cochlea and the cochlear nuclei and the connections among central nervous system regions are initially formed in a highly precise topographic manner, with little error or "exuberance." While acoustic stimulation may be needed for the maintenance and some subtle refinement of these connections, we must look elsewhere for the signals responsible for their establishment. Finally, careful comparison of the specific results and general trends in Chapters 5, 6, and 7 reveals that dramatic developmental changes take place in the growth of almost all auditory regions of the brain well after the ages where existing physiological studies suggest that the system is responding in a mature fashion. Although these changes are unlikely to involve the fundamental organization of principal connections, they almost certainly involve a massive elaboration of dendritic and axonal processes as well as changes in glial organization. They may also involve the establishment of new modulatory and regulatory circuits. It is disappointing, and thereby an important opportunity for the future, that we are largely ignorant of their physiological and behavioral consequences.

1.4 Conclusion

The purpose of this volume is to bring together some of the literature and important concepts that have emerged from research on the normal development of the auditory system. Each chapter not only provides the reader with the *facts* as we know them today but also provides, explicitly or implicitly, a set of questions begging to be investigated by our ever-expanding arsenal of biological and behavioral methods. As emphasized throughout this introduction, a plethora of information is available on normative development of the vertebrate auditory system. No doubt, much more is needed, particularly in areas involving the perception of complex ethologically relevant stimuli, biophysical and pharmacological properties of developing cells, and the expression of developmentally important molecules such as transmembrane receptors, matrix molecules, and transcriptional regulators. This information will undoubtedly emerge as it is needed; most of the tools and many of the reagents are readily available.

On the other hand, knowledge and understanding of the cascades of inductive and regulative events governing biological and behavioral development of the auditory system is largely absent. Discovery of these processes provides the major challenge for the next generation of auditory scientists.

Today's ignorance is tomorrow's opportunity!

References

Bonini NM, Cho K-W (1995) Early decisions in *drosophilia* eye morphogenesis. Curr Opin Gene Dev 5:507-515.

Cotanche DA (1987) Regeneration of hair cell stereociliary bundles in the chick cochlea following severe acoustic trauma. Hear Res 30:181-195.

Corti A (1851) Recherches sur l'orane de l'ouie des mammiferes. Wiss Zool 8:109-169.

Cruz RM, Lambert PR, Rubel EW (1987) Light microscopic evidence of hair cell regeneration after gentamicin toxicity in chick cochlea. Arch Otolaryngol Head Neck Surg 113:1058-1062.

Dallos P, Popper AN, Ray RR (eds) (1996) The Cochlea. New York: Springer-Verlag.

Gottlieb G (1988) Development of species identification in ducklings. XV. Individual auditory recognition. Dev Psychbiol 21:509-522.

Groves AK, Anderson DJ (1996) Role of environmental signals and transcriptional reulators in neural crest development. Dev Genet 18:64-72.

Kelley MW, Xu XM, Wagner MA, Warchol ME, Corwin JT (1993) Supernumerary hair cells in response to exogenous retinoic acid in culture. Development 119:1041-1053.

Kraus N, McGee TJ, Carrell TD, Zecker SG, Nicol TG, Koch DB (1996) Auditory neurophysiologic responses and discrimination deficits in children with learning problems. Science 273:971-973.

Kuhl PK, Williams KA, Lacerda F, Stevens KN, Lindblom B (1992) Linguistic experience alters phonetic perception in infants by 6 months of age. Science 255:606-608.

MacDougall LK, Waterfield MD (1966) To sevenless, a daughter. Curr Biol 6:1250-1253.

Merzenich MM, Jenkins WM, Johnston P, Schreiner C, Miller SL, Tallal P (1996) Temporal processing deficits of language-learning impaired children ameliorated by training. Science 271:77-81.

Popper AN, Fay RR (eds) (1992) The Mammalian Auditory Pathway: Neurophysiology. New York: Springer-Verlag.

Rubel EW (1978) Ontogeny of structure and function in the vertebrate auditory system. In: Jacobson M (ed) Handbook of Sensory Physiology. Vol. IX. Development of Sensory Systems. New York: Springer-Verlag, chap. 5, pp. 135-237.

Tallal P, Miller SL, Bedi G, Byma G, Wang X, Nagarajan SS, Schreiner C, Jenkins WM, Merzenich MM (1996) Language comprehension in language-learning impaired children improved with acoustically modified speech. Science 271:81–84.

Webster DB, Popper AN, Fay RR (eds) (1992) The Mammalian Auditory Pathways: Neuroanatomy. New York: Springer-Verlag.

Yost WA, Popper AN, Fay RR (eds) (1993) Human Psychophysics. New York: Springer-Verlag.

2
Behavioral Studies of Hearing Development

Lynne A. Werner and Lincoln Gray

1. Introduction

This chapter reviews progress that points to the exciting potential in studies of behavioral development. The goal is twofold: to review common trends in psychoacoustic data from different species of newborn vertebrates and to indicate the many interesting questions that remain to be answered. The purpose of the introduction is to state and to justify the assumptions that underlie that goal.[1]

1.1 What Does the Study of Auditory Behavioral Development Tell Us About the Development of Hearing?

1.1.1 Why Study Behavioral Development?

Hearing scientists generally agree on the importance of studying auditory behavior. Ultimately, hearing is only defined by the behavior of an organism. Psychoacoustics, in particular, is an important link among various aspects of hearing research, allowing comparison of humans and nonhumans and providing a link between cellular and behavioral changes (Brindley 1970; Werner 1992). From a clinical standpoint, behaviors define deafness and communicative disorders because behavioral problems are what lead people to seek professional assistance. And, of course, both psychoacoustic and clinical results are usually stable, showing little variability within and among individuals and across decades of investigation

---

[1] As used here, the terms "infant" and "child" refer only to humans. "Neonate," "newborns," and "birth" refer to all vertebrates. The terms "neonate" and "newborn" are intended to indicate that it has not been long since the subject was born. This could mean within the first postnatal year for humans but within a few days of hatching for a chick.

(Stevens and Newman 1936; Green 1976; Moore 1989). Such reliability is sometimes seen as the hallmark of "real science."

Hearing scientists have much greater difficulty in accepting the study of behavior as an approach to the study of development. There are at least two reasons for this difficulty. First, it is clear to all that differences between neonatal and mature responses to sound can stem from many sources. These include both sensory factors, representing primary sensory processes, and nonsensory factors such as attention or memory. In the view of many, nonsensory factors thus interfere with the isolation of the primary sensory-processing immaturities that limit true sensitivity during development. Because nonsensory factors can often not be controlled in immature organisms in this view, auditory behavior cannot provide much information about the development of hearing. Second, variability rather than stability seems to be the hallmark of developmental behavioral data. To some, this variability makes the techniques for studying the development of auditory behavior suspect. In any case, variability increases the difficulty of identifying developmental trends, leading to the conclusion that some other approach to the study of hearing development would be more useful. In the course of reviewing the literature on the development of auditory behavior, we hope to convince the reader that the contributions of nonsensory factors to age differences in auditory behavior and the variability associated with developmental behavioral data reflect important developmental processes that cannot be studied with nonbehavioral approaches.

All behavioral responses are the result of an interaction between sensory and nonsensory factors (Gray 1992b), even if it is difficult to determine their independent effects. Moreover, nonsensory factors limit sensitivity as much as sensory factors do under nearly all circumstances, even among mature listeners. In the unusual case of simple detections or discriminations by mature listeners, the effects of nonsensory factors may be minimized given the appropriate psychophysical technique and sufficient practice. Nonsensory factors become very important, however, under the complex and uncertain conditions in which organisms ordinarily listen (Green 1983; Hall, Haggard, and Fernandes 1984; Yost and Watson 1987; Neff and Callaghan 1988; Yost 1991). Sensitivity in such cases is limited not by primary sensory processing but by the attentional and memory processes that allow listeners to construct an auditory representation of the world. If age-related change in the behavioral response to all kinds of sounds reflects measurable effects of both sensory and nonsensory processes, the study of behavioral development provides a unique opportunity to understand hearing and its development.

An influential idea in the study of development has been that variability in developmental data reflects real variability in the developmental process rather than the effects of uncontrolled nuisance variables (Lerner, Perkins, and Jacobson 1993). For many characteristics, the end point of development for all individuals is the same: an efficient kidney or a finely tuned basilar membrane. Variability in these characteristics is small when they are measured in mature individuals. If a characteristic is examined in a cross section taken at a given age during a period of development, however, variability will inevitably be high because individuals develop at somewhat different times and different rates. Such observations point to the importance of longitudinal studies but also suggest that variability is an important aspect of development. Several recent studies have shown that variability during development can actually be informative with respect to the mechanisms underlying sensory maturation (Werner, Folsom, and Mancl 1993; Werner, Folsom, and Mancl 1994; Peterzell, Werner, and Kaplan 1995). It is also important to note that variability is characteristic of complex and uncertain perception, even when listeners are highly trained with the most rigorous psychophysical techniques (e.g., Green 1983; Neff and Callaghan 1988). To the extent that immature nonsensory processes are important factors in neonatal audition, variability is to be expected. In sum, the variability that characterizes developmental behavioral data, although not always helpful, does not necessarily indicate a flaw in the behavioral approach.

Finally, it is important to recognize that perception of even simple sounds cannot be completely described in terms of detections and discriminations. Sounds also have attributes, perceived patterns or properties revealed by judgments (Stevens 1975). Pitch and loudness are examples of attributes. There is renewed interest in the dichotomy between attributes and detection or discrimination in mature hearing (Stebbins 1993). This is an interesting coincidence because the pattern of development of perceptual attributes is evidently different from that of detections and discriminations (e.g., Gray 1987a). In both mature and immature individuals, auditory behavior remains the most straightforward way to study perceptual attributes.

1.1.2 Comment on the Methods Used to Assess Auditory Behavior

Notwithstanding the convincing arguments for the theoretical importance of variability during development, the methods used to assess neonates' behavioral response to sound may still be insensitive or unreliable. Substantial progress has been made, however, on sensitive and unbiased methods for evaluating the hearing of neonates.

Infants make a variety of responses to sound (e.g., Watrous et al. 1975). These responses represent a general orientation to stimulation rather than a unique response to sound. Perhaps as a consequence, the responses tend to habituate rather quickly (e.g., Bridger 1961). To encourage infants to continue responding long enough to estimate a threshold, a conditioning paradigm is frequently used. The first successful technique of this type was visual reinforcement audiometry (Moore, Thompson, and Thompson 1975), so-called because a turn of the infant's head toward a sound source is reinforced by the presentation of an interesting visual display. This technique works well for infants between ~6 and 24 months of age. The observer-based psychoacoustic procedure (OPP; Olsho et al. 1987; Werner 1995) was developed to test infants younger than 6 months of age. In OPP, an observer judges on each trial whether or not a signal occurred on the basis of the infant's response. Both signal and no-signal trials are randomly presented, and the observer has no prior knowledge of trial type. If the observer is able to correctly identify a signal trial, the infant is visually reinforced for responding. Any of the many responses that an infant might make, changes in motor activity, eye movements, or head turns, can be conditioned and provide the basis of the observer's decision. Thus, even though young infants do not make directed, short-latency head turns to sound sources, they can be successfully tested with OPP. For the most part, OPP gives equivalent results to VRA among older infants (Olsho et al. 1987) and thus has the advantage that it can be applied to infants throughout the first postnatal year. Starting around 3 years of age, children are able to perform more or less standard psychophysical procedures, which are now usually disguised as video games (e.g., Wightman et al. 1989; Hall and Grose 1991).

An unbiased procedure for testing the hearing of newborn chicks has also been developed. Isolated chicks normally peep incessantly but momentarily delay their ongoing vocalizations when they hear a novel sound. Kerr, Ostapoff, and Rubel (1979) originally used this pause in peeping to study frequency generalization gradients in chicks. Gray (1987b) subsequently showed that the duration of silence during stimulus and control trials can be taken as a measure of confidence that a stimulus occurred. These "confidence ratings" can then be used to generate receiver operating characteristics (ROC), or isosensitivity, curves. Analysis of the properties of ROC derived from peep-suppression data show them to be a sensitive measure of auditory responsiveness. Classic psychophysical measures, such as thresholds and difference limens, can be rapidly and reliably measured in subjects between a few hours and 1 week of age (Gray and Rubel 1985a; Gray 1992b).

There is no way to know that a future methodological improvement will not yield lower thresholds. For that matter, we will never know the "left-hand limit" of adults' (let alone neonates') psychometric function. In the age ranges in which the development of auditory behavior has been studied most extensively, however, the consistency of recent records, obtained with somewhat different techniques, encourages the belief that the records reflect a stable process (Werner 1992). Moreover, the reasonable and consistent dependence of behavioral measures on the intensity, frequency, or azimuth of the stimulus suggests that it is hearing that is being measured (Gray 1992b; Werner 1992). In the final analysis, age-related change in the relationship between stimulus and response may well be more important to understanding auditory development than is the precise value of the threshold (Banks and Dannemiller 1987; Werner and Bargones 1992).

Currently available measurement techniques are clearly sensitive and reliable enough to describe such relationships.

1.2 What Do Comparisons Between Species Tell Us About the Development of Hearing?

There are clear pragmatic reasons for studying behavioral development in nonhuman species. Human infants are not ideal subjects for many studies. It is difficult and time consuming to attract large numbers of infants for extended psychoacoustic testing. In contrast, large numbers of, say, newly hatched chickens can easily be recruited to this task. Behavioral trends can be related more easily to physiological, anatomic, and biochemical data in nonhumans. The apparent effects of enriched, deprived, or traumatic early acoustic environments on human perceptual development (Besing, Koehnke, and Goulet 1993; Philbin, Balweg, and Gray 1994; Wilmington, Gray, and Jahrsdorfer 1994) can also be studied experimentally in nonhumans.

The theoretical justification for using comparisons between humans and nonhumans to understand auditory development is less apparent but still compelling. Virtually all land-dwelling vertebrates share the same fundamental problems in generating an auditory representation of the world. Parallel developmental trends in different species suggest general principles. These conservative trends likely reflect the most important determinants of perceptual development. Conversely, if differences between species can be related to the unique specializations of each species, then we glean important insight into how structural and functional characteristics are related.

Because the authors happen to study auditory behavior in humans and chickens, the review is biased toward these species. Several casual observations led to the expectation that comparisons between these species may be particularly interesting. The auditory system of chicks and humans is at a comparable stage, partially but not fully developed, at birth (Rubel 1978). Both species start hearing about two-thirds of the way through gestation (Jackson and Rubel 1978; Birnholz and Benacerraf 1983). Both species are highly responsive to naturalistic stimuli (DeCasper and Fifer 1980; Gottlieb 1985). Both chicks and infants fail to respond perfectly even to apparently audible sounds (Werner and Gillenwater 1990; Gray 1992a), and stimuli must be presented at appropriate times relative to the neonates' changing behavioral states to elicit a response (Wilson and Thompson 1984; Gray 1990a).

More important, the available data indicate that the normal development of human and nonhuman hearing is parallel in many ways. Furthermore, similar effects of abnormal experience have been identified in many species. These similarities, reviewed in Section 2, provide the strongest justification for pursuing comparisons between species as a means to understanding auditory development.

TABLE 2.1. Summary of studies of the development of auditory behavior.

Behavior	Humans	Other mammals	Birds	Results
Intensity processing				
Absolute sensitivity	1–16	17–21	22–25	*
Intensity discrimination	5, 26–33			* humans ? nonhumans
Loudness and dynamic range	34–40		41	?
Frequency Processing				
Frequency resolution				
Critical ratio	6, 42–44	45	46	*
Critical band	47			* humans ? nonhumans
Auditory filter width	48–52		53	*
Frequency discrimination	5, 27, 32, 54–59		60, 61	*
Frequency representation		62	63	?
Temporal Processing				
Gap detection	64–67			* humans ? nonhumans
Amplitude modulation	68, 69			?
Duration discrimination	27, 70, 71			?
Temporal integration	5, 72–75		76	*
Frequency modulation	77, 78			?
Complex Sound Processing				
Pitch	79–82			?
Spectral shape discrimination, timbre	83, 84			?
Comodulation masking release	67, 85			?
Music	86–110			* humans ? nonhumans
Discrimination of species-typical vocal productions (e.g., speech)	See Table 2.2		111–113	?
Cross-language perception	114–123			*
Localization and Binaural Processing				
Masking level difference	124–132			* humans ? nonhumans
Location identification	133–145	146–150	151, 152	*
Minimum audible angle	137, 141, 153–158			* humans ? nonhumans
Interaural time discrimination	159–161			?
Interaural intensity discrimination	159, 160			?

(Continued)

TABLE 2.1. (*Continued*)

Behavior	Humans	Other mammals	Birds	Results
Attention				
Responsiveness and preference	162–187		112, 113, 188–206	*
Habituation	207–220		221	*
Distraction and selective attention	222–232		233, 234	*
Representation				
Categories	235–249			?
Scales and perceptual maps	250–252		253–255	*

*Developmental trend is evident; ?, developmental trend is not clear. 1, Berg and Smith 1983; 2, Eisele, Berry, and Shriner 1975; 3, Elliott and Katz 1980; 4, Hoversten and Moncur 1969; 5, Maxon and Hochberg 1982; 6, Nozza and Wilson 1984; 7, Olsho et al. 1988; 8,Schneider, Trehub, and Bull. 1980; 9, Schneider et al. 1986; 10, Sinnott, Pisoni, and Aslin 1983; 11, Trehub, Schneider, and Endman 1980; 12, Trehub et al. 1988; 13, Weir 1976; 14, Weir 1979; 15, Werner and Gillenwater 1990; 16, Werner and Mancl 1993; 17, Ehret 1976; 18, Ehret and Romand 1981; 19, Pilz, Schnitzler, and Menne 1987; 20, Sheets, Dean, and Reiter 1988; 21, Zimmermann 1993; 22, Gray 1987b; 23, Gray 1992a; 24, Gray and Rubel 1985a; 25, Gray and Rubel 1985b; 26, Bull, Eilers, and Oller 1984; 27, Jensen and Neff 1993; 28, Moffitt 1973; 29, Schneider, Bull, and Trehub 1988; 30, Sinnott and Aslin 1985; 31, Steinschneider, Lipton, and Richmond 1966; 32, Stratton and Connolly 1973; 33, Tarquinio, Zelazo, and Weiss 1990; 34, Bartushuk 1964; 35, Bond and Stevens 1969; 36, Collins and Gescheider 1989; 37, Dorfman and Megling 1966; 38, Kawell, Kopun, and Stelmachowicz 1988; 39, MacPherson et al. 1991; 40, Stuart, Durieux-Smith, and Stenstrom 1991; 41, Gray and Rubel 1981; 42, Allen and Wightman 1994; 43, Schneider, Bull, and Trehub 1988; 44, Schneider, et al. 1989; 45, Ehret 1977; 46, Gray 1993a; 47, Schneider, Morrongiello, and Trehub 1990; 48, Allen et al. 1989; 49, Hall and Grose 1991; 50, Irwin, Stillman,and Schade 1986; 51, Olsho 1985; 52, Spetner and Olsho 1990; 53, Gray 1993b; 54, Leavitt et al. 1976; 55, Olsho et al. 1982a; 56, Olsho et al. 1982b; 57, Olsho 1984; 58, Olsho et al. 1987; 59, Wormith, Moffitt, and Pankhurst 1975; 60, Gray and Rubel 1985a; 61, Kerr, Ostapoff, and Rubel 1979; 62, Hyson and Rudy 1987; 63, Gray, unpublished data; 64, Irwin et al. 1985; 65, Werner et al. 1992; 66, Wightman et al. 1989; 67, Trehub, Schneider, and Henderson 1995; 68, Grose, Hall, and Gibbs 1993; 69, Hall and Grose 1994; 70, Elfenbein, Small, and Davis 1993; 71, Morrongiello and Trehub 1987; 72, Berg 1991; 73, Berg 1993; 74, Blumenthal, Avenando, and Berg 1987; 75, Thorpe and Schneider 1987; 76, Gray 1990b; 77, Aslin 1989; 78, Colombo and Horowitz 1986; 79, Bundy, Colombo, and Singer 1982; 80, Clarkson and Clifton 1985; 81, Clarkson and Clifton 1995; 82, Clarkson and Rogers 1995; 83, Trehub, Endman, and Thorpe 1990; 84, Clarkson, Clifton, and Perris 1988; 85, Veloso, Hall, and Grose 1990; 86, Drake and Gerard 1989; 87, Cohen, Thorpe, and Trehub 1987; 88, Krumhansl and Keil 1982; 89, Thorpe et al. 1988; 90, Trainor and Trehub 1992; 91, Trehub and Unyk 1992; 92, Trehub, Thorpe, and Trainor 1990; 93, Bartlett and Dowling 1980; 94, Chang and Trehub 1977; 95, Demany 1977; 96, Demany 1982; 97, Demany and Armand 1984; 98, Drake 1993; 99, Ferland and Mendelson 1989; 100, Jusczyk and Krumhansl 1993; 101, Lynch et al. 1990; 102, Morrongiello, Endman & Thrope 1985; 103, Pick et al. 1993; 104, Trainor and Trehub 1993; 105, Trehub 1989; 106, Trehub 1990; 107, Trehub, Bull, and Thorpe 1984; 108, Trehub, Thorpe, and Morrongiello 1985; 109, Trehub et al. 1986; 110, Trehub, Thorpe, and Morrongiello 1987; 111, Gray and Jahrsdoerfer 1986; 112, Gottlieb 1974; 113, Dooling and Searcy 1980; 114, Best, McRoberts, and Sithole 1988; 115, Eilers, Gavin, and Wilson 1979; Critique, Jusczyk, Shea, and Aslin 1984; Reply, Eilers, Gavin,

(*Continued*)

TABLE 2.1. (*Continued*)

and Wilson 1980; 116, Werker and Polka 1993; 117, Aslin et al. 1981; 118, Eilers, Gavin, and Oller 1982; Critique, Aslin and Pisoni 1980; Reply, Eilers et al. 1984b; 119, Lasky, Syrdal-Lasky, and Klein 1975; 120, Oller and Eilers 1983; 121, Streeter 1976; 122, Werker and Tees 1983; 123, Werker et al. 1981; 124, Werker and Tees 1984; 125, Hall and Derlacki 1988; 126, Hall, Grose, and Pillsbury 1990; 127, Hall and Grose 1990; 128, Moore, Hutchings, and Meyer 1991; 129, Nozza 1987; 130, Nozza, Wagner, and Crandell 1988; 131, Pillsbury, Grose, and Hall 1991; 132, Roush and Tait 1984; 133, Schneider, Bull, and Trehub 1988; 134, Clarkson, Clifton, and Morrongiello 1985; 135, Clifton, Morrongiello, and Dowd 1984; 136, Clifton et al. 1981; 137, Hillier, Hewitt, and Morrongiello 1992; 138, Litovsky and Macmillan 1994; 139, Litovsky and Clifton 1992; 140, Morrongiello and Clifton 1984; 141, Morrongiello, Hewitt, and Gotowiec 1991; 142, Morrongiello, Fenwick, and Chance 1990; 143, Morrongiello and Rocca 1987a; 144, Muir, Clifton, and Clarkson 1989; 145, Perris and Clifton 1988; 146, Wilmington, Gray, and Jahrsdorfer, 1994; 147, Clements and Kelly 1978a; 148, Clements and Kelly 1978b; 149, Kelly and Potash 1986; 150, Kelly, Judge, and Fraser 1987; 151, Kelly 1986; 152, Knudsen, Knudsen, and Esterly 1982; 153, Knudsen, Esterly, and Knudson 1984; 154, Ashmead, Clifton, and Perris 1987; 155, Ashmead et al. 1991; 156, Morrongiello 1988; 157, Morrongiello and Rocca 1987b; 158, Morrongiello and Rocca 1987c; 159, Morrongiello and Rocca 1990; 160, Ashmead et al. 1991; 161, Bundy 1980; 162, Kaga 1992; 163, Berg, Berg, and Graham 1971; 164, Clifton and Meyers 1969; 165, Ewing and Ewing 1944; 166, Heron and Jacobs 1969; 167, Leavitt et al. 1976; 168, Orchik and Rintelman 1978; 169, Rewey 1973; 170, Bohlin, Lindhagen, and Nagekull 1981; 171, Brown 1979; 172, Clarkson and Berg 1983; 173, Colombo 1985; 174, Colombo and Bundy 1981; 175, DeCasper and Fifer 1980; 176, DeCasper and Prescott 1984; 177, DeCasper and Spence 1986; 178, Fernald 1985; 179, Fernald and Kuhl 1987; 180, Flexer and Gans 1985; 181, Hutt et al. 1968; Critique, Bench 1973; Reply, Hutt 1973; 182, Johansson and Salmivalli 1983; 183, Mehler et al. 1978; 184, Mendel 1968; 185, Cooper and Aslin 1990; 186, Panneton and DeCasper 1984; 187, Segall 1972; 188, Standley and Madsen 1990; 189, Gottlieb 1971; 190, Gottlieb 1975a; 191, Gottlieb 1975b; 192, Gottlieb 1975c; 193, Gottlieb 1978; 194, Gottlieb 1979; 195, Gottlieb 1980a; 196, Gottlieb 1980b; 197, Gottlieb 1981; 198, Gottlieb 1982; 199, Gottlieb 1983; 200, Gottlieb 1984; 201, Gottlieb 1985; 202, Gottlieb 1987; 203, Gottlieb 1988; 204, Gottlieb 1991a; 205, Gray and Jahrsdoerfer 1986; 206, Miller and Gottlieb 1981; 207, Miller 1980; 208, Bartushuk 1962; 209, Berg 1972; 210, Brody, Zelazo, and Chaika 1984; 211, Clifton, Graham, and Hatton 1968; 212, Field et al. 1979; 213, Graham, Clifton, and Hatton 1968; 214, Hepper and Shahidullah 1992; 215, Kinney and Kagan 1976; 216, O'Connor 1980; 217, O'Connor, Cohen, and Parmalee 1984; 218, Segall 1972; 219, Tarquinio et al. 1991; 220, Zelazo, Brody, and Chaika 1984; 221, Zelazo et al. 1989; 222, Philbin, Balweg, and Gray 1994; 223, Bargones and Werner 1994; 224, Greenberg, Bray, and Beasley 1970; 225, Olsho 1985; 226, Pearson and Lane 1991; 227, Werner and Bargones 1991; 228, Doyle 1973; 229, Geffen and Sexton 1978; 230, Hagen 1967; 231, Maccoby and Konrad 1966; 232, Pearson and Lane 1991; 233, Hallahan, Kauffman, and Ball 1974; 234, Gray 1993b; 235, Gray 1993a; 236, Jusczyk and Thompson 1978; 237, Jusczyk, Copan, and Thompson 1978; 238, Ferland and Mendelson 1989; 239, Fodor, Garrett, and Brill 1975; 240, Bertoncini et al. 1988; 241, Grieser and Kuhl 1989; 242, Hillenbrand 1983; 243, Hillenbrand 1984; Critique, Moroff 1985; Reply, Hillenbrand 1985; 244, Kuhl 1979; 245, Kuhl 1983; 246, Marean, Werner, and Kuhl 1992; 247, Miller et al. 1983; 248, Miller, Younger, and Morse 1982; 249, Morrongiello 1986; 250, Trehub and Thorpe 1989; 251, Demany and Armand 1984; 252, Kuhl 1991; 253, Kuhl et al. 1992; 254, Gray 1987a; 255, Gray 1991; 256, Schneider and Gray 1991.

2. Trends in Behavioral Development

The discussion that follows is organized around Table 2.1. This is our attempt to summarize the literature on the development of auditory behavior, a cross between Rubel (1978) and Fay (1988). Table 2.1 lists the major aspects of audition and the behavioral studies that have examined their development. Studies of humans, other mammals, and birds are listed separately. Table 1 also indicates, by empty cells, some potentially important abilities that have been measured in adults but not yet in neonates. At least two properties of this literature should be readily apparent. First, many more studies of behavioral development have been completed with humans than with any other species. This is a major limitation in our understanding of the general mechanisms underlying early development: there is basically no species for which detailed studies of behavior, physiology, and structure have been completed over the entire period of development. Second, many more developmental studies have been done in birds than in nonhuman mammals. As discussed in Section 1.2, there are several common characteristics of auditory development in humans and birds that suggest it is meaningful to compare these species. On the other hand, it would also be helpful to compare humans with other mammals, and additional work on behavioral development in nonhuman mammals is certainly needed.

In the Results column of Table 2.1, we have indicated whether the literature in this area presents to us a consistent picture of the course of development. In this context, we suggest that "a consistent picture" means that some similar developmental changes seem to have been observed. That a result is consistent does not mean that we know why or how the developmental trend occurs, only that an age-related change occurs or does not occur. A question mark is used to indicate that we do not believe that there are sufficient data at this time to make a statement about development or its course.

In the Sections 2.1–2.7, the trends (or lack of trends) summarized in Table 1 will be discussed. The intent is to describe the trends that have been demonstrated, to suggest mechanisms that could be responsible for these trends, and to point out issues that have yet to be resolved.

2.1 Intensity Processing

Four measures of intensity processing are included in Table 2.1: absolute sensitivity, intensity discrimination, loudness, and dynamic range. Of the four, absolute sensitivity is the only one that has been examined extensively and the only one that has been examined to any extent in nonhumans. Although absolute sensitivity is an important parameter describing any sensory system, it is unfortunate that so little is known about the supra-threshold processing of intensity during development.

2.1.1 Absolute Sensitivity

Absolute thresholds are available for humans from the neonatal period through childhood and adolescence and into adulthood. A schematic illustrating the progression of absolute thresholds is shown in Figure 2.1. A few days after birth, humans tend to have thresholds that are 30–70 dB higher than those of adults (Weir 1976, 1979). The audibility curve is more or less flat at this age (Eisele, Berry, and Shriner 1975; Weir 1979). Adults' thresholds tend to get lower with increasing frequency, at least up to ~4 kHz, and as a result, neonates' thresholds are more mature at low than at high frequencies. Thresholds improve progressively during infancy. During the first 6 postnatal months, the improvement is greater at high than at low frequencies. Both 1 and 3 month olds still have more adultlike thresholds at lower frequencies (Olsho et al. 1988; Werner and Gillenwater 1990; Werner and Mancl 1993). By 6 months, thresholds above ~4 kHz are actually

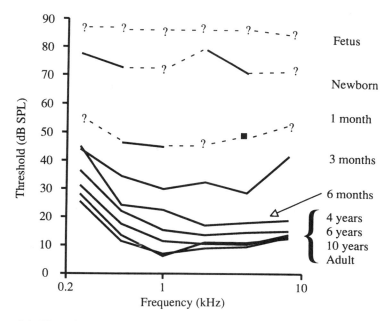

FIGURE 2.1. Hypothetical audibility curves during human auditory development (based on the results of several studies; see text). Frequency regions where data are not available are plotted as *question marks* and *dashed lines; filled symbol* and *solid lines* are based on a summary of available data. SPL, sound pressure level. Adult data are taken from Olsho et al. (1988) with permission. From Werner and Marean 1996. Reprinted by permission of Westview Press.

closer to those of adults than are thresholds at lower frequencies (Trehub, Schneider, and Endman 1980; Nozza and Wilson 1984; Olsho et al. 1988). Schneider, Trehub, and Bull (1980), in fact, have shown that very high frequency thresholds (i.e., 10 and 19 kHz) approach adult values by 24 months of age. The "high-frequency-first" pattern of development continues through the remainder of childhood: mature thresholds are observed at progressively lower frequencies as a child grows, until ~ 10 years of age when thresholds are mature across the frequency range (Elliott and Katz 1980; Schneider et al. 1986; Trehub et al. 1988).

The data from nonhuman mammals are limited to a rather short period after the onset of hearing. Of course, the developmental period of other mammals is generally shorter than that of humans, but the fact remains that nonhuman threshold development has yet to be followed into adulthood by any investigator. Moreover, thresholds reported for young animals never reach the values reported in standard comparative psychophysical experiments with adults. Ehret (1976), for example, followed threshold development for mice between 11 and 19 days (summarized in Figure 2.2). Initially, thresholds are quite high and the audibility curve is relatively flat. With age, sensitivity progressively improves, with greater improvement occurring in the frequencies above 10 kHz. Around 18 postnatal days, the rate of improvement slows; however, at the oldest age tested by Ehret (1976), thresholds are still some 10 dB higher than those reported by Ehret (1974) for young adult mice.

A similar pattern of threshold development is reported for cats (Ehret and Romand 1981) and for chickens (Saunders and Salvi 1993; see Figure

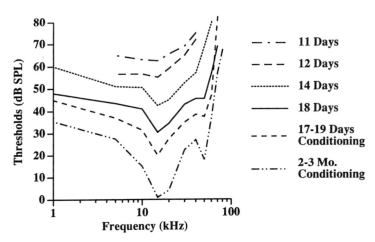

FIGURE 2.2. Audibility curves of mouse pups in days after the onset of response to sound. Thresholds for 11–18 days were based on an unconditioned response; thresholds for 17–19 days and 2–3 months were obtained in a conditioning procedure. From Ehret (1976) with permission.

2.3). The pattern reported for tree shrews (Zimmermann 1993) is slightly different in that the first change is increasing sensitivity to relatively low frequencies. But soon improvement at high frequencies accelerates, surpassing the initial gains at low frequencies to reach adult level first, leaving the final improvement at low frequencies. The progression reported for nonhumans is, thus, similar to that observed before 6 months postnatal age in humans.

One aspect of early development in nonhuman species that has not been observed in humans is the expansion of the frequency range over which a response can be observed. As Rubel (1978) noted, there is a general tendency for vertebrates to respond initially only to frequencies in the low or middle frequencies of the adult range of hearing. There is no reason to believe that this tendency is not present in humans, but, by most accounts, one would expect to see the clearest evidence for its existence during what is normally the human prenatal period. There are obvious difficulties in stimulus calibration and response recording *in utero* and other problems in the testing of prematurely born humans that have prevented determining whether humans also show an early frequency-range expansion.

There are basically two common accounts of the pattern of threshold development just described. The first and most prevalent is that improvement in behavioral thresholds is a direct reflection of cochlear development, which is said to be complete in mice, for example, at ~18 days postnatal age (see discussion by Walsh and McGee 1986). Any remaining difference between neonates and adults beyond the period of cochlear maturation is accounted for by unspecified nonsensory factors (Ehret 1976; Gray 1992b).

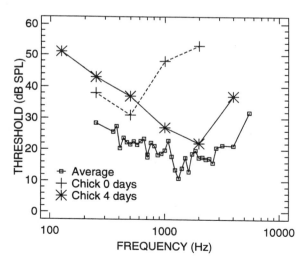

FIGURE 2.3. Audibility curves from 0-day-old, 4-day-old, and adult ("average") chickens. Chick data from Gray and Rubel (1985a); adult data from Saunders and Salvi (1993). Reprinted with permission.

This account is supported by observations that thresholds of single units from the cochlear nucleus to the auditory cortex follow the same course of development as those in the auditory nerve (Brugge, Reale, and Wilson 1988).

The second account of absolute threshold development holds that there are several functional improvements within the primary auditory system that are responsible for the age-related change in sensitivity. One of these factors is the conductive apparatus of the ear. The resonant frequency of the external ear, for example, changes as the pinna and ear canal grow (Carlile 1991; Keefe et al. 1994). The acoustic power transmitted by the middle ear of chicks (Saunders et al. 1986), small mammals (e.g., Cohen et al. 1993), and humans (Keefe et al. 1993) continues to increase with age well beyond the time that the cochlea is considered mature. In humans, middle ear development could account for threshold development after ~6 years of age (Okabe et al. 1988). Other factors are less well established. Schneider et al. (1989), for example, have proposed that the auditory neural response grows at a slower rate with increasing intensity in infants and children and that the difference in response growth is responsible for immature thresholds. Shallow rate-intensity functions have been reported for nonhumans in the auditory nerve and brain stem just after the onset of cochlear response (Sanes and Walsh, Chapter 6). In support of this neural contribution to human threshold development, Schneider et al. (1989) cite the general correspondence between the developmental time courses of thresholds and of some evoked potential measurements (e.g., Eggermont 1985). In fact, Werner, Folsom, and Mancl (1993, 1994) recently demonstrated correlations between ABR (auditory brain stem response) threshold or interpeak latency and behavioral threshold in 3-month-old infants, but these correlations are no longer significant by 6 months of age. This pattern of results suggests that brain stem immaturity may contribute to immature thresholds early in infancy but that other factors must be responsible for immature thresholds after 6 months of age. At this point, there are no data that directly relate the immaturity of structures central to the auditory brain stem to threshold development.

Most investigators seem to agree that nonsensory processes such as attention or motivation make a contribution to early threshold development. Several attempts have been made to model the effects of a simple sort of inattentiveness (e.g., Green 1990; Viemeister and Schlauch 1992; Werner 1992; Wightman and Allen 1992). These models assume that neonates are inattentive on a certain proportion of trials and that they effectively guess whether or not a sound occurred on those trials. Although such models are consistent with the slope and upper asymptote of the psychometric function of young organisms (Gray 1992a; Allen and Wightman 1994; Bargones, Werner, and Marean 1995), among 6-month-old humans, for example, they can only account for ~3 dB of a 15-dB threshold immaturity. Other, more specific models of auditory attention, however, may be able to do a better job of explaining immature detection performance (see Section 2.2.1).

In summary, external and middle ear, cochlear, neural, and probably attentional development all contribute to absolute threshold development. In the period immediately after the onset of hearing, cochlear maturation may dominate the process, although middle ear development is probably also reflected in threshold development during this period (at least in animals that begin to hear in air). But thresholds continue to develop beyond the time when people consider the cochlea mature. The middle ear is not mature until almost adolescence in humans. Middle ear immaturity accounts for as much as 10 dB of the difference between young human infants and adults and $\sim$3 dB to the threshold difference between 10 year olds and adults. Data from hamsters suggest that middle ear maturation contributes 10–15 dB to threshold development below 5,000 Hz (Relkin and Saunders 1980). Before 6 months of age in humans, it is clear that neural maturation plays a role in threshold development. We currently do not know what the neural structural and physiological correlates of threshold development are, whether neural development at more central parts of the auditory nervous system continues to contribute to threshold maturation in later development, or the extent to which neural contributions to threshold development will be demonstrable in nonhumans. Finally, the development of auditory attention may play a major role in threshold development. However, simple models of inattention have not succeeded in accounting for neonates' insensitivity to sound.

2.1.2 Intensity Discrimination

Although several studies have addressed the development of intensity discrimination in humans, nothing is known about its development in nonhumans. Human adults can discriminate about a 1-dB change in the intensity of a pure tone whether the tone is barely audible or at a high intensity (e.g., Viemeister 1988). Nonhuman adults generally discriminate changes on the order of a few decibels, with the best species and individuals approaching the performance of well-trained human listeners (summarized by Fay 1988).

The results of several studies of pure-tone intensity discrimination among developing humans are summarized in Figure 2.4. They suggest a definite improvement in intensity discrimination between infancy and middle childhood, but the amount of improvement and its timing are less certain. There is but one pure-tone data point for infants, that for 6 month olds tested by Sinnott and Aslin (1985). The infants are definitely and significantly poorer in intensity discrimination than adults tested in the same study. The data of Maxon and Hochberg (1982) from 4 year olds show the intensity DL (difference limen) improving from $\sim$6 dB to $\sim$3 dB between 6 months and 4 years of age. Thus there appears to be agreement that the intensity DL

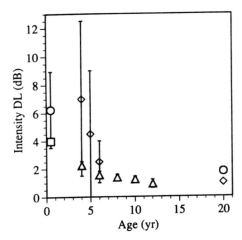

FIGURE 2.4. Intensity difference limens (DL; $10 \log [I + \Delta I]/I$) as a function of age from four studies. Error bars represent $\pm$ SE; in some cases the error bars are smaller than the plot symbols. *Squares:* Bull, Eilers, and Oller (1984); *circles:* Sinnott and Aslin (1985); *triangles:* Maxon and Hochberg (1982); *diamonds:* Jensen and Neff (1993). Data replotted with permission.

approaches adult values around 6 years of age in humans. What is in question is the amount and rate of improvement during infancy and early childhood, and, of course, given that this topic has not been explored in nonhumans, it is too early to conclude that the postnatal maturation of intensity discrimination constitutes a developmental trend.

A few studies have examined infants' ability to discriminate intensity changes in nonsense syllables or words and report lower intensity DLs for these stimuli than have been reported for pure tones (Bull, Eilers, and Oller 1984; Tarquinio, Zelazo, and Weiss 1990; see Figure 2.4). The difference is greater than would be expected to result from increasing the bandwidth of the stimulus (Raab and Goldberg 1975). The suggestion is that one might demonstrate better intensity discrimination, at least among infants, if one were to choose a more "attractive" stimulus (see Section 2.4.5). This would be another example of how nonsensory or attentional factors interact with sensory changes to determine the behavior of neonates.

2.1.3 Loudness and Dynamic Range

The least studied aspects of the development of intensity coding are loudness and dynamic range. By loudness, we mean the perceived magnitude of sound; by dynamic range, we mean the range of intensities over which the perceived magnitude changes.

Despite the fact that intensity discrimination seems to undergo maturation during infancy and early childhood and that intensity discrimination is related to the growth of loudness in some fashion (e.g., Schlauch and Wier 1987), what little evidence there is fails to show postnatal change in the perception of loudness. First, an early and unreplicated report by Bartushuk (1964) claimed that the degree to which a newborn infant's heart rate accelerated in response to a tone was a power function of the intensity of the tone with the same exponent as that seen for adults' numerical magnitude estimates (Stevens 1956). Second, Bond and Stevens (1969) reported that the growth of loudness was adultlike in 5-year-old children when it was measured by cross-modality matching of light brightness to tone loudness. More recently, Collins and Gescheider (1989) found that the growth of loudness in a group of children ranging from 4 to 7 years of age did not differ from that in adults. A notable feature of the Collins and Gescheider study is that they used different methods of scaling loudness, finding similar results in all cases.

The results of Gray and Rubel (1981) however, hint at early changes in the growth of loudness in chicks. Changing the intensity of a tone from 70 to 90 dB produced about the same increase in the duration of peep suppression to tones at three frequencies in 4-day-old chicks. In 1-day-old chicks, the same intensity change produced twice as great a response when the frequency was 900 Hz as when the frequency was higher or lower. The rapid rate of loudness growth for the 900-Hz tone is interesting for two reasons. First, it was not the case that 1-day-old infants' absolute thresholds were better at 900 Hz. Second, of the three frequencies tested, only 900 Hz is within the frequency range of the species' maternal assembly call. The implication is that attention and other nonsensory factors play a role in the development of loudness. We remind the reader who feels that such an effect does not truly reflect a change in *loudness* that attention does appear to influence loudness, measured in traditional psychophysical paradigms, among adult listeners (Schlauch 1992).

Dynamic range has not been assessed in neonates, so it is not clear whether it undergoes early development. A few studies have attempted to measure the upper limit of the dynamic range in children by estimating the intensity at which children report discomfort (Kawell, Kopun, and Stelmachowicz 1988; MacPherson et al. 1991). No differences between children and adults have been reported.

It should be evident that no clear developmental trend in loudness or dynamic range can be identified on the basis of the currently available information. All of the published human data are from listeners > 4 years of age. Thus recent methodological advances in the measurement of loudness in nonverbal organisms may have great impact when applied to this area.

2.2 Frequency Processing

Frequency processing holds a pivotal position in auditory theory and research, so it is not surprising that its development has been well studied. The major subdivisions of this topic are frequency resolution (the ability to selectively process a single component of a complex sound), frequency discrimination (the ability to distinguish between sounds of different frequency), and frequency representation (the stability of a given frequency's perceptual identity). These three processes are related through a common dependence on the width of the so-called auditory filter, the psychophysical reflection of the frequency analysis accomplished at the cochlea and maintained in some pathways throughout the auditory nervous system. However, each also depends on other mechanisms. For example, measures of frequency resolution are also influenced by intensity coding (e.g., Patterson 1974), and frequency discrimination also depends on the temporal frequency code (e.g., Moore 1974).

2.2.1 Frequency Resolution

Table 1 lists three general classes of psychophysical paradigms that are used to assess frequency resolution. The critical ratio is the signal-to-noise ratio at threshold for a narrow-band signal in a broadband background noise. By making some assumptions about the listener's detection criterion, or efficiency, it is possible to estimate the bandwidth of the auditory filter (Fletcher 1940). The accuracy of the estimate one obtains, of course, depends on the validity of the original assumptions. Estimates of the critical band are a more direct measure of the auditory filter width in that they estimate a bandwidth at which a change in perception (e.g., an increase in masked threshold) occurs. Finally, the family of paradigms that we have labeled "auditory filter width" includes techniques that are able to provide some information with respect to the shape of the auditory filter as well as to separate the effects of detection efficiency from those of filter width. These are the currently preferred techniques. A comparison among the results obtained with these three types of procedure provides an important lesson for those interested in the roles of sensory and nonsensory maturation in auditory development. Because nonsensory factors have different effects in different procedures, mistakes in interpretation are possible when a single method is considered. Furthermore, this is a case in which there are some clear trends across species.

Critical ratios have been estimated for infants and children, mouse pups, and chicks. For example, Schneider et al. (1989) followed the development of thresholds for octave-band noises masked by broad-band noise in infants, children, and young adults. At each of five octave-band center frequencies, masked thresholds improved by ~ 15 dB between 6 months and adulthood, only approaching adult values at around 10 years of age. Ehret (1977) measured mouse pups' thresholds for tones masked by broad-band noise. Masked thresholds declined by ~20 dB across the frequency range of 2–80 kHz between 10 and 18 days of age. Gray (1993b) estimated chicks' thresholds for tones in broad-band noise; 1-day-old chicks had thresholds that were ~5 dB higher than those of 4-day-old chicks. Thus there is a consistent trend for masked thresholds, and thus critical ratios, to decline with age.

An important observation in these studies is that in many respects, neonates' masked thresholds are qualitatively adultlike. The shape of neonates' masked audibility curve is not dramatically different from that of adults, and increasing the masker level has the same effect on neonates' masked thresholds as it does on adults' (Ehret 1977; Schneider et al. 1989; Gray 1993b).

One interpretation of these findings is that the auditory filter is maturing, becoming increasingly narrow with age. There are good reasons, however, to question this interpretation. As Nozza and Wilson (1984) and Gray (1993b) noted, estimates of the bandwidth of the auditory filter based on critical ratios obtained from neonates are unreasonable. In chicks, for example, the estimated critical bandwidth is greater than the chicken's range of hearing! Furthermore, it is hard to reconcile neonates' abilities to discriminate among species-specific vocalizations, with a deficit in frequency resolution of the magnitude that their critical ratios suggest.

In fact, critical band and auditory filter width studies of neonates uniformly indicate that the width of the auditory filter is mature quite early in postnatal life. Olsho (1985) was the first to report that psychophysical tuning curves, the behavioral analog to single-unit rate-tuning curves, were adultlike in width by 6 months of age. Schneider, Morrongiello, and Trehub (1990) found that critical bandwidths were adultlike among 6-month-old infants. Although Spetner and Olsho (1990) found some immaturity of high-frequency filter widths among 3 month olds, that immaturity had disappeared by 6 months. Gray (1993a) reported that simultaneous masking patterns were adultlike in shape among 0-day-old chicks.

Although Irwin, Stillman, and Schade (1986) and Allen et al. (1989) found that filter widths estimated with notched noise maskers were broader among 4 to 6-year-old children, Hall and Grose (1991) showed that in a slightly different procedure, 4 year olds had mature auditory filter widths. Their data suggest that young children's decision strategies were responsible for the immature filter widths in previous studies. Specifically, it appears that 4 year olds only attempt to detect relatively loud tones.

The most parsimonious explanation of this body of results, then, is that auditory filter widths mature early in postnatal life. It is worth noting that the physiological measures of frequency resolution that have been examined to date (otoacoustic emissions, Bargones and Burns 1988; ABR, Folsom and Wynne 1987; Abdala and Folsom 1995) also appear to mature by 6 months of age. Single-unit recordings (Manley et al. 1991) and compound action potentials in chicks (e.g., Rebillard and Rubel 1981) also indicate early maturation of frequency resolution. Under current models of frequency resolution (e.g., Patterson et al. 1982), if the filter width remains constant, then elevated masked thresholds are accounted for by the efficiency of the filter or the signal-to-noise ratio required for detection.

There are several possible explanations for neonates' inefficiency. First, the neural response to the signal may be highly variable during development, making it more difficult for young listeners to distinguish signal from noise (Schneider et al. 1989). Higher neonatal intensity DLs (Maxon and Hochberg 1982; Sinnott and Aslin 1985; Jensen and Neff 1993) and higher variability in neonatal psychoacoustic data (Werner and Bargones 1991; Gray 1993b) are consistent with this hypothesis. Second, the neural response to the signal may grow at a slower rate during development so that a higher intensity is required to achieve a given response magnitude. That infants and young children have higher intensity DLs than adults is also consistent with this alternative, but observations that the growth in loudness is adultlike in young children are inconsistent with it (e.g., Collins and Gescheider 1989; see Section 2.1.3). Physiological data from nonhumans indicate that both response variability and shallow response growth characterize the immature auditory nervous system (Sanes and Walsh, Chapter 6).

It is also possible that the immaturity is the way that infants and children use the information provided by the sensory system. As is the case for absolute sensitivity, however, only a small part of the age difference in masked threshold can be accounted for by assuming that the neonate simply tunes out and guesses on some trials (see Section 2.1.1). Another notion about early auditory attention begins with the observation that adults selectively monitor the auditory filter centered on the signal frequency when they are detecting a tone in noise (e.g., Greenberg and Larkin 1968). If listeners monitor the wrong filter or multiple filters, masked threshold will increase by as much as 7 or 8 dB depending on the assumptions (e.g., Dai, Scharf and Buus 1991). This effect is not hard to understand: monitoring filters beside the one with the signal in it adds noise to the observation. At least one study suggests that 6-year-old children can selectively monitor the filter centered on the signal frequency (Greenberg, Bray and Beasley 1970). Six-month-old infants do not appear to be selective in the filter that they monitor (Bargones and Werner 1994), and they can readily be distracted from the signal by energy in remote spectral regions (Werner and Bargones 1991). These results suggest that the development of listening strategies contributes to the development of masked thresholds after ~6 months of age in humans. To the extent that similar patterns of development are observed in other species (as in Gray 1993a), the same mechanism may be involved.

2.2.2 Frequency Discrimination

Frequency discrimination is believed to depend on auditory filter width at high frequencies but on a temporal frequency code, or phase locking, at low frequencies (e.g., Moore 1974; Freyman and Nelson 1986). Nonuniformity in the development of frequency discrimination across frequencies could mean that frequency resolution and temporal frequency coding develop along different courses.

There appear to be two phases in the development of frequency discrimination (Figure 2.5). Early in the postnatal period, neonates are rather poor at frequency discrimination, but they are particularly poor at high frequencies. For example, Olsho, Koch, & Halpin (1987) found that 3-month-old

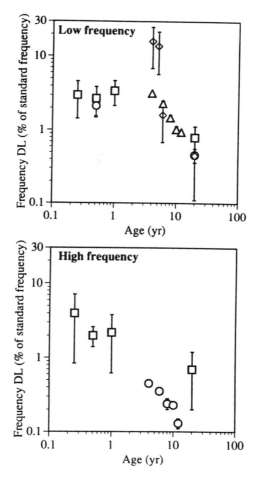

FIGURE 2.5. Frequency DL as a function of age from four studies. Error bars represent ±SE; in some cases, the error bars are smaller than the plot symbols. Top panel shows results for standard frequencies of 440 Hz (Jensen and Neff 1993; *diamonds*), 500 Hz (Olsho et al. 1987; *squares*; Maxon and Hochberg, 1982; *triangles*), and 1,000 Hz (Sinnott and Aslin 1985; *circles*). Bottom panel shows results at a standard frequency of 4,000 Hz (Olsho et al. 1987; *squares*; Maxon and Hochberg 1982; *circles*). Data used with permission.

infants could discriminate a 3% change in frequency at 500 Hz but that they needed more than a 4% change to discriminate between tones around 4,000 Hz. By 6 months of age, frequency DLs at 500 Hz had changed very little, but those at 4,000 Hz had decreased to ~2%. In fact, high-frequency DLs have been reported to be adultlike by 6 months (Olsho 1984; Olsho, Koch, & Halpin 1987). Maxon and Hochberg (1982) also found that 4 year olds could discriminate changes well under 1% at 4,000 Hz, comparable to adult performance. Similarly, Gray and Rubel (1985b) reported a greater improvement in frequency discrimination at 2,800 Hz than at either 400 or 800 Hz between 1 and 4 days of age in chicks. Thus this initial phase is characterized by immature frequency DLs, but rapid improvement in high-frequency discrimination.

The second phase of development may continue until 8 or 10 years of age in humans, but it is not clear whether or when it occurs in nonhumans. During this period, the low-frequency DL progressively improves to adult values. Several studies have reported that the frequency DL at 500 or 1,000 Hz, for example, is ~2–3% among 6-month-old infants but <1% among adults tested in a similar procedure (Olsho 1984; Sinnott and Aslin 1985; Olsho, Koch & Halpin 1987). Maxon and Hochberg (1982) reported that between 4 and 10 years of age, frequency DLs decrease from 3 to 1% at 500 Hz and from 1 to 0.4% at 1,000 Hz. Jensen and Neff (1993) found that frequency DLs at 440 Hz declined from an average of 70 Hz, or ~16%, at 4 years to an average of ~5 Hz, or just over 1% and still not quite adultlike, at 6 years. Although there is, again, some inconsistency in the precise value of the frequency DL, and when it improves, the pattern in humans is nonetheless consistent with the notion that high-frequency discrimination approaches maturity early but that low-frequency discrimination continues to develop well into childhood. This pattern is evident even when the tones are presented at equal sensation level (SL) for each subject and over a range of intensities (Maxon and Hochberg 1982; Olsho et al. 1987).

Again, there are several explanations for this pattern of development, although as yet none is widely accepted. The nonsensory hypotheses have focused on the changes in low-frequency DLs after 6 months of age. They are based on the idea that if low-frequency discrimination makes greater cognitive demands on a listener than does high-frequency discrimination, then infants and children may be especially handicapped at these frequencies because of cognitive immaturities. Olsho, Koch, & Carter (1988) explored the possibility that poor low-frequency discrimination resulted from a differential requirement for training at low frequencies, as suggested by a comparison between little-trained (Harris 1952) and well-trained listeners (Wier, Jesteadt and Green 1977). For some adults, Olsho et al. (1988) found that training effects were more pronounced at low frequencies, but a reanalysis of Olsho's (1984) infant frequency DLs suggested that training effects could not entirely account for the age differences observed in the latter study.

Sensory explanations have also been advanced. Werner (1992) suggested that the development of frequency discrimination reflected differential rates of development in frequency resolution and in temporal frequency coding, consistent with models of mature frequency discrimination. Werner noted that the improvement observed in high-frequency discrimination in the early postnatal months coincides with the final phase of maturation in frequency resolution at high frequencies (Spetner and Olsho 1990). There is evidence that phase locking is a rather late developing capacity in other mammals (Kettner, Feng and Brugge 1985), and there is one electrophysiological study suggesting that phase locking is immature in 1-month-old infants (Levi, Folsom, and Dobie 1993). These studies are consistent with the hypothesis that the maturation of phase locking is responsible for the late development of low-frequency discrimination, but further study of the development of temporal frequency coding is clearly needed.

2.2.3 Frequency Representation

One of the most interesting discoveries about audition in recent years was that the cochlear tonotopic map changes during development (Lippe and Rübsamen, Chapter 5). Ryals and Rubel (1985) and Lippe and Rubel (1985) presented data from chickens consistent with the hypothesis that a given cochlear position codes progressively higher frequencies with increasing age. Subsequently, this basic result has been confirmed and refined by observations in a variety of mammals (e.g., Arjmand, Harris, and Dallos 1988; Robertson and Irvine 1989; Rübsamen, Neuweiler, and Marimuthu 1989; Sanes, Merickel, and Rubel 1989).

It would be logical to ask how this change in tonotopy might be reflected in behavior during development. There is no direct evidence that the processes underlying the shift are also responsible for improved high-frequency sensitivity and frequency resolution in the early postnatal period. Hyson and Rudy (1987), however, demonstrated that the so-called place-code shift could have a direct impact on perception. They conditioned 15-day-old rat pups to suppress activity in response to an 8-kHz pure tone. Immediately after this training, the rat pups suppressed activity when they heard the 8-kHz tone but not when they heard tones of other frequencies. Three days later, Hyson and Rudy reexamined the pups' responses to tones of various frequencies. This time the pups suppressed activity when they heard a 12-kHz tone but not when they heard the 8-kHz tone on which they had originally been trained. This suggests not only that the tonotopic map had changed but also that, at these frequencies, the code for frequency among rat pups is based on the place in the nervous system that responds to the stimulus.

Other implications of the place-code shift have yet to be explored. For example, how is early experience with sound translated into later behavior, given this change in the organization of the auditory nervous system? This area is one of the potentially fascinating areas of behavioral development.

2.3 Temporal Processing

Many auditory capacities depend on accurate temporal coding, including localization, pitch perception, and sound-source segregation. Two different categories of neural information fall under the rubric of "temporal processing," and the distinction will become important to the discussion of the mechanisms underlying its development. First, for low frequencies, the input to the auditory nervous system is phase locked, and the timing of action potentials provides information about the frequency of the stimulus. Second, at all frequencies, auditory nerve responses are also phase locked to any amplitude modulation of the stimulus (e.g., Wang and Sachs 1992). In jargon, we would say that the phase-locked response to the carrier frequency encodes the fine structure of the stimulus, whereas the phase-locked response to the modulation frequency encodes the amplitude envelope.

All of the behavioral studies of auditory temporal development have examined the development of envelope coding. The data are few but are in agreement that substantial improvement occurs in envelope coding with age, at least in humans. The question of underlying mechanisms has not been answered satisfactorily, and the answer may prove to be complicated.

2.3.1 Envelope Processing: Gap Detection, Amplitude Modulation, and Duration Discrimination

Measures of temporal resolution are intended to determine the smallest change in the amplitude envelope that a listener can detect. In the case of gap detection, the task is to identify a brief interruption in the stimulus; the shortest detectable gap is called the gap threshold. The studies that have examined gap thresholds in infants and children have all reported age differences. Werner et al. (1992) found that 3-, 6-, and 12-month-old infants' gap thresholds were ~60 ms, in contrast to adults' gap thresholds of ~5 ms. There was little difference among infants at different ages, although variability was high among 12 month olds and some of these infants had gap thresholds that were close to adult values. Gray (unpublished observations) found only marginal responsiveness to gaps in white noise as long as 40 ms in chicks, suggesting that poor temporal resolution will be found in neonates of other species.

The results of Irwin et al. (1985) and Wightman et al. (1989) are in agreement insofar as they found that children also had immature gap thresholds. They disagree on the age at which gap thresholds mature: Irwin et al. (1985) found that gap threshold was not mature until 10–12 years, whereas Wightman et al. (1989) obtained adultlike gap thresholds among 5–7 year olds. Both Werner et al. (1992) and Wightman et al. (1989) found that the effect of stimulus frequency on gap detection was qualitatively similar at all ages; Irwin et al. (1985) found that there were greater gap detection immaturities among 6 year olds for a low frequency.

The convergence of results from different paradigms is an important tool in understanding auditory development, particularly when different paradigms may involve differential contributions of sensory and nonsensory mechanisms (see Section 2.2.1; Hall and Grose 1991). Because gap detection is a special case of amplitude modulation detection, one would predict from the gap detection results that infants and children would have difficulty processing amplitude modulation. Grose, Hall, and Gibbs (1993) have examined sensitivity to amplitude modulation in 4- to 10-year-old children using a masking-period pattern paradigm. Listeners detected a 500- or a 2,000-Hz tone when it was masked by a band-pass noise that was either modulated by a 10-Hz square wave or unmodulated. The difference between the detection thresholds for the tone in the modulated and unmodulated masking conditions is the measure of temporal resolution. All of the children had poorer temporal resolution than the adults at 500 Hz; at 2,000 Hz, the 4–5 year olds were poorer than adults. Hall and Grose (1994) subsequently examined the detection of amplitude modulation in a broadband noise. They found that 4–5 year olds required a greater modulation depth to detect modulation than adults did but that the effects of modulation rate were similar for children and adults. The latter result indicates that it is not temporal resolution per se but the representation of the intensity change that makes it difficult for infants and young children to detect changes in the amplitude envelope.

Duration discrimination has also been reported to undergo considerable development in the human postnatal period. Morrongiello and Trehub (1987) reported that 6 month olds responded to a change in the duration of a repeated 200-ms noise burst only when the duration changed by ~20 ms; 5 year olds responded when the duration changed by ~15 ms; adults responded to changes as small as 10 ms. Two other studies of duration discrimination among children agree that 4 year olds are immature on this measure, whereas by 6 years many children perform in an adultlike manner (Elfenbein, Small, and Davis 1993; Jensen and Neff 1993). Whether there are frequency gradients in the development of duration discrimination has not been explored.

It is difficult to speculate about the mechanisms that are responsible for age-related change in amplitude-envelope processing. First, all we know about this aspect of development is that infants are worse than adults. Second, there is no corresponding information on other species. Third, the processes underlying amplitude-envelope coding have not been well studied developmentally in any species.

At least at the levels of the auditory nerve and cochlear nucleus, phase locking to frequencies as low as those present in amplitude envelopes appears to be developed rather early (Kettner, Feng, and Brugge 1985). One electrophysiological study of 1-month-old humans indicates immaturity in the processing of amplitude modulation (Levi, Folsom, and Dobie 1993). A recent study by Trehub, Schneider, and Henderson (1995) reports that gap detection performance is better, if not adultlike, when infants detect gaps between two tone pips than has been reported for continuous noise. Trehub et al. suggest that the difference results from a greater susceptibility to adaptation in younger listeners. This suggestion is consistent with the results of some evoked'potential studies (e.g., Lasky 1984; Donaldson and Rubel 1990) and with the results of one behavioral study of forward masking in infants (Werner 1996).

Whenever the data indicate a simple effect of age on performance, it is difficult to separate the development of sensory processes from that of nonsensory processes. In the case of temporal resolution, the issue has yet to be addressed directly. However, there are models of mature temporal processing that posit an explicit role for higher order mechanisms (e.g., Jones and Boltz 1989; Viemeister and Wakefield 1991; see Section 2.3.2), and these could provide a starting point.

2.3.2 Temporal Integration

Traditionally, measures such as gap detection have been described as measures of the minimum integration time of the auditory system, whereas temporal-integration measures address the maximum integration time of the system (Green 1985). The typical temporal-integration experiment estimates detection threshold as a function of stimulus duration; the typical result for mature listeners is that threshold improves with duration at a rate of nearly 10 dB per decade up to ~200–300 ms (e.g., Watson and Gengel 1969). The exact value of the time constant of integration depends on several factors (reviewed by Gelfand 1990).

There is a clear trend in the development of temporal integration: compared with adults, neonates' detection improves at a faster rate with increasing duration and continues to improve to longer durations. Thorpe and Schneider (1987) first reported this tendency for 6 month olds detecting a 4-kHz octave-band noise, and Gray (1990b) made the same observation for 1-day-old chicks. Blumenthal, Avenando, and Berg (1987), studying newborn infants, Berg (1991), studying 6 month olds, and Werner and Marean (1991), studying 3 and 6 month olds, concur. Maxon and Hochberg (1982) are the only investigators to have examined temporal integration in older children; they observed little change between 4 and 12 years, consistent with the idea that temporal integration is mature by this time.[2] Some recent findings by Berg (1991, 1993) indicate that the slope of infants' temporal-integration function also depends on frequency, bandwidth, and background noise in a complex way. These results will have to be taken into account by any model that accounts for the development of temporal integration.

---

[2]Maxon and Hochberg (1982) did not test adult listeners and their results are a little different from what one might expect of a mature well-trained listener. However, these minor differences between studies and laboratories are difficult to interpret.

Some models of mature temporal integration hold that there are auditory channels with different time constants and that a listener selects a channel depending on the requirements of the task at hand (e.g., Penner 1978; Green 1985). This idea, along with limited physiological observations of mature auditory systems (Gersuni, Baru, and Hutchinson-Clutter 1971), have led some to suggest that channels that process long-duration stimuli mature before those that process transient stimuli (e.g., Berg 1991), resulting in the steeper temporal-integration function in neonates. This is consistent with data from chicks (Gray 1990b).

A sensory explanation of the age-related change in the slope of the temporal integration function that has not been considered in the behavioral-development literature is based on age differences in the growth of neural excitation with increases in intensity. As Fay and Coombs (1983) point out, if excitation grows at a slower rate with increases in intensity, then the temporal-integration function will be steeper. It has been suggested that such a reduced rate of excitation growth is responsible for age-related improvements in absolute thresholds, masked thresholds, and intensity discrimination (e.g., Schneider et al. 1989). There are data suggesting that auditory nerve and brain stem rate-intensity functions steepen just after the onset of cochlear response for nonhumans (Sanes and Walsh, Chapter 6) and during early human infancy (Cornacchia, Martini, and Morra 1983; Durieux-Smith et al. 1985).

Another possibility is suggested by the "multiple-looks" model of temporal processing of Viemeister and Wakefield (1991), which posits two stages of temporal processing. The first stage is an integrator with a time constant similar to that estimated as the minimum time constant in gap detection or amplitude modulation detection experiments. This is followed by a memory-like stage in which the short "looks" provided by the integrator can be accumulated and combined in an intelligent fashion to form the basis of a detection decision. In this model, threshold would improve for longer duration stimuli because the larger the number of "looks," the greater the probability that the detection criterion will be exceeded. One hypothesis with respect to the development of temporal processing is that the short time-constant integrator matures early, how early is not clear, but that neonates do not combine information over time as adults do. This model could also account for the fact that measures like gap detection seem to take longer to mature than measures of temporal integration: even if the integrator provides the same input to the decision process, some decision strategies may be easier than others.

2.3.3 Frequency Modulation

Frequency modulation is the oddball in this section because it refers to changes in a sound's frequency, rather than intensity, with time. The perception of frequency modulation has also not been widely studied in mature listeners (see recent review by Moore and Sek 1992) despite the fact that aspects of frequency modulation appear to be important cues to the discrimination and identification of many species-specific vocalizations (e.g., Gottlieb 1985). Because of neonates' apparently precocious ability to perceive species-specific vocalizations and because in at least one species (see Section 2.6.1) frequency modulation is believed to play a key role in determining neonates' preference for the vocalizations of their own species, we include two studies that have systematically studied it during development in Table 2.1. Colombo and Horowitz (1986) showed that 4-month-old infants could discriminate between two tones that were frequency modulated to different extents. The infants did not show more of a response to one sweep than the other, a fact that has some significance for understanding an infant's preference for certain types of speech (e.g., Fernald and Kuhl 1987; see Section 2.6.1). Aslin (1989) conducted several experiments addressing 6 month olds' processing of frequency modulation of a 1-kHz tone, with a view toward understanding how infants of this age may process the frequency transitions characteristic of speech sounds. The major findings of this investigation were that (1) 6 month olds do process the spectral change occurring during the frequency transition; (2) 6 month olds require larger transitions to detect a frequency change than adults do; and (3) changes in the transition, such as changing its duration or appending it to a steady-state tone affect 6-month-olds' and adults' detection of transitions in much the same way. Thus the current data suggest that infants are able to process frequency modulation in some respects as adults do.

2.4 Complex Sound Processing

There are many reasons why the development of complex sound processing is of interest. All other motivations aside, it is of interest to understand complex perception because that is the perceptual activity in which we are most frequently engaged. There is also the matter of concatenated complexities: if neonates have immature frequency or intensity or temporal resolution, then their perception of complex sounds may be affected by unpredictable combinations and interactions of these processes. Complex sound processing is usually held to involve more than the initial sensory processing. For example, after a complex is analyzed and temporally coded, some additional processing is deemed necessary to actually extract a pitch (e.g., Moore 1989). The maturation of these later processing stages may be an important aspect of auditory development.

There are also those who strongly believe that the psychophysical approach of using simple stimuli to probe the workings of the auditory system is an inappropriate way to study developing organisms (e.g., Turkewitz, Birch, and Cooper 1972; Gottlieb 1985). In this view, pure tones and noises are unlikely to elicit responses from neonates because organisms tend to be born selectively responsive to stimuli that are of the greatest significance for survival: avian maternal assembly calls or human speech, for example. To the extent that any complex stimulus is more "naturalistic" than a pure tone, one might predict a greater degree of responsiveness and perhaps more mature processing in neonates' processing of complexes. There is some evidence, in fact, that newborn birds achieve lower detection thresholds when they are tested with species-appropriate stimuli (Gray and Jahrsdoerfer 1986). Developmental psychophysical data would require reinterpretation if studies using complex stimuli indicated a much greater state of auditory maturity. One of the major issues addressed in Sections 2.4.1–2.4.5, consequently, will be the extent to which the complex-processing data are consistent with results from studies of simple sounds during development.

2.4.1 Pitch Discrimination

The perception of pitch, the psychological correlate of acoustic frequency, is an important component of complex perception. Contemporary models of pitch perception (e.g., Srulovicz and Goldstein 1983; Rosen and Fourcin 1986; Moore 1989) recognize the influence of two peripheral processes, frequency analysis and temporal coding. A central processor must also use the information provided by peripheral processing to assign a pitch to the stimulus. Based on this very general model, one might predict that pitch perception would be immature if either frequency resolution, temporal coding, or the central pitch processor was immature.

Those immaturities of frequency resolution that have been described in 3-month-old infants (Spetner and Olsho 1990), for example, are probably not sufficient to have a major impact on pitch perception, particularly because they occur only at high frequencies. Should larger, low-frequency immaturities be found at younger ages, some impact might be expected. By 6 months, frequency resolution appears to be mature (see Section 2.2.1). Physiological data from neonates (e.g., Kettner, Feng, and Brugge 1985) indicate that phase locking to low-frequency tones continues to develop for some time during the postnatal period, and there is some indication that envelope coding may be immature among 1-month-old infants (Levi, Folsom, and Dobie 1993). Such immaturities would be expected to limit the precision with which pitch could be assigned (also see Section 2.2.2). It is difficult to make predictions about the development of the central processor without being more specific about what it does, but, of course, any model that posits a role of experience with sound (e.g., Terhardt 1974) would predict that developmental effects would be observed.

The data on the development of pitch perception are sparse. When adults are presented with a harmonic complex with the fundamental component missing, they still match the pitch of the complex to the fundamental frequency. This phenomenon is referred to as perceiving the missing fundamental or as "low pitch." Bundy, Colombo, and Singer (1982) asked whether 4-month-old infants discriminated a change in the order of the notes in a repeated three-note sequence. Whether the fundamental was present and whether the harmonics in the complex changed from trial to trial were varied among subjects. Only infants who heard the fundamental and the same three harmonics on each presentation showed evidence of discrimination. This result provides no evidence that 4 month olds perceive complex pitch.

Studies of older infants suggest that pitch perception is at least qualitatively adultlike by 7 months. Clarkson and her colleagues have been responsible for all of this interesting work. Clarkson and Clifton (1985) first demonstrated that infants hear the missing fundamental: once infants had learned to respond to a change in pitch in complexes with the fundamental, they generalized the response to complexes without it. Clarkson and Clifton (1995) showed that infants' pitch discrimination performance was poorer for inharmonic than for harmonic complexes, as is the case for adults (e.g., Schouten, Ritsma, and Cardozo 1962). Finally, Clarkson and Rogers (1995) showed that infants perform better when a complex contains only low-frequency harmonics than they do when the complex contains only high-frequency harmonics, suggesting that the "existence region" of low pitch is similar to that of adults (Ritsma 1967).

There are clearly some questions remaining in this area. When does low pitch develop in nonhumans, where physiological data may help inform theories of mature pitch processing? Can infants younger than 7 months of age synthesize low pitch? Finally, what are the limits of pitch processing among 7 month olds? Do their complex-frequency DLs develop in parallel with their pure-tone frequency DLs?

2.4.2 Spectral Shape Discrimination, Timbre

The term timbre is generally defined, rather vaguely, as sound quality, but, in essence, it refers to whatever is left after pitch and loudness are accounted for. Spectral shape, or the relative amplitudes of different components of a complex, is an important determinant of timbre, but other parameters such as onset characteristics are also involved (e.g., see discussion by Bregman 1990). Data on the development of spectral shape discrimination have implications for our understanding of the other aspects of auditory development such as intensity discrimination and the identification of many naturally occurring sounds such as vowels. In addition, spectral shape discrimination has some implications for the development of pitch perception: infants' ability to extract pitch from spectrally varying complexes would certainly be less impressive if they were unable to discriminate the spectral changes.

Two studies have examined timbre perception in 7-month-old infants. Clarkson, Clifton, and Perris (1988) used stimuli very similar to the complexes used in their pitch discrimination experiment except that the pitch of the stimuli was constant and the infants were conditioned to respond when they heard a change in the spectral shape (harmonic content). Infants quickly learned the task whether or not the fundamental component of the complex was present. Trehub, Endman, and Thorpe (1990) extended this result by showing that 7–8 month olds could distinguish between two spectral shapes even when the frequency, duration, and intensity of the complex were varied.

It seems clear that by 7 months human infants are capable of processing spectral shape. As was the case for pitch perception, however, it is not known whether younger infants possess the same capabilities or whether other species are as sensitive to spectral shape changes early in development. Furthermore, although we know that infants can discriminate rather gross changes in spectral shape, the limits of that ability have yet to be established.

2.4.3 Comodulation Masking Release

Pitch perception and spectral shape discrimination both indicate that mature listeners make comparisons across frequency regions in processing complex sounds. Comodulation masking release (CMR) and related phenomena (McFadden 1978; Yost and Sheft 1989) are other examples of how comparisons across frequency may be important. In the CMR paradigm, listeners are able to use the similarity in the amplitude envelope of a masking noise to the amplitude envelope of noise in other spectral regions ("flanking noise") to perceptually separate a tone from the noise (Hall et al. 1984). The improvement in threshold that results from this process is CMR. This is an interesting task developmentally because it depends on at least three developing capacities: temporal resolution, combining information across frequency regions, and selective attention to sound at a certain point in time (see Sections 2.3.1, 2.3.2, and 2.6)

CMR has not been studied in infants, nor has it been examined in nonhuman neonates. Veloso, Hall, and Grose (1990) showed that 6-year-old children had adultlike CMR in at least some conditions. Grose, Hall, and Gibbs (1993) examined thresholds for 500-Hz tones masked by modulated noise bands in 4- to 10-year-old children and in adults. Children of all ages obtained as much CMR as adults when the modulated noise extended into frequency regions away from the signal. Thus, even though children's internal representation of modulation is immature (see Section 2.3.1), they are able to improve their detection by combining information across frequency. The superiority of broadband over narrow-band sound processing in neonates may constitute an important developmental trend (see Section 2.2.1).

2.4.4 Music

The perception of music certainly involves the processes of frequency discrimination and temporal resolution. On first blush one might be tempted to hypothesize that infants' and children's music perception would be limited by any immaturity of these processes, but the changes in frequency and timing that occur in music are well above threshold even for the immature auditory system (e.g., Trehub et al. 1986; Olsho et al. 1987). What is interesting about music perception is that it involves the perception of auditory patterns, and it is reasonable to ask how the ability to perceive such patterns develops.

In fact, infants as young as 5 months of age appear to perceive the pattern of frequency changes in a melody (i.e., the melodic contour), recognize the same contour when it is transposed to a different key (e.g., Chang and Trehub 1977; Ferland and Mendelson 1989), and detect transpositions (Trehub, Bull, and Thorpe 1984). Infants are sensitive to the temporal properties, or rhythm, of music (Trehub and Thorpe 1989), although young children may not be as accurate as adults at reproducing any but simple rhythms (Drake 1993). Jusczyk and Krumhansl (1993) showed that 4.5-month-old infants were sensitive to the cues (e.g., note duration and pitch contour) that signal the end of a musical phrase in much the same way that slightly older infants have been shown to be sensitive to the phrase structure of speech (e.g., Jusczyk et al. 1992). Thus, in many respects, it appears that the basics of this type of auditory pattern perception are in place within a few months of birth, although the performance of infants in experiments involving music discrimination tends to be mediocre. The age at which these capacities are first demonstrated is not clear.

Adult listeners consistently judge certain musical patterns as sounding better than others, and they may be more accurate in detecting changes to the better sounding patterns (e.g., Cuddy, Cohen, and Miller 1979). Because many aspects of what is considered a good musical pattern vary from culture to culture, it is held that music perception is largely determined by experience with the music of a specific culture (e.g., Krumhansl 1990). In fact, North American infants and adults appear to be affected differently by manipulations of culture-specific musical relations. For example, a change in musical scale that makes melody discriminations more difficult for adults or older children appear to make little difference to the discrimination performance of 6- to 11-month-old infants (Cuddy, Cohen, and Mewhort 1981; Trehub et al. 1986; Lynch et al. 1990). In at least one case, however, an effect in adults attributed to acculturation (Bartlett and Dowling 1980) has also been demonstrated in infants of this age (Trainor and Trehub 1993).

Findings that infants demonstrate certain effects in music perception that were believed to result from musical acculturation has been taken to mean that these effects may actually represent "natural proclivities" in auditory perception (e.g., Trainor and Trehub 1993). The infants tested in these studies, however, are at least 5 months old and usually 8 or more months old. Kuhl et al. (1992) have shown that 6-month-old infants discriminate between vowels in a way that reflects their experience with their native language, and in the case of visual patterns, young infants may recognize the prototypical aspects of a pattern after a very limited exposure (Walton and Bower 1993). Thus it is not out of the question that infants' superior performance on some musical patterns reflects their postnatal experience with Western music.

As a final note, research into the development of music perception frequently parallels studies of the development of speech perception. For example, just as infants are initially capable of discriminating speech sounds from nonnative languages, they are capable of discriminating changes in melodies based on nonnative musical scales. In both cases, older children and adults are incapable or less capable of making the discrimination. Such findings would suggest that general perceptual processes, rather than specific speech or music perceptual processes, are developing. Unfortunately, there are instances in which such parallels break down. For example, although variation around a prototypical vowel is less discriminable than variation around a bad example of a vowel (Kuhl 1991), discrimination is better when prototypical musical patterns are involved. Furthermore, the perceptual correlates of a space formed by the formant frequencies of vowels are generally accepted as the space within which vowel prototypes exist. The dimensions of the "culture-specific schemata" (Trainor and Trehub 1993) for music have not been specified. Thus whether similar organizational processes are operating in music and speech perception is uncertain.

2.4.5 Discrimination of Species-Typical Vocal Productions

Developing organisms show strong tendencies to attend to or to approach the sources of species-typical vocalizations (e.g., Hutt et al. 1968; Gottlieb 1985). An interesting question is how these tendencies in developing animals are related to, or interact with, the developing auditory system.

In the case of humans, the most important species-typical vocalization is speech. The development of speech perception has been so widely studied that a summary of that research requires a table of its own (Table 2.2). A variety of issues peculiar to the study of speech and language have been addressed. These issues are yet to be settled, and a full treatment of them would require more space than available here. The studies in Table 2.2 are also categorized according to whether their results showed infants or children to be mature ("Positive or Adultlike") or immature ("Negative or Not Adultlike"). For many of the speech perception topics that have been examined developmentally, there are many more entries under "Positive or Adultlike" than there are under "Negative or Not Adultlike." Moreover, many of these positive reports involve very young infants. Many speech sounds, for example, can be discriminated by infants within a few days of birth, and both native language and mother's voice are apparently recognized as well. Such observations have led some to refer to the "surprising"

TABLE 2.2. Summary of speech discrimination developmental studies.

Topic	Positive or Adultlike	Negative or Not Adultlike	Results
Phonetic Discrimination			
Yes/no	1–35	22, 22, 24, 36	*
Accuracy	37–46	37–39, 41–43, 46–48	?
Cues for discrimination	4, 7, 22, 36, 49–53	50, 54–57	?
Identification			
Boundaries	37, 45, 50, 58–60	37, 60, 61	?
Function slope	44, 45, 60, 61	50, 58–61	?
Isolated Features or Nonspeech Analogues	2, 26, 28, 32, 62	63, 64	?
Stress and Prosody	11, 27, 32, 65–71		*
Voice discrimination	66, 72–75	72	*

Positive, study shows that neonates can discriminate along the given dimension; negative, it does not; adultlike, study shows that neonates can discriminate along the given dimension as adults can; not adultlike, it does not; *, developmental trend in this type of discrimination is evident; ?, there is no clear trend. 1, Eilers et al. 1989; 2, Eimas 1974; 3, Eimas 1975; 4, Eimas and Miller 1980a; 5, Eimas and Miller 1992; 6, Eimas et al. 1971; 7, Fowler, Best and McRoberts 1990; 8, Karzon 1985; 9, Kuhl 1979; 10, Kuhl 1983; 11, Kuhl and Miller 1982; 12, Marean, Werner, and Kuhl 1992; 13, Moon and Fifer 1990; 14, Murphy, Shea, and Aslin 1989; 15, Swoboda, Morse, and Leavitt 1976; 16, Trehub 1973; 17, Walley, Pisoni, and Aslin 1984; 18, Abbs and Minifie 1969; 19, Bertoncini and Mehler 1981; 20, Bertoncini et al. 1987; 21, Eilers and Minifie 1975; 22, Eilers, Wilson, and Moore 1977; 23, Eimas and Miller 1980b; 24, Hillenbrand, Minifie, and Edwards 1979; 25, Holmberg, Morgan, and Kuhl 1977; 26, Jusczyk et al. 1977; 27, Jusczyk and Thompson 1978; 28, Jusczyk et al. 1989; 29, Leavitt et al. 1976; 30, Moffitt 1971; 31, Moon, Bever, and Fifer 1992; 32, Morse 1972; 33, Swoboda, Morse, and Leavitt 1976; 34, Swoboda et al. 1978; 35, Trehub and Rabinovitch 1972; 36, Eilers 1977; 37, Elliott et al. 1986; 38, Menary, Trehub, and McNutt 1982; 39, Abbs and Minifie 1969; 40, Barton 1980; 41, Eilers and Oller 1976; 42, Graham and House 1970; 43, Shvachkin 1973; 44, Strange and Broen 1980; 45, Wolf 1973; 46, Elliott 1986; 47, Velleman 1988; 48, Aslin et al. 1981; 49, Greenlee 1980; 50, Nittrouer and Studdert-Kennedy 1987; 51, Levitt et al. 1988; 52, Miller et al. 1983; 53, De Weirdt 1987; 54, Allen and Norwood 1988; 55, Morrongiello and Robson 1984; 56, Eilers et al. 1984a; 57, Walley and Carrell 1983; 58, Burnham, Earnshaw, and Clark 1991; 59, Elliott et al. 1981b; 60, Zlatin and Koenigsnecht 1975; 61, Simon and Fourcin 1978; 62, Jusczyk et al. 1983; 63, Elliott et al. 1989; 64, Jusczyk et al. 1980; 65, Bull, Eilers, and Oller 1984; 66, Jusczyk et al. 1992; 67, Karzon and Nicholas 1989; 68, Culp and Boyd 1974; 69, Hirsh-Pasek et al. 1987; 70, Kemler Nelson et al. 1989; 71, Spring and Dale 1977; 72, Mehler et al. 1978; 73, Miller et al. 1982; 74, Miller 1983; 75, Mills and Melhuish, 1974.

or "highly developed" speech-processing capacities of infants (Eimas, Miller, and Jusczyk 1987; Kuhl 1990). If, as described in Sections 2.1–2.3, newborns have elevated detection thresholds, poor temporal resolution, and perhaps immature frequency resolution, how can they be so good at discriminating speech sounds?

There are two common answers to this question. The first is based on the idea that infants attend more to speech than they do to boring tones and noises. By this account, infants will appear to be more sensitive when tested with speech than with nonspeech. Thus the immaturity in detection threshold or in temporal resolution results from a failure of attention rather than immature sensory processing. Enhanced sensitivity to species-typical stimuli has been demonstrated in neonatal birds (Gray and Jahrsdoerfer 1986). Several studies of infants and children, however, find that detection thresholds are no lower for speech sounds than for tones (Elliott et al. 1981a; Nozza, Wagner, and Crandell 1988). Moreover, Nozza et al. (1991) and Elliott et al. (1981a) reported that infants and children only achieved adultlike speech discrimination performance when the level of the speech sounds was adjusted to compensate for age differences in absolute sensitivity.

The second answer to the question is that psychoacoustic measures reflect immaturities of sensory processing but that speech perception measures used with young infants are not sensitive to these immaturities. When testing speech discrimination in infant subjects, the issue is whether or not a discrimination can be made with a relatively large difference between stimuli ("Phonetic discrimination, Yes/no" in Table 2.2). These studies do not ask *how well* infants can make the discrimination. One exception is the study by Aslin et al. (1981), which showed that infants' threshold for discriminating a change in voice onset time was poorer than that of adults. Once children reach the age when methods are available for establishing threshold or another measure of sensitivity, it is common to find reports of immature speech discrimination, as indicated in Table 2.2 ("Phonetic discrimination, Accuracy"). Thus an argument can be made for the position that processing of sound, including speech, improves with age. At the same time, young infants' fuzzy representations of speech sounds are clearly sufficient to allow them to discriminate among many speech sounds.

There are at least three hypothetical ways in which auditory constraints on speech perception could be exhibited. First, the timing of age-related changes in speech perception could depend on the maturation of basic capacities such as frequency and temporal resolution. Changes in the perception of native and nonnative speech sounds occur at apparently regular ages during infancy (Best 1993; Kuhl 1993; Werker and Polka 1993), and there are interesting parallels in the timing of maturation of basic auditory capacities. Second, infants or young children could use different acoustic cues than adults to identify speech sounds and thus compensate for immaturities in auditory processing. Although there are indications that even young infants can use some of the same acoustic cues that adults use to discriminate speech sounds, there is evidence that the relative importance of various cues is different for older children than it is for adults ("Phonetic discrimination, Cues for discrimination" in Table 2.2). Third, the exaggerated prosodic features typical of speech directed toward infants (e.g., Fernald 1985) may compensate for auditory limitations. In fact, there is some indication that infants' very early sensitivity to changes in stress within words, prosodic features such as intonation contour, and differences among voices depends on the prosodic exaggerations that occur in infant-directed speech (e.g., Mehler et al. 1978).

There is a troubling paucity of rigorous psychoacoustic data quantifying developmental changes in the perception of naturalistic stimuli in nonhumans. Gray and Jahrsdoerfer (1986) used trial-by-trial data from an adaptive procedure to show that newborn ducks had a steeper psychometric function than newborn chicks for detecting a ducklike sound, but the stimulus was a pulsed filtered noise, not truly natural. The probability of a correct head turn toward a naturalistic stimulus presented from speakers separated by 180° has been quantitied (e.g., Kelly, Judge, and Fraser 1987). Such data would allow the calculation of traditional psychoacoustic measures such as percent correct, false alarms, and fits to a psychometric function. To date, however, we know of no studies that have used this approach to quantify the limits of neonatal responsiveness to naturalistic stimuli. A complexity of social communication comparable to that in mammals is evident in song birds, and the acquisition of bird song is similar in many ways to the acquisition of human speech. Many investigators (reviewed in Fay 1988) have applied traditional psychoacoustic techniques to song birds, but the perceptual studies have only been done on mature birds (but see Dooling and Searcy 1980). Clearly, this area makes fertile territory for important future research.

2.5 Localization and Binaural Processing

The development of localization and binaural hearing have received attention for many reasons. Probably the major reason is that these are basic auditory processes in which the brain plays an obvious role. It is natural that researchers would be interested in studying a process that is likely to show age-related change, and the brain appears to undergo more extensive postnatal development than the ear. In addition, binaural hearing represents the auditory parallel to binocular vision, which develops substantially in the postnatal period and depends on normal input to develop normally (see Shimojo 1993 for a review).

2.5.1 Binaural Masking Level Difference

The masking-level difference (MLD) refers to the improvement in the detection of a signal under dichotic listening conditions. Its development has been examined in humans but not in nonhumans. Nozza and his colleagues (Nozza 1987; Nozza, Wagner, and Crandell 1988) demonstrated that 6-month-old infants derive less benefit than adults from interaural phase differences in the signal. Schneider et al. (1988), using a somewhat different paradigm, found that 12 month olds similarly derived less benefit from interaural differences. By about 5 years of age, the MLD appears to be mature (Nozza, Wagner, and Crandell 1988; Hall and Grose 1990). Abnormal experience with sound, caused by otitis media or another conductive disorder, appears to affect the size of the MLD (Hall and Derlacki 1988; Hall, Grose and Pillsbury 1990; Moore, Hutchings, and Meyer 1991; Pillsbury, Grose and Hall 1991; Wilmington, Gray, and Jahrsdorfer 1994).

2.5.2 Localization

Even newborn humans are capable of making at least crude localizations of a sound source. Muir and Field (1979) were the first to convincingly demonstrate that, within a few days of birth, infants would make slow long-latency head turns to the left or right toward a sound source. Whether the response is present in preterm infants is not known. Subsequent investigations (summarized by Clarkson and Clifton 1991) showed that this response depends on the duration and repetition rate of the stimulus and that only certain stimuli (e.g., rattle sounds, speech) elicit the response reliably. It is interesting that the response tends to "disappear" around 2 months of age, and when it reappears, it is quicker and shorter in latency (Muir, Clifton, and Clarkson 1989). Morrongiello and Rocca (1987a) found that the accuracy with which infants turned to face a sound source when only auditory cues were available increases progressively until ~18 months of age. Even though the development of distance perception has not been widely studied, it appears that 7 month olds can use intensity cues to judge whether a sound source is within reach or out of reach (Clifton, Perris, and Bullinger 1991).

Young rats (Potash and Kelly 1980) will approach a social call but not a distress call or noise played at a moderate level. Gerbils (Kelly and Potash 1986) and guinea pigs (Clements and Kelly 1978a) show similar unconditioned approach responses. Cats can be trained to approach an attractive sound for milk (Clements and Kelly 1978b; Olmstead and Villablanca 1980). Approach responses are similar to the early head turns observed in infants in that they depend on the stimulus and they disappear as the subjects age. It is also notable that young dogs, beginning at ~16 days of age, and rat pups, beginning on day 14, show consistent head turns toward a sound opposite one ear (Ashmead, Clifton, and Reese 1986; Kelly, Judge, and Fraser 1987)[3].

Several studies of infants have examined the development of the minimum audible angle (MAA), or the threshold for detection of a change in a sound source location. Morrongiello and her colleagues (e.g., Morrongiello

---

[3]An important research program on sound localization in owls (Knudsen, Knudsen, and Esterly 1982; Knudsen, Esterly, and Knudsen 1984; Knudsen 1988) is not reviewed here. Although owls provide exciting details about the roles of early experience, the original emergence of the binaural behavioral responses has not been studied.

1988; Morrongiello, Fenwick, and Chance 1990; Morrongiello and Rocca 1990) have followed the development of the MAA over the longest age range, finding a progressive decrease from 2 months until sometime between 18 months and 5 years of age. The MAAs measured in other laboratories (summarized by Clifton 1992; Fig. 2.6) agree well with those estimated in these studies. Thus the development of the MAA seems to parallel that of the MLD. Although detecting changes in the elevation of a sound source does not depend on interaural differences, it should be noted that Morrongiello and Rocca (1987c) reported that the MAA for elevation changes also improves during infancy but appears adultlike by ~18 months.

An obvious question is whether the development of the MAA is the result of an improvement in the coding of the binaural cues for localization: interaural intensity and interaural time differences (IIDs and ITDs, respectively). Improvements in interaural cue discriminations have been reported among infants (Bundy 1980) and into childhood (Kaga 1992). Ashmead et al. (1991), however, found that ITD threshold did not change between 16 and 28 weeks of age, an age period during which the MAA improves substantially. Furthermore, infants' MAAs are far worse than their ITD discrimination performance would predict. This implies that an age-related change in the MAA cannot be accounted for by an age-related change in ITD coding. If infants are poor at discriminating IIDs and nonetheless depend on IIDs to localize sound in azimuth, their MAAs would be poor. Parallels in IID discrimination and MAAs during development have not yet been examined.

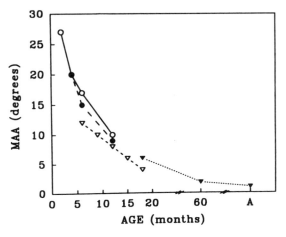

FIGURE 2.6. Minimum audible angle (MAA) as a function of age from several studies. *Filled circles:* Ashmead et al. (1991); *filled triangles:* Litovsky and Macmillan (1994); *open triangles:* Morrongiello (1988); *open circles:* Morrongiello, Fenwick, and Chance (1990). Reprinted from Clifton (1992) with permission.

Another potential contributor to immature localization is the way in which interaural differences are mapped onto positions in space. Because head size increases with age, the interaural cues that specify a particular location change as well. Thus the map of auditory space might be in a state of flux during infancy and early childhood, a situation that is unlikely to promote accurate localization. Gray (unpublished observations) applied both signal-detection and multidimensional-scaling analyses to the responses of 0- and 4-day-old chicks to changes in the source of noise bursts from one speaker to another in a pentagonal array of speakers. The signal-detection analysis showed that the newborn chicks were as sensitive as the older chicks to small changes in sound source location. But although young chicks responded for the same duration to large and small location changes, the responses of older chicks were graded with the size of the location change. Multidimensional scaling of the duration of chicks' responses to different-size location changes was used to construct a perceptual "map" of the pentagonal speaker array (see Section 2.7.2). These maps are shown in Figure 2.7, where it is evident that it was not until 4 days of age that the chicks' responses reflected a realistic representation of the speaker positions. It is not the perception of the cues that is changing but rather the organization of a subjective map of auditory space. This aspect of development may reflect age-related changes in the neural map of

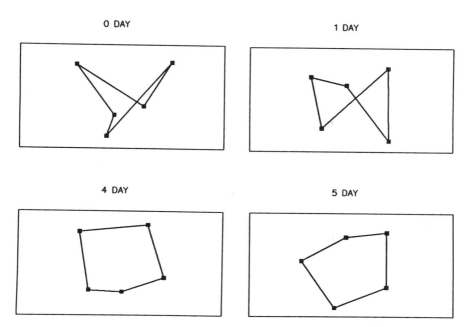

FIGURE 2.7. "Maps" of auditory space derived from chicks' response durations with multiple dimensional scaling.

auditory-visual space, as in the superior colliculus (Withington-Wray, Binns, and Keating 1990). Whether infants' MAAs are constrained by disorganized auditory spatial maps remains an interesting but unanswered question.

2.6 Attention

We use the term attention to mean a selective processing of a sound or a dimension of sound. Immaturity of attention has figured prominently in explanations for immature thresholds of all kinds, and the specific effects of inattention on thresholds are beginning to be described (see Sections 2.1, 2.2, 2.3, and 2.4). Studies explicitly addressing the development of auditory attention suggest that neonates show little evidence of "listening" in the sense of directing attention toward an expected frequency or spectral pattern, a skill at which adults excel (e.g., Schlauch and Hafter 1991). Although school-age children may be able to accomplish such selectivity for simple tasks, they appear to have difficulty with more complex listening situations.

There is a literature characterizing infants' and children's attention to sound in various ways that is typically not considered by those of us who use psychoacoustic methods to study development. Because studies in this category strongly influence developmental psychologists' views of neonatal attention, they are reviewed before more conventional studies of attention. These studies suggest that neonates are not entirely unselective in their listening behavior in that they prefer or show differential responsiveness to some sounds over others. Furthermore, normal newborns quickly learn to ignore repeated signals but redirect attention when the stimulus has changed.

2.6.1 Responsiveness and Preference

As mentioned in Section 2.4.5, infants are more likely to respond to some sounds than others, and if given a choice between sounds, they will chose to listen to some sounds over others. Responsiveness and preference are often taken as evidence that a sound has captured or kept infants' attention.

The acoustic parameters that regulate infants' responsiveness and preferences have not been extensively studied. Although bandwidth, intensity, and frequency have all been suggested as important factors (e.g., Hutt et al. 1968; Mendel 1968; Turkewitz, Birch, and Cooper 1972; Flexer and Gans 1985), in many studies supporting these suggestions, the stimuli used confounded two or more variables or the stimuli were not specified sufficiently to eliminate other possibilities (e.g., Hutt et al. 1968; Eisenberg 1976). Pulsed sounds seem to produce a greater response than continuous sounds among infants younger than 4 months but not among older infants (Mendel 1968; Bohlin, Lindhagen, and Nagekull 1981; Clarkson and Berg

1983). The one thing on which there is a consensus is that infants are more responsive to speech than to other sounds (Hutt et al. 1968; Colombo and Bundy 1981; Standley and Madsen 1990). Note that "more responsive" does not necessarily mean "more sensitive" (see Section 2.4.5).

There are numerous studies of infants' preferences for a particular type of speech, so-called "motherese" or infant-directed speech. When adults speak to infants, they tend to increase the extent of both amplitude and frequency modulation of the fundamental frequency and the duration of syllables (e.g., Fernald and Kuhl 1987). They speak more clearly, they use simple sentences, and they tend to repeat themselves (reviewed by Cooper 1993). If looking at a visual display is reinforced by the presentation of speech, even neonates will look longer at a display producing infant-directed speech than at one producing adult-directed speech (Cooper and Aslin 1990). The extent of fundamental frequency modulation appears to be an important determinant of the preference, but the aspects of infant-directed speech responsible for the preference may change during early infancy (Fernald and Kuhl 1987; Cooper 1993). Considerable attention has been given to the idea that infant-directed speech may help the infant to segment running speech into sentences, phrases, and words. Although there is some evidence for this idea (e.g., Fernald and Mazzie 1991), it remains controversial (e.g., Aslin 1993). Little attention has been given to the idea that these changes in speech may represent a parental adaptation to immature sensory processing, particularly for young infants. There is at present no empirical support for the latter idea.

A final factor that is known to influence infant preferences is familiarity. Within a few days of birth, infants show preferences for their own mother's voice over the voice of another infant's mother (DeCasper and Fifer 1980). That this preference results from prenatal experience is suggested by the finding that newborns prefer to hear a recording of one highly inflected story over another if their mothers read that story aloud during the pregnancy (DeCasper and Spence 1986). The preference for the mother's voice may depend on whether the mother uses infant-directed speech (Mehler et al. 1978). Newborn infants do not show a preference for their father's voice over that of someone else's father, although they appear to be able to discriminate the voices (DeCasper and Prescott 1984). Familiarity also affects infants' responses to nonspeech sounds (e.g., Zelazo and Komer 1971).

A similar pattern of preferences is exhibited by chicks and ducklings. Gottlieb and his colleagues have conducted an extensive series of experiments demonstrating that ducklings will approach a maternal call typical of their own species but not calls of other species (summarized by Gottlieb 1985). Mallard ducks use the repetition rate to identify the maternal call of their species (Miller 1980), and wood ducks use frequency modulation (Gottlieb 1974). Early experience is crucial in the development of appropriate preferences. Hearing the subject's own vocalization is sufficient for mallards (Gottlieb 1980a); hearing a sibling is necessary for wood ducks (Gottlieb 1983). Ducklings exposed to the maternal call of a chicken prenatally will approach the chicken maternal call after hatching (Gottlieb 1991b). Fledgling song birds also show preferential responses to conspecific songs, even before they produce songs themselves, but the role of experience in establishing this preference is uncertain (Dooling and Searcy 1980).

In summary, it is clear that neonates of several species exhibit acoustic preferences. The nature of the factors that regulate this behavior in humans is not well understood. Are neonates born with a built-in "neonate-directed vocalization detector"? Are neonates actually more sensitive to sounds with the acoustic characteristics of neonate-directed vocalizations? What sorts of prenatal experience are necessary for the development of these preferences?

2.6.2 Habituation

Habituation is a ubiquitous phenomenon in behavior. Subjects normally decrease responsiveness when a moderate stimulus is repeatedly presented. Just as responsiveness is taken as a sign of a neonates' attention, habituation is taken as an indication of neonates' inattention.

Neonatal habituation is believed to reflect some internal representation of the stimulus, although there is no consensus about the mechanisms involved (e.g., Dannemiller and Banks 1983; Slater and Morison 1985; Dannemiller and Banks 1986; Malcuit, Pomerleau, and Lamarre 1988; McCall 1988; Ackles and Karrer 1991). There have been few studies directed at the nature of habituation among neonates, but many studies have used habituation to one stimulus and a recovery of responsiveness when the stimulus is changed as evidence of discrimination between the original and changed stimuli. Rovee-Collier (1987) provides a complete description of the uses of habituation in infant research. In any case, habituation is normal. Rapid habituation seems to be a "cost of doing business" in neonatal psychoacoustic studies: normal mature subjects may respond sensitively at both ends of a long testing session, but neonates typically do not.

Although the assumption is that habituation reflects the same process regardless of the modality of stimulation, studies of the process of habituation and its development have tended to use visual stimuli (discussed by Horowitz 1974). Moreover, it has been noted that patterns of habituation to auditory and visual stimuli can be quite different (McCall 1979). The available studies of auditory habituation identify a few factors that are known to affect the rate of habituation and the extent to which the response will be reinstated when the stimulus is changed. Stimulus-variable effects have been assessed in a few studies. Clifton, Graham, and Hatton (1968) showed that the rate of infant habituation is inversely related to stimulus duration. Berg (1972) similarly showed that infants habituate more rapidly to sounds with long rise times than they do to sounds with short rise times. The infant's initial familiarity with the stimulus as well as the relationship between the stimulus and stimuli that the infant has heard previously both affect habituation and response recovery (e.g., Kinney and Kagan 1976; McCall and McGhee 1977; McCall 1988). Familiarity also affects habituation in newborn chicks (Gray 1992a).

A common question asked about habituation has been whether the rate of habituation or response recovery during infancy is predictive of later cognitive or intellectual abilities (e.g., O'Connor 1980; O'Connor, Cohen and Parmelee 1984; Bornstein 1985; see review by Rovee-Collier 1987). It has been demonstrated that high-risk and premature infants differ from normal infants in both the rate of habituation and extent of response recovery to change (Segall 1972; O'Connor 1980; Brazelton 1984; Zelazo et al. 1989), and atypical postnatal experience among high-risk infants has been suggested as a cause (e.g., Segall 1972). Atypical early experience also appears to affect the rate of habituation in nonhuman neonates: chicks that are exposed to noise[4] for 4 days just after hatching fail to habituate to repeated presentations of brief noise bursts in a quiet testing chamber (Philbin, Balweg, and Gray 1994). Whether high-risk infants and noise-exposed chicks have trouble forming an adequate representation of the sound or inhibiting their response is not clear. It has also been suggested that early visual experience affects visual attention (e.g., Movshon and Van Sluyters 1981). One wonders how many of the reported effects of early experience (e.g., otitis media) on speech and language development (e.g., Kavanagh 1986) may be mediated by deficits or delays in the development of attention.

2.6.3 Distraction and Selective Attention

Distraction, or attentional, masking was first described in infants by Werner and Bargones (1991). The task was to detect a pure tone in the presence of a noise so different in frequency from the signal that it could not possibly cause peripheral interference. Such noise increases thresholds in neonates but not adults. Similar processes appear to affect the responsiveness of chicks by about the same amount (Gray 1993b). The characteristics of the distractor (e.g., bandwidth) may be an important determinant for the degree of distraction in neonates (Gray 1993a,b).

Among humans, the development of resistance to distraction appears to continue well into childhood, although the paradigms used to assess children are quite different from that of Werner and Bargones (1991). Studies of children typically use a dichotic listening task in which different sounds are presented to the two ears and the subject is instructed to process the sound arriving at one (attended) ear in some way but to ignore the sound arriving at the other (unattended) ear. In general, the results of these studies

---

[4]The chicks were exposed to noise recorded in a human neonatal intensive care nursery.

show that, with age, there is a progressive improvement in processing sounds from the attended ear and a progressive decrease in the number of "intrusions" from the unattended ear. The exact age at which performance becomes adultlike is not clear; estimates range from 10 to sometime after 3 years. Similar effects are observed in still other paradigms with visual stimuli (see review by Lane and Pearson 1982).

Questions about the nature of the development of resistance to distraction remain. Lane and Pearson (1982) list several processes that could be involved. Infants and children may encode sounds more holistically than adults (see Section 2.7), be less selective in choosing stimuli for processing, and/or be unable to inhibit a response to an irrelevant sound.

A few studies support the contention that selectivity in the choice of stimuli develops between infancy and childhood. Bargones and Werner (1994), for example, showed that although adults detect a tone at an expected frequency better than they detect tones at unexpected frequencies, 7- to 9-month-old infants detect equal sensation-level expected and unexpected tones equally well. Greenberg, Bray, and Beasley (1970) found that by 6 years of age children listen as selectively as adults in this paradigm. Dichotic listening studies, however, show that 7- or 8-year-old children still have difficulty "directing attention"; they are worse than older children or adults in switching attention between ears (Pearson and Lane 1991) and in dividing attention between two ears.

It is not entirely clear how we can model this lack of selective listening. One possibility has to do with the way that attention is directed. Neonates may listen in a "broadband mode"; that is, they may monitor the output of many auditory filters even when it would be beneficial to monitor only one. Alternatively, neonates may not actively listen at all but wait for a sound to capture their attention. More salient stimuli (broader bandwidths, longer durations, and species-specific vocalizations) are more effective at capturing attention and thus receive preferential processing. Both of these models are consistent with the characteristics of neonates' psychometric function in detection (e.g., Dai, Scharf, and Buus 1991; Gray 1992a; Hubner 1993; Allen and Wightman 1994; Bargones, Werner, and Marean 1995).

2.7 Representation

Throughout the discussion so far, we have referred to cases in which the internal representation of sound may play an important role in the development of auditory behavior. One explanation for immature localization is the detail available in an internal "map" of auditory space (see Section 2.5), and the detail in representations may be a limiting factor for selective attention in early life (see Section 2.6). This section discusses two types of studies of the development of auditory representation: categorization, because sounds cannot be categorized along a dimension that is not represented; and perceptual maps, which involve the way that the representations of sounds are related to each other in perceptual space.

2.7.1 Categories

It is abundantly clear that neonates categorize sounds as well as sights (see, e.g., Quinn and Eimas 1986). Fodor, Garrett, and Brill (1975) first demonstrated that 14- to 18-week-old infants learned to respond to two syllables and not to a third more readily if the two syllables shared a common initial consonant than if they did not. In other words, infants are sensitive to the fact that the syllables with common consonants shared some feature; they place them within the same category. Subsequent to that study there have been demonstrations that infants can also categorize vowels (e.g., Kuhl 1979; Marean, Werner, and Kuhl 1992), voice gender (Miller 1983), musical contours (Ferland and Mendelson 1989), and musical rhythms (Trehub and Thorpe 1989).

Because infants are categorizing sounds that vary along some adult-defined dimension, the assumption is often made that infants are using the same physical dimensions as adults to form categories and, hence, must represent sound in an adultlike way. The validity of this assumption has not been explored extensively. Hillenbrand (1983, 1984) showed that infants' categorization of speech sounds is probably not accomplished by simply "memorizing" which sounds go in which group. Infants must be recognizing some acoustic similarity between the members of a speech category. Miller, Younger, and Morse (1982) showed that even though 7 month olds correctly classified male and female voices, they did not learn to form voice categories on the basis of fundamental frequency alone. This suggests that some other dimension (e.g., spectral shape or timbre) is the basis of the infants' categorization of voices.

There has been some debate as to whether infants represent speech as a string of phonetic segments or in a more "holistic" form such as syllables (e.g., Jusczyk and Krumhansl 1993; Kuhl 1993). Although infants must clearly represent the acoustic properties of speech that signal a change in a phonetic segment to be able to perform the many speech discriminations of which they are capable, it is quite possible that they do not represent those acoustic properties as a unit separate from the rest of a syllable or multisyllable utterance. Two varieties of evidence support the idea that neonates' speech representations are more holistic than those of adults. The results of several studies (e.g., Jusczyk and Derrah 1987; Bertoncini et al. 1988) find that infants do not respond to changes in all the phonetic segments in a syllable any differently than they do to a change in a single phonetic segment. Furthermore, preliterate children and adults tend to have difficulty in speech tasks that involve manipulations of individual phonetic segments, suggesting that the representation of words as a string of phonemes is the result of learning letter-sound correspondences in reading (reviewed by Walley 1993). Whether the representation of words or syllables is "linguistic" or "acoustic" is difficult to determine, but Bertoncini et al. (1988) present interesting data indicating that infants' representations of speech sounds depend on the context in which the sounds are heard and that 2 month olds may form more detailed representations than do newborns.

2.7.2 Perceptual Maps

All of the topics considered so far have been concerned with the neonates' behavior in detecting and discriminating sounds. There are important aspects of perception that cannot be addressed in such tasks. How similar do these two sounds seem? How much higher is this tone than another? The perceptions addressed by the latter questions involve attributes, judgments about the relationship between stimuli. One way to think about attributes is that they reflect the distance between the representations of sounds in a perceptual space such that closely spaced sounds are heard as more similar. This space is defined by stimulus dimensions, but how sounds are spaced along these dimensions is determined by perceptual processes. We refer to such a space as a "perceptual map." A one-dimensional perceptual map is a "scale." The generation of a perceptual map involves making assumptions about the relationship between behavior and perceptual distance: the frequency of confusions will be greater or the response latency will be longer when two sounds to be discriminated are close together in the perceptual map. Multidimensional scaling is the statistical procedure that is commonly used to find the best fitting map for a particular set of data.

Gray (1991) and Schneider and Gray (1991) used the duration of chicks' peep suppression to a frequency change to describe their pitch scale under the assumption that the duration of the response is related to the magnitude of the frequency change. No reliable scale could be derived for newborn chicks, but by 1 day of age, chicks exhibited a consistent frequency scale, and by 4 days of age, the scale was as consistent as that derived from the reaction times of mature humans to similar stimuli. This analysis suggests that a pitch scale emerges during development. Similar patterns of development have been found for the loudness scale (Gray 1987a) and for the map of auditory space (Gray 1992b). As suggested above, it may be the emergence of the map of auditory space that is responsible for much of the age-related change in localization behavior.

Perceptual maps appear to be an important aspect of many types of perception. For example, Grieser and Kuhl (1989) showed that vowels in a perceptual map, defined by a transform of the frequencies of their first two formants, were spaced in a complicated way. Vowels in the vicinity of the "best example," or prototype, of a particular vowel category tended to be closely spaced, but vowels at some distance from the prototype tended to be spaced further apart. Kuhl (1991) referred to this as a "perceptual magnet effect": the prototype seems to pull nearby exemplars toward it. Six-month-old infants demonstrate this effect, but it depends on experience with speech: American infants only show the effect around a good exemplar of an English vowel; Swedish infants only show the effect around a prototypical Swedish vowel (Kuhl et al. 1992). When this organization of vowel space first appears is not known.

The similarities in development across content areas and species strongly suggest a general developmental trend: neonates appear able to discriminate between sounds along an acoustic dimension before they develop a consistent representation of the perceptual relationship between stimuli in the space defined by that dimension. Normal experience with sound appears to be important for the development of perceptual maps in humans and birds (Schneider and Gray 1991; Besing, Koehnke, and Goulet 1993; Wilmington, Gray, and Jahrsdorfer 1994), suggesting another general trend that is still to be firmly established. The relationship between the accuracy of discrimination along a dimension and the consistency of the corresponding perceptual map also remain to be examined in detail.

3. Summary and Conclusions

A summary of the established general trends in auditory development is neither long nor complicated: (1) Absolute thresholds and frequency discrimination mature first at high and then at low frequencies. (2) Masked thresholds develop in parallel across the frequency range. (3) Auditory filter width and critical bandwidth mature early but last at high frequencies. (4) The temporal integration function becomes shallower over the course of development. (5) Neonates are capable of locating sound sources in at least a crude fashion, but normal binaural experience with sound is required to maintain localization capability. (6) Neonates respond selectively to sound, showing particular responsiveness to certain species-specific vocalizations, but the ability to selectively process one sound among many or one dimension of a multidimensional sound develops much later. (7) During development, the ability to discriminate between sounds along some dimension precedes the ability to place sounds in consistent positions in a perceptual map defined by that dimension.

Some trends are evident among humans but have not been established among nonhumans. These include the improvement in intensity discrimination during infancy and early childhood, improvements in gap detection and duration discrimination, the early appearance of auditory pattern discrimination in music, the perceptual reorganization that accompanies native language learning, and improvements in binaural hearing. For both humans and nonhumans, little is known about the growth of loudness, implications of shifting tonotopic maps, sensitivity to interaural cues underlying binaural hearing, and perception of complex sounds.

The gaps in our understanding of the processes underlying age-related change in auditory behavior are at this point larger than the gaps in the descriptions of the trends. An overriding theme in this review has been the complexity of the sensory and nonsensory maturation that gives rise to the development of auditory behavior. The cases in which we can identify specific sensory processes responsible for behavioral development are few. It is clear that the cochlea is a primary limit on early sensitivity and that the middle ear continues to limit sensitivity over a longer time period. The limits imposed on hearing by primary neural immaturity and the important implications of immature attentional and representational processes for auditory sensitivity are just beginning to be appreciated. If any conclusion is justified, it is that the sources of development in auditory behavior are neither trivial nor easily understood. The exciting aspect of the field is that powerful methods and approaches to understanding the development of auditory behavior are in hand and waiting to be applied to many interesting questions.

Acknowledgements. Data on chicks were collected with support from Grant DC-00253 to Lincoln Gray. Preparation of the manuscript was supported by National Institute of Child Health and Human Development Grant HD-28261 to Lincoln Gray and National Institute on Deafness and Other Communication Disorders Grants DC-00520 and DC-00396 to Lynne A. Werner.

References

Abbs MS, Minifie FD (1969) Effect of acoustic cues in fricatives on perceptual confusions in preschool children. J Acoust Soc Am 46:1535–1542.

Abdala C, Folsom R (1995) Frequency contribution to the click-evoked ABR in human adults and infants. J Acoust Soc Am 97:2394–2404.

Ackles PK, Karrer R (1991) A critique of the Dannemiller and Banks (1983) neuronal fatigue (selective adaptation) hypothesis of young infant habituation. Merrill-Palmer Q 37:325–334.

Allen GD, Norwood JA (1988) Cues for intervocalic /t/ and /d/ in children and adults. J Acoust Soc Am 84:868–875.

Allen P, Wightman F (1994) Psychometric functions for children's detection of tones in noise. J Speech Hear Res 37:205–215.

Allen P, Wightman F, Kistler D, Dolan T (1989) Frequency resolution in children. J Speech Hear Res 32:317–322.

Arjmand E, Harris D, Dallos P (1988) Developmental changes in frequency mapping of the gerbil cochlea: comparisons of two cochlear locations. Hear Res 32:93–96.

Ashmead D, Clifton RK, Reese EP (1986) Development of auditory localization in dogs: single source and precedence effect sounds. Dev Psychobiol 19:91–103.

Ashmead D, Davis D, Whalen T, Odom R (1991) Sound localization and sensitivity to interaural time differences in human infants. Child Dev 62:1211–1226.

Ashmead DH, Clifton RK, Perris EE (1987) Precision of auditory localization in human infants. Dev Psychol 23:641–647.

Aslin RN (1989) Discrimination of frequency transitions by human infants. J Acoust Soc Am 86:582–590.

Aslin RN (1993) Segmentation of fluent speech into words: learning models and the role of maternal input. In: de Boysson-Bardies B, de Schonen S, Jusczyk P, McNeilage P (eds). Developmental Neurocognition: Speech and Face Processing in the First Year of Life. Boston, MA: Kluwer Academic Publishers, pp. 305–315.

Aslin RN, Pisoni DB (1980) Effects of early linguistic experience on speech discrimination by infants: a critique of Eilers, Gavin, and Wilson (1979). Child Dev 51:107–112.

Aslin RN, Pisoni DB, Hennessey BL, Perey AJ (1981) Discrimination of voice onset time by human infants: new findings and implications for the effects of early experience. Child Dev 52:1135–1145.

Banks MS, Dannemiller JL (1987) Infant visual psychophysics. In: Salapatek P, Cohen LB (eds). Handbook of Infant Perception: From Sensation to Perception. New York: Academic Press, pp. 115–184.

Bargones JY, Burns EM (1988) Suppression tuning curves for spontaneous otoacoustic emissions in infants and adults. J Acoust Soc Am 83:1809–1816.

Bargones JY, Werner LA (1994) Adults listen selectively; infants do not. Psychol Sci 5:170–174.

Bargones JY, Werner LA, Marean GC (1995) Infant psychometric functions for detection: mechanisms of immature sensitivity. J Acoust Soc Am 98:99–111.

Bartlett JC, Dowling WJ (1980) Recognition of transposed melodies: a key-distance effect in developmental perspective. J Exp Psychol Hum Percept Perform 6:501–515.

Barton D (1980) Phonemic perception in children. In: Yeni-Komshian GH, Kavanagh JF, Ferguson CA (eds). Child Phonology. Vol. 2. Perception. New York: Academic Press, pp. 97–116.

Bartushuk AK (1962) Human neonatal cardiac acceleration to sound: habituation and dishabituation. Percept Mot Skills 15:15–27.

Bartushuk AK (1964) Human neonatal cardiac responses to sound: a power function. Psychon Sci 1:151–152.

Bench J (1973) "Square-wave" stimuli and neonatal auditory behavior: some comments on Ashton (1971), Hutt et al. (1968) and Lenard et al. (1969). J Exp Child Psychol 16:521–527.

Berg KM (1991) Auditory temporal summation in infants and adults: effects of stimulus bandwidth and masking noise. Percept Psychophys 50:314–320.

Berg KM (1993) A comparison of thresholds for 1/3-octave filtered clicks and noise bursts in infants and adults. Percept Psychophys 54:365–369.

Berg KM, Smith MC (1983) Behavioral thresholds for tones during infancy. J Exp Child Psychol 35:409–425.

Berg KM, Berg WK, Graham FK (1971) Infant heart rate response as a function of stimulus and state. Psychophysiology 8:30–44.

Berg WK (1972) Habituation and dishabituation of cardiac responses in 4-month-old, alert infants. J Exp Child Psychol 14:92–107.

Bertoncini J, Mehler J (1981) Syllables as units in infant speech perception. Infant Behav Dev 4:247–260.

Bertoncini J, Bijeljac-Babic R, Blumstein SE, Mehler J (1987) Discrimination in neonates of very short CVs. J Acoust Soc Am 82:31–37.

Bertoncini J, Bijeljac-Babic R, Jusczyk PW, Kennedy LJ, Mehler J (1988) An investigation of young infants' perceptual representations of speech sounds. J Exp Psychol Gen 117:21–33.

Besing J, Koehnke J, Goulet C (1993) Binaural performance associated with a history of otitis media in children. Abstr Assoc Res Otolaryngol 16:57.

Best CT (1993) Emergence of language-specific constraints in perception of non-native speech: a window on early phonological development. In: de Boysson-Bardies B, de Schonen S, Jusczyk P, McNeilage P, Morton J (eds) Developmental Neurocognition: Speech and Face Processing in the First Year of Life. Boston, MA: Kluwer Academic Publishers, pp. 289–304.

Best CT, McRoberts GW, Sithole NM (1988) Examination of perceptual reorganization for nonnative speech contrasts: Zulu click discrimination by English-speaking adults and infants. J Exp Psychol Hum Percept Perform 14:345–360.

Birnholz JC, Benacerraf BR (1983) The development of human fetal hearing. Science 222:516–518.

Blumenthal TD, Avenando A, Berg WK (1987) The startle response and auditory temporal summation in neonates. J Exp Child Psychol 44:64–79.

Bohlin G, Lindhagen K, Nagekull B (1981) Cardiac orienting to pulsed and continuous auditory stimulation: a developmental study. Psychophysiology 18:440–446.

Bond B, Stevens SS (1969) Cross-modality matching of brightness to loudness by 5-year-olds. Percept Psychophys 6:337–339.

Bornstein MH (1985) Habituation of attention as a measure of visual information processing in human infants: summary, systematization, and synthesis. In: Gottlieb G, Krasnegor NA (eds). Measurement of Audition and Vision in the First Year of Postnatal Life: A Methodological Overview. Norwood, NJ: Ablex Publishing, pp. 3–30.

Brazelton TB (1984) Neonatal Behavioral Assessment Scale. London: Spastics International Medical Publications with JB Lippincott.

Bregman AS (1990) Auditory Scene Analysis: The Perceptual Organization of Sound. Cambridge, MA: MIT Press.

Bridger WH (1961) Sensory habituation and discrimination in the human neonate. Am J Psychiatry 117:991–996.

Brindley GS (1970) Physiology of the Retina and Visual Pathway. London: E. Arnold.

Brody L, Zelazo PR, Chaika H (1984) Habituation-dishabituation to speech in the newborn. Dev Psychol 20:114–119.

Brown CJ (1979) Reactions of infants to their parents' voices. Infant Behav Dev 2:295–300.

Brugge JF, Reale RA, Wilson GF (1988) Sensitivity of auditory cortical neurons of kittens to monaural and binaural high frequency sound. Hear Res 34:127–140.

Bull D, Eilers RE, Oller DK (1984) Infants' discrimination of intensity variation in multisyllabic stimuli. J Acoust Soc Am 76:13–17.

Bundy R (1980) Discrimination of sound localization cues in young infants. Child Dev 51:292–294.

Bundy R, Colombo J, Singer J (1982) Pitch perception in young infants. Dev Psychol 18:10–14.

Burnham DK, Earnshaw LJ, Clark JE (1991) Development of categorical identification of native and non-native bilabial stops: infants, children and adults. J Child Lang 18:231–260.

Carlile S (1991) The auditory periphery of the ferret: postnatal development of acoustic properties. Hear Res 51:265–278.

Chang HW, Trehub SE (1977) Auditory processing of relational information by young infants. J Exp Child Psychol 24:324–331.

Clarkson MG, Berg WK (1983) Cardiac orienting and vowel discrimination in newborns: crucial stimulus parameters. Child Dev 54:162–171.

Clarkson MG, Clifton RK (1985) Infant pitch perception: evidence for responding to pitch categories and the missing fundamental. J Acoust Soc Am 77:1521–1528.

Clarkson MG, Clifton RK (1991) Acoustic determinants of newborn orienting. In: Weiss MJS, Zelazo PR (eds). Newborn Attention: Biological Constraints and the Influence of Experience. Norwood, NJ: Ablex Publishing, pp. 99–119.

Clarkson MG, Clifton RK (1995) Infants' pitch perception: inharmonic tonal complexes. J Acoust Soc Am 98:1372–1379.

Clarkson MG, Rogers EC (1995) Infants require low-frequency energy to hear the pitch of the missing fundamental. J Acoust Soc Am, 98:148–154.

Clarkson MG, Clifton RK, Morrongiello BA (1985) The effects of sound duration on newborns' head orientation. J Exp Child Psychol 39:20–36.

Clarkson MG, Clifton RK, Perris EE (1988) Infant timbre perception: discrimination of spectral envelopes. Percept Psychophys 43:15–20.

Clements M, Kelly JB (1978a) Auditory spatial responses of young guinea pigs (*Gavia porcellus*) during and after ear blocking. J Comp Physiol Psychol 92:34–44.

Clements M, Kelly JB (1978b) Directional responses by kittens to an auditory stimulus. Dev Psychobiol 11:505–511.

Clifton RK (1992) The development of spatial hearing in human infants. In: Werner LA, Rubel EW (eds). Developmental Psychoacoustics. Washington, DC: American Psychological Association, pp. 135–157.

Clifton RK, Meyers WJ (1969) The heart-rate response of four-month-old infants to auditory stimuli. J Exp Child Psychol 7:122–135.

Clifton RK, Graham FK, Hatton HM (1968) Newborn heart-rate response and response habituation as a function of stimulus duration. J Exp Child Psychol 6:265–278.

Clifton RK, Morrongiello B, Kulig J, Dowd J (1981) Auditory localization of the newborn infant: its relevance for cortical development. Child Dev 52:833–838.

Clifton RK, Morrongiello B, Dowd J (1984) A developmental look at an auditory illusion: the precedence effect. Dev Psychobiol 17:519–536.

Clifton RK, Perris EE, Bullinger A (1991) Infants' perception of auditory space. Dev Psychol 27:187–197.

Cohen AJ, Thorpe LA, Trehub SE (1987) Infants' perception of musical relations in short transposed tone sequences. Can J Psychol 41:33–47.

Cohen YA, Doan DE, Rubin DM, Saunders JC (1993) Middle ear development V: development of umbo sensitivity in the gerbil. Am J Otolaryngol 14:191–198.

Collins AA, Gescheider GA (1989) The measurement of loudness in individual children and adults by absolute magnitude estimation and cross-modality matching. J Acoust Soc Am 85:2012–2021.

Colombo J (1985) Spectral complexity and infant attention. J Gen Psychol 146:519–526.

Colombo J, Bundy RS (1981) A method for the measurement of infant auditory selectivity. Infant Behav Dev 4:219–223.

Colombo J, Horowitz FD (1986) Infants' attentional responses to frequency modulated sweeps. Child Dev 57:287–291.

Cooper RP (1993) The effect of prosody on young infants' speech perception. In: Rovee-Collier C, Lipsitt LP (eds). Advances in Infancy Research. Vol. 8. Norwood, NJ: Ablex Publishing, pp. 137–167.

Cooper RP, Aslin RN (1990) Preference for infant-directed speech in the first month after birth. Child Dev 61:1584–1595.

Cornacchia L, Martini A, Morra B (1983) Air and bone conduction brain stem responses in adults and infants. Audiology 22:430–437.

Cuddy LL, Cohen AJ, Miller J (1979) Melody recognition: the experimental application of musical rules. Can J Psychol 33:148–156.

Cuddy LL, Cohen AJ, Mewhort DJK (1981) Perception of structure in short melodic sequences. J Exp Psychol Hum Percept Perform 7:869–883.

Culp RE, Boyd EF (1974) Visual fixation and the effect of voice quality differences in 2-month-old infants. Monogr Soc Res Child Dev 39:78–91.

Dai H, Scharf B, Buus S (1991) Effective attenuation of signals in noise under focused attention. J Acoust Soc Am 88:2837–2842.

Dannemiller JL, Banks MS (1983) Can selective adaptation account for early infant habituation? Merrill-Palmer Q 29:151–158.

Dannemiller JL, Banks MS (1986) Testing models of early infant habituation: a reply to Slater and Morison. Merrill-Palmer Q 32:87–91.

DeCasper AJ, Fifer WP (1980) Of human bonding: newborns prefer their mothers' voices. Science 208:1174–1176.

DeCasper AJ, Prescott PA (1984) Human newborns' perception of male voices: preference, discrimination, and reinforcing value. Dev Psychobiol 17:481–491.

DeCasper AJ, Spence MJ (1986) Prenatal maternal speech influences newborns' perception of speech sounds. Infant Behav Dev 9:133–150.

Demany L (1977) Rhythm perception in early infancy. Nature 266:718–719.

Demany L (1982) Auditory stream segregation in infancy. Infant Behav Dev 5:261–276.

Demany L, Armand F (1984) The perceptual reality of tone chroma in early infancy. J Acoust Soc Am 76:57–66.

De Weirdt W (1987) Age differences in place-of-articulation phoneme boundary. Percept Psychophys 42:101–103.

Donaldson GS, Rubel EW (1990) Effects of stimulus repetition rate on ABR threshold, amplitude and latency in neonatal and adult Mongolian gerbils. Electroencephalogr Clin Neurophysiol 77:458–470.

Dooling R, Searcy M (1980) Early perceptual selectivity in the swamp sparrow. Dev Psychobiol 13:499–506.

Dorfman DD, Megling R (1966) Comparison of magnitude estimation of loudness in children and adults. Percept Psychophys 1:239–241.

Doyle A-B (1973) Listening to distraction: a developmental study of selective attention. J Exp Child Psychol 15:100–115.

Drake C (1993) Reproduction of musical rhythms by children, adult musicians and adult nonmusicians. Percept Psychophys 53:25–33.

Drake C, Gerard C (1989) A psychological pulse train: how young children use this cognitive framework to structure simple rhythms. Psychol Res 51:16–22.

Durieux-Smith A, Edwards CG, Picton TW, McMurray B (1985) Auditory brainstem responses to clicks in neonates. J Otolaryngol 14:12–18.

Eggermont JJ (1985) Evoked potentials as indicators of auditory maturation. Acta Otolaryngol 421:41–47.

Ehret G (1974) Age-dependent hearing loss in normal hearing mice. Naturwissenschaften 61:506.

Ehret G (1976) Development of absolute auditory thresholds in the house mouse (*Mus musculus*). J Am Audiol Soc 1:179–184.

Ehret G (1977) Postnatal development in the acoustic system of the house mouse in the light of developing masked thresholds. J Acoust Soc Am 62:143–148.

Ehret G, Romand R (1981) Postnatal development of absolute auditory thresholds in kittens. J Comp Physiol Psychol 95:304–311.

Eilers RE (1977) Context-sensitive perception of naturally produced stop and fricative consonants by infants. J Acoust Soc Am 61:1321–1336.

Eilers RE, Minifie FD (1975) Fricative discrimination in early infancy. J Speech Hear Res 18:158–167.

Eilers RE, Oller KD (1976) The role of speech discrimination in developmental sound substitutions. J Child Lang 3:319–329.

Eilers RE, Wilson WR, Moore JM (1977) Developmental changes in speech discrimination in infants. J Speech Hear Res 20:766–780.

Eilers RE, Gavin W, Wilson WR (1979) Linguistic experience and phonemic perception in infancy: a crosslinguistic study. Child Dev 50:14–18.

Eilers RE, Gavin WJ, Wilson WR (1980) Effects of early linguistic experience on speech discrimination by infants: a reply. Child Dev 51:113–117.

Eilers RE, Gavin WJ, Oller DK (1982) Cross-linguistic perception in infancy: early effects of linguistic experience. J Child Lang 9:289–302.

Eilers RE, Bull DH, Oller DK, Lewis DC (1984a) The discrimination of vowel duration by infants. J Acoust Soc Am 75:1213–1218.

Eilers RE, Oller DK, Bull DH, Gavin WJ (1984b) Linguistic experience and infant speech perception: a reply to Jusczyk, Shea and Aslin (1984). J Child Lang 11:467–475.

Eilers RE, Oller DK, Urbano R, Moroff D (1989) Conflicting and cooperating cues: perception of cues to final consonant voicing by infants and adults. J Speech Hear Res 32:307–316.

Eimas PD (1974) Auditory and linguistic processing of cues for place of articulation by infants. Percept Psychophys 16:513–521.

Eimas PD (1975) Speech perception in early infancy. In: Cohen LB, Salapatek P (eds). Infant Perception: From Sensation to Cognition. New York: Academic Press, pp. 193–231.

Eimas PD, Miller JL (1980a) Contextual effects in infant speech perception. Science 209:1140–1141.

Eimas PD, Miller JL (1980b) Discrimination of information for manner of articulation. Infant Behav Dev 3:367–375.

Eimas PD, Miller JL (1992) Organization in the perception of speech by young infants. Psychol Sci 3:340–345.

Eimas PD, Miller JL, Jusczyk (1987) On infant speech perception and the acquisition of language. In: Harnad S (ed). Categorical Perception: The Groundwork of Cognition. New York: Cambridge University Press, pp. 161–195.

Eimas PD, Siqueland ER, Jusczyk P, Vigorito J (1971) Speech perception in infants. Science 171:303–306.

Eisele WA, Berry RC, Shriner TA (1975) Infant sucking response patterns as a conjugate function of changes in the sound pressure level of auditory stimuli. J Speech Hear Res 18:296–307.

Eisenberg R (1976) Auditory Competence in Early Life. Baltimore, MD: University Park Press.

Elfenbein JL, Small AM, Davis M (1993) Developmental patterns of duration discrimination. J Speech Hear Res 36:842–849.

Elliott LL (1986) Discrimination and response bias for CV syllables differing in voice onset time among children and adults. J Acoust Soc Am 80:1250–1255.

Elliott LL, Katz DR (1980) Children's pure-tone detection. J Acoust Soc Am 67:343–344.

Elliott LL, Longinotti C, Clifton L, Meyer D (1981a) Detection and identification thresholds for consonant-vowel syllables. Percept Psychophys 30:411–416.

Elliott LL, Longinotti C, Meyer D, Raz I, Zucker K (1981b) Developmental differences in identifying and discriminating CV syllables. J Acoust Soc Am 70:669–677.

Elliott LL, Busse LA, Partridge R, Rupert J, DeGraaff R (1986) Adult and child discrimination of CV syllables differing in voicing onset time. Child Dev 57:628–635.

Elliott LL, Hammer MA, Scholl ME, Wasowicz JM (1989) Age differences in discrimination of simulated single-formant frequency transitions. Percept Psychophys 46:181–186.

Ewing IR, Ewing AWG (1944) The ascertainment of deafness in infancy and early childhood. J Laryngol Otol 59:309–333.

Fay RR (1988) Hearing in Vertebrates: A Psychophysics Databook. Winnetka, IL: Hill-Fay Associates.

Fay RR, Coombs S (1983) Neural mechanisms in sound detection and temporal summation. Hear Res 10:69–92.

Ferland MB, Mendelson MJ (1989) Infants' categorization of melodic contour. Infant Behav Dev 12:341–355.

Fernald A (1985) Four-month-old infants prefer to listen to motherese. Infant Behav Dev 8:181–195.

Fernald A, Kuhl P (1987) Acoustic determinants of infant perception for motherese speech. Infant Behav Dev 10:279–293.

Fernald A, Mazzie C (1991) Prosody and focus in speech to infants and adults. Dev Psychol 27:209–221.

Field TM, Dempsey JR, Hatch J, Ting G, Clifton RK (1979) Cardiac and behavioral responses to repeated tactile and auditory stimulation in preterm and term neonates. Dev Psychol 15:406–416.

Fletcher H (1940) Auditory patterns. Rev Mod Phys 12:47–65.

Flexer C, Gans DP (1985) Comparative evaluation of the auditory responsiveness of normal infants and profoundly multihandicapped children. J Speech Hear Res 28:163–168.

Fodor JA, Garrett MF, Brill SL (1975) Pi ka pu: the perception of speech sounds by pre-linguistic infants. Percept Psychophys 18:74–78.

Folsom RC, Wynne MK (1987) Auditory brain-stem responses from human adults and infants: wave V tuning curves. J Acoust Soc Am 81:412–417.

Fowler CA, Best CT, McRoberts GW (1990) Young infants' perception of liquid coarticulatory influences on following stop consonants. Percept Psychophys 48:559–570.

Freyman RL, Nelson DA (1986) Frequency discrimination as a function of tonal duration and excitation-pattern slopes in normal and hearing-impaired listeners. J Acoust Soc Am 79:1034–1044.

Geffen G, Sexton MA (1978) The development of auditory strategies of attention. Dev Psychol 14:11–17.

Gelfand SA (1990) Hearing: An Introduction to Psychological and Physiological Acoustics. New York: Marcel Dekker.

Gersuni G, Baru VKH, Hutchinson-Clutter M (1971) Effects of temporal lobe lesions on perception of sounds of short duration. In: Gersuni GV (ed). Sensory Processes at the Neuronal and Behavioral Levels. New York: Academic Press, pp. 287–300.

Gottlieb G (1971) Development of Species Identification in Birds: An Inquiry into the Prenatal Determinants of Perception. Chicago, IL: University of Chicago Press.

Gottlieb G (1974) On the acoustic basis of species identification in wood ducklings (*Aix sponsa*). J Comp Physiol Psychol 87:1038–1048.

Gottlieb G (1975a) Development of species identification in ducklings. I. Nature of perceptual deficit caused by embryonic auditory deprivation. J Comp Physiol Psychol 89:387–399.

Gottlieb G (1975b) Development of species identification in ducklings. II. Experiential prevention of perceptual deficit caused by embryonic auditory deprivation. J Comp Physiol Psychol 89:675–684.

Gottlieb G (1975c) Development of species identification in ducklings. III. Maturational rectification of perceptual deficit caused by auditory deprivation. J Comp Physiol Psychol 89:899–812.

Gottlieb G (1978) Development of species identification in ducklings. IV. Changes in species-specific perception caused by auditory deprivation. J Comp Physiol Psychol 92:375–387.

Gottlieb G (1979) Development of species identification in ducklings. V. Perceptual differentiation in the embryo. J Comp Physiol Psychol 93:831–854.

Gottlieb G (1980a) Development of species identification in ducklings. VI. Specific embryonic experience required to maintain species-typical perception in Peking ducklings. J Comp Physiol Psychol 94:579–587.

Gottlieb G (1980b) Development of species identification in ducklings. VII. Highly specific early experience fosters species-specific perception in wood ducklings. J Comp Physiol Psychol 94:1019–1027.

Gottlieb G (1981) Development of species identification in ducklings. VIII. Embryonic versus postnatal critical period for the maintenance of species-typical perception. J Comp Physiol Psychol 93:831–854.

Gottlieb G (1982) Development of species identification in ducklings. IX. The necessity of experiencing normal variations in embryonic auditory stimulation. Dev Psychobiol 15:517–517.

Gottlieb G (1983) Development of species identification in ducklings. X. Perceptual specificity in the wood duck embryo requires sib stimulation for maintenance. Dev Psychobiol 16:323–333.

Gottlieb G (1984) Development of species identification in ducklings. XII. Ineffectiveness of auditory self-stimulation. J Comp Psychol 98:137–141.

Gottlieb G (1985) On discovering the significant acoustic dimensions of auditory stimulation for infants. In: Gottlieb G, Krasnegor N (eds). Measurement of Audition and Vision in the First Year of Postnatal Life: A Methodological Overview. Norwood, NJ: Ablex Publishing, pp. 3–29.

Gottlieb G (1987) Development of species identification in ducklings. XIV. Malleability of species-specific perception. J Comp Psychol 101:178–182.

Gottlieb G (1988) Development of species identification in ducklings. XV. Individual auditory recognition. Dev Psychobiol 21:509–522.

Gottlieb G (1991a) Experiential canalization of behavioral development: Results. Dev Psychol 27:35–39.

Gottlieb G (1991b) Social induction of malleability in ducklings. Anim Behav 41:953–962.

Graham FK, Clifton RK, Hatton HM (1968) Habituation of heart rate response to repeated auditory stimulation during the first five days of life. Child Dev 39:35–52.

Graham LW, House AS (1970) Phonological oppositions in children: a perceptual study. J Acoust Soc Am 49:559–566.

Gray L (1987a) Multidimensional perceptual development: consistency of responses to frequency and intensity in young chickens. Dev Psychobiol 20:299–312.

Gray L (1987b) Signal detection analyses of delays in neonates' vocalizations. J Acoust Soc Am 82:1608–1614.

Gray L (1990a) Activity level and auditory responsiveness in neonatal chickens. Dev Psychobiol 23:297–308.

Gray L (1990b) Development of temporal integration in newborn chickens. Hear Res 45:169–177.

Gray L (1991) Development of a frequency dimension in chickens (Gallus gallus). J Comp Physiol 105:85–88.

Gray L (1992a) An auditory psychometric function from newborn chicks. J Acoust Soc Am 91:1608–1615.

Gray L (1992b) Interactions between sensory and nonsensory factors in the responses of newborn birds to sound. In: Werner LA, Rubel EW (eds). Developmental Psychoacoustics. Washington, DC: American Psychological Association, pp. 89–112.

Gray L (1993a) Developmental changes in chickens' masked thresholds. Dev Psychobiol 26:447–457.

Gray L (1993b) Simultaneous masking in newborn chickens. Hear Res 69:83–90.

Gray L, Jahrsdoerfer R (1986) Naturalistic psychophysics: thresholds of ducklings (Ana platyrynchos) and chicks (Gallus gallus) to tones that resemble mallard calls. J Comp Psychol 100:91–94.

Gray L, Rubel EW (1981) Development of responsiveness to suprathreshold acoustic stimulation in chickens. J Comp Physiol Psychol 95:188–198.

Gray L, Rubel EW (1985a) Development of absolute thresholds in chickens. J Acoust Soc Am 77:1162–1172.

Gray L, Rubel EW (1985b) Development of auditory thresholds and frequency difference limens in chicks. In: Gottlieb G, Krasnegor NA (eds). Measurement of Audition and Vision in the First Year of Postnatal Life: A Methodological Overview. Norwood, NJ: Ablex Publishing, pp. 145–166.

Green DM (1976) An Introduction to Hearing. Hillsdale, NJ: Lawrence Erlbaum Associates.

Green DM (1983) Profile analysis: a different view of auditory intensity discrimination. Am Psychol 38:133–142.

Green DM (1985) Temporal factors in psychoacoustics. In: Michelson A (ed) Time Resolution in Auditory Systems. New York: Springer-Verlag, pp. 122–140.

Green DM (1990) Stimulus selection in adaptive psychophysical procedures. J Acoust Soc Am 87:2662–2674.

Greenberg GZ, Larkin WD (1968) The frequency response characteristic of auditory observers detecting signals of a single frequency in noise: the probe-signal method. J Acoust Soc Am 44:1513–1523.

Greenberg GZ, Bray NW, Beasley DS (1970) Children's frequency-selective detection of signals in noise. Percept Psychophys 8:173–175.

Greenlee M (1980) Learning the phonetic cues to the voiced-voiceless distinction: a comparison of child and adult speech perception. J Child Lang 7:459–468.

Grieser D, Kuhl PK (1989) Categorization of speech by infants: support for speech-sound prototypes. Dev Psychol 25:577–588.

Grose JH, Hall JW III, Gibbs C (1993) Temporal analysis in children. J Speech Hear Res 36:351–356.

Hagen JW (1967) The effect of distraction on selective attention. Child Dev 38:685–694.

Hall JW III, Derlacki EL (1988) Binaural hearing after middle ear surgery. Masking level difference for interaural time and amplitude cues. Audiology 27:89–98.

Hall JW III, Grose JH (1990) The masking level difference in children. J Am Acad Audiol 1:81–88.

Hall JW III, Grose JH (1991) Notched-noise measures of frequency selectivity in adults and children using fixed-masker-level and fixed-signal-level presentation. J Speech Hear Res 34:651–660.

Hall JW III, Grose JH (1994) Development of temporal resolution in children as measured by the temporal modulation transfer function. J Acoust Soc Am 96:150–154.

Hall JW III, Haggard MP, Fernandes MA (1984) Detection in noise by spectro-temporal pattern analysis. J Acoust Soc Am 76:50–56.

Hall JW III, Grose JH, Pillsbury HC (1990) Predicting binaural hearing after stapedectomy from presurgery results. Arch Otolaryngol Head Neck Surg 116:946–950.

Hallahan DP, Kauffman JM, Ball DW (1974) Developmental trends in recall of central and incidental auditory material. J Exp Child Psychol 17:409–421.

Harris JD (1952) Pitch discrimination. J Acoust Soc Am 24:750–755.

Hepper PG, Shahidullah S (1992) Habituation in normal and Down's syndrome fetuses. Q J Exp Psychol B Comp Physiol Psychol 44:305–317.

Heron TG, Jacobs R (1969) Respiratory curve responses of the neonate to auditory stimulation. Int Audiol 8:77–84.

Hillenbrand J (1983) Perceptual organization of speech sounds by infants. J Speech Hear Res 26:268–282.

Hillenbrand J (1984) Speech perception by infants: categorization based on nasal consonant place of articulation. J Acoust Soc Am 75:1613–1622.

Hillenbrand J (1985) Perception of feature similarities by infants. J Speech Hear Res 28:317–318.

Hillenbrand J, Minifie FD, Edwards TJ (1979) Tempo of spectrum change as a cue in speech-sound discrimination by infants. J Speech Hear Res 22:147–165.

Hillier L, Hewitt KL, Morrongiello BA (1992) Infants' perception of illusions in sound localization: reaching to sounds in the dark. J Exp Child Psychol 53:159–179.

Hirsh-Pasek K, Kemler Nelson DG, Jusczyk PW, Wright Cassidy K, Druss B, Kennedy L (1987) Clauses are perceptual units for young infants. Cognition 26:269–286.

Holmberg TL, Morgan KA, Kuhl PK (1977) Speech perception in early infancy: discrimination of fricative consonants. J Acoust Soc Am 62:S99.

Horowitz FD (ed) (1974) Visual attention, auditory stimulation, and language discrimination in young infants. Monographs of the Society for Research in Child

Development 39(5–6); whole number.

Hoversten GH, Moncur JP (1969) Stimuli and intensity factors in testing infants. J Speech Hear Res 12:677–686.

Hubner R (1993) On possible models of attention in signal detection. J Math Psychol 37:266–281.

Hutt SJ (1973) Square-wave stimuli and neonatal auditory behavior: reply to Bench. J Exp Child Psychol 16:530–533.

Hutt SJ, Hutt C, Lenard HG, von Bernuth H, Muntjewerff WJ (1968) Auditory responsivity in the human neonate. Nature 218:888–890.

Hyson RL, Rudy JW (1987) Ontogenetic change in the analysis of sound frequency in the infant rat. Dev Psychobiol 20:189–207.

Irwin RJ, Stillman JA, Schade A (1986) The width of the auditory filter in children. J Exp Child Psychol 41:429–442.

Irwin RJ, Ball AKR, Kay N, Stillman JA, Rosser J (1985) The development of temporal acuity in children. Child Dev 56:614–620.

Jackson H, Rubel EW (1978) Ontogeny of behavioral responsiveness to sound in the chick embryo as indicated by electrical recordings of motility. J Comp Physiol Psychol 92:682–696.

Jensen JK, Neff DL (1993) Development of basic auditory discrimination in preschool children. Psychol Sci 4:104–107.

Johansson RK, Salmivalli A (1983) Arousing effect of sounds for testing infants' hearing ability. Audiology 22:417–420.

Jones MR, Boltz M (1989) Dynamic attending and responses to time. Psychol Rev 96:459–491.

Jusczyk PW, Derrah C (1987) Representation of speech sounds by young infants. Dev Psychol 23:648–654.

Jusczyk PW, Krumhansl CL (1993) Pitch and rhythmic patterns affecting infants' sensitivity to musical phrase structure. J Exp Psychol Hum Percept Perform 19:627–640.

Jusczyk PW, Thompson E (1978) Perception of a phonetic contrast in multisyllabic utterances by 2-month-old infants. Percept Psychophys 23:105–109.

Jusczyk PW, Rosner BS, Cutting JE, Foard CF, Smith LB (1977) Categorical perception of nonspeech sounds by 2-month-old infants. Percept Psychophys 21:50–54.

Jusczyk PW, Copan H, Thompson E (1978) Perception by 2-month-old infants of glide contrasts in multisyllabic utterances. Percept Psychophys 24:515–520.

Jusczyk PW, Pisoni DB, Walley A, Murray J (1980) Discrimination of relative time of two-component tones by infants. J Acoust Soc Am 67:262–270.

Jusczyk PW, Pisoni DB, Reed MA, Fernald A, Myers M (1983) Infants' discrimination of the duration of a rapid spectrum change in nonspeech signals. Science 222:175–176.

Jusczyk PW, Shea SL, Aslin RN (1984) Linguistic experience and infant speech perception: a re-examination of Eilers, Gavin and Oller (1982). J Child Lang 11:453–466.

Jusczyk PW, Rosner BS, Reed MA, Kennedy LJ (1989) Could temporal order differences underlie 2-month-olds' discrimination of English voicing contrasts? J Acoust Soc Am 85:1741–1749.

Jusczyk PW, Hirsh-Pasek K, Nelson DG, Kennedy LJ, Woodward A, Piwoz J (1992) Perception of acoustic correlates of major phrasal units by young infants. Cognit Psychol 24:252–293.

Kaga M (1992) Development of sound localization. Acta Paediatr Jpn 34:134–138.

Karzon RG (1985) Discrimination of polysyllabic sequences by one-to four-month-old infants. J Exp Child Psychol 39:326–342.

Karzon RG, Nicholas JG (1989) Syllabic pitch perception in 2- to 3-month-old infants. Percept Psychophys 45:10–14.

Kavanagh JF (ed) (1986) Otitis Media and Child Development. Parkton, MD: York Press.

Kawell ME, Kopun JG, Stelmachowicz PG (1988) Loudness discomfort levels in children. Ear Hear 9:133–136.

Keefe DH, Bulen JC, Arehart KH, Burns EM (1993) Ear-canal impedance and reflection coefficient in human infants and adults. J Acoust Soc Am 94:2617–2638.

Keefe DH, Burns EM, Bulen JC, Campbell SL (1994) Pressure transfer function from the diffuse field to the human infant ear canal. J Acoust Soc Am 95:355–371.

Kelly JB (1986) The development of sound localization and auditory processing in mammals. In: Aslin RN (ed) Advances in Neural and Behavioral Development. Norwood, NJ: Ablex Publishing, pp. 205–234.

Kelly JB, Potash M (1986) Directional responses to sounds in young gerbils (*Meriones unguiculatus*). J Comp Psychol 100:37–45.

Kelly JB, Judge PW, Fraser IH (1987) Development of the auditory orientation response in the albino rat (*Rattus norvegicus*). J Comp Psychol 101:60–66.

Kemler Nelson DG, Hirsch-Pasek K, Jusczyk PW, Wright Cassidy K (1989) How the prosodic cues in motherese might assist language learning. J Child Lang 16:55–68.

Kerr LM, Ostapoff EM, Rubel EW (1979) Influence of acoustic experience on the ontogeny of frequency generalization gradients in the chicken. J Exp Psychol Anim Behav Processes 5:97–115.

Kettner RE, Feng J-Z, Brugge JF (1985) Postnatal development of the phase-locked response to low frequency tones of the auditory nerve fibers in the cat. J Neurosci 5:275–283.

Kinney DK, Kagan J (1976) Infant attention to auditory discrepancy. Child Dev 47:155–164.

Knudsen EI (1988) Experience shapes sound localization and auditory unit properties during development in the barn owl. In: Edelman GM, Gall WE, Cowan WM (eds) Auditory Function: Neurobiological Bases of Hearing. New York: John Wiley and Sons, pp. 137–149.

Knudsen EI, Knudsen PF, Esterly SD (1982) Early auditory experience modifies sound localization in barn owls. Nature 295:238–240.

Knudsen EI, Esterly SD, Knudsen PF (1984) Monaural occlusion alters sound localization during a sensitive period in the barn owl. J Neurosci 4:1001–1011.

Krumhansl CL (1990) Cognitive Foundations of Musical Pitch. New York: Oxford University Press.

Krumhansl CL, Keil FC (1982) Acquisition of the hierarchy of tonal functions in music. Mem Cognit 10:243–251.

Kuhl PK (1979) Speech perception in early infancy: perceptual constancy for spectrally dissimilar vowel categories. J Acoust Soc Am 66:1668–1679.

Kuhl PK (1983) Perception of auditory equivalence classes for speech in early infancy. Infant Behav Dev 6:263–285.

Kuhl PK (1990) Auditory perception and the ontogeny and phylogeny of human speech. Semin Speech Lang 11:77–91.

Kuhl PK (1991) Human adults and infants exhibit a "perceptual magnet effect" for speech sounds, monkeys do not. Percept Psychophys 50:93–107.

Kuhl PK (1993) Innate predispositions and the effects of experience in speech perception: the native language magnet theory. In: de Boysson-Bardies B, de Schonen S, Jusczyk P, McNeilage P, Morton J (eds) Developmental Neurocognition: Speech and Face Processing in the First Year of Life. Boston, MA: Kluwer Academic Publishers, pp. 259–274.

Kuhl PK, Miller JL (1982) Discrimination of auditory target dimensions in the presence or absence of variation in a second dimension by infants. Percept Psychophys 31:279–292.

Kuhl PK, Williams KA, Lacerda F, Stevens KN, Lindblom B (1992) Linguistic experience alters phonetic perception in infants by 6 months of age. Science 255:606–608.

Lane DM, Pearson DA (1982) The development of selective attention. Merrill-Palmer Q 28:317–337.

Lasky RE (1984) A developmental study on the effect of stimulus rate on the auditory evoked brain-stem response. Electroencephalogr Clin Neurophysiol 59:411–419.

Lasky RE, Syrdal-Lasky A, Klein RE (1975) VOT discrimination by four to six and a half month old infants from Spanish environments. J Exp Child Psychol 20:215–225.

Leavitt LA, Brown JW, Morse PA, Graham FK (1976) Cardiac orienting and auditory discrimination in 6-week old infants. Dev Psychol 12:514–523.

Lerner RM, Perkins DF, Jacobson LP (1993) Timing process, and the diversity of developmental trajectories in human life: a developmental contextual perspective. In: Turkewitz G, Devenny DA (eds) Developmental Time and Timing. Hillsdale, NJ: Lawrence Erlbaum Associates, pp. 41–60.

Levi EC, Folsom RC, Dobie RA (1993) Amplitude-modulation following response (AMFR): effects of modulation rate, carrier frequency, age, and state. Hear Res 68:42–52.

Levitt A, Jusczyk PW, Murray J, Carden G (1988) Context effects in two-month-old infants' perception of labiodental/interdental fricative contrasts. J Exp Psychol Hum Percept Perform 14:361–368.

Lippe WR, Rubel EW (1985) Ontogeny of tonotopic organization of brain stem auditory nuclei in the chicken: implications for the development of the place principle. J Comp Neurol 237:273–289.

Litovsky RY, Clifton RK (1992) Use of sound-pressure level in auditory distance discrimination by 6-month-old infants and adults. J Acoust Soc Am 92:794–802.

Litovsky RY, Macmillan NA (1994) Sound localization precision under conditions of the precedence effect: effects of azimuth and standard stimuli. J Acoust Soc Am 96:752–758.

Lynch MP, Eilers RE, Oller DK, Urbano RC (1990) Innateness, experience, and music perception. Psychol Sci 1:272–276.

Maccoby EE, Konrad KW (1966) Age trends in selective listening. J Exp Child Psychol 3:113–122.

MacPherson BJ, Elfenbein JL, Schum RL, Bentler RA (1991) Thresholds of discomfort in young children. Ear Hear 12:184–190.

Malcuit G, Pomerleau A, Lamarre G (1988) Habituation, visual fixation and cognitive activity in infants: a critical analysis and attempt at a new formulation.

Eur Bull Cognit Psychol 8:415–440.

Manley GA, Kaiser A, Brix J, Gleich O (1991) Activity patterns of primary auditory-nerve fibres in chickens: development of fundamental properties. Hear Res 57:1–15.

Marean GC, Werner LA, Kuhl PK (1992) Vowel categorization by very young infants. Dev Psychol 28:396–405.

Maxon AB, Hochberg I (1982) Development of psychoacoustic behavior: sensitivity and discrimination. Ear Hear 3:301–308.

McCall RB (1979) Individual differences in the pattern of habituation at 5 and 10 months of age. Dev Psychol 15:559–569.

McCall RB (1988) Habituation, response to new stimuli, and information processing in human infants. Eur Bull Cognit Psychol 8:481–488.

McCall RB, McGhee PE (1977) The discrepancy hypothesis of attention and affect in infants. In: Uzgiris IC, Weizmann F (eds) The Structuring of Experience. New York: Plenum Press, pp. 179–210.

McFadden D (1987) Comodulation detection differences using noise-band signals. J Acoust Soc Am 81:1519–1527.

Mehler J, Bertoncini J, Barriere M, Jassik-Gerschenfeld D (1978) Infant recognition of mother's voice. Perception 7:491–497.

Menary S, Trehub SE, McNutt J (1982) Speech discrimination in preschool children: a comparison of two tasks. J Speech Hear Res 25:202–207.

Mendel ML (1968) Infant responses to recorded sounds. J Speech Hear Res 11:811–816.

Miller CL (1983) Developmental changes in male/female voice classification by infants. Infant Behav Dev 6:313–330.

Miller CL, Younger BA, Morse PA (1982) The categorization of male and female voices in infancy. Infant Behav Dev 5:143–159.

Miller D (1980) Maternal vocal control of behavioral inhibition in mallard ducklings (*Ana platyrhychos*). J Comp Physiol Psychol 94:606–623.

Miller DB, Gottlieb G (1981) Effects of domestication on production and perception of mallard maternal calls: developmental lag in behavioral arousal. J Comp Physiol Psychol 95:205–219.

Miller JL, Connine CM, Schermer TM, Kluender KR (1983) A possible auditory basis for internal structure of phonetic categories. J Acoust Soc Am 73:2124–2133.

Mills M, Melhuish E (1974) Recognition of mother's voice in early infancy. Science 252:123–124.

Moffitt AR (1971) Consonant cue perception by twenty- to twenty-four-week-old infants. Child Dev 42:717–731.

Moffitt AR (1973) Intensity discrimination and cardiac reaction in young infants. Dev Psychol 8:357–359.

Moon C, Fifer WP (1990) Syllables as signals for 2-day-old infants. Infant Behav Dev 13:377–390.

Moon C, Bever TG, Fifer WP (1992) Canonical and non-canonical syllable discrimination by two-day-old infants. J Child Lang 19:1–17.

Moore BC, Sek A (1992) Detection of combined frequency and amplitude modulation. J Acoust Soc Am 92:3119–3131.

Moore BCJ (1974) Relation between the critical bandwidth and the frequency difference limen. J Acoust Soc Am 55:359.

Moore BCJ (1989) Introduction to the Psychology of Hearing. New York:

Academic Press.

Moore DR, Hutchings ME, Meyer SE (1991) Binaural masking level differences in children with a history of otitis media. Audiology 30:91–101.

Moore JM, Thompson G, Thompson M (1975) Auditory localization of infants as a function of reinforcement conditions. J Speech Hear Disord 40:29–34.

Moroff D (1985) Do infants perceive similarity within feature classes? A critique of Hillenbrand. J Speech Hear Res 28:316–318.

Morrongiello BA (1986) Infants' perception of multiple-group auditory patterns. Infant Behav Dev 9:307–319.

Morrongiello BA (1988) Infants' localization of sounds in the horizontal plane: estimates of minimum audible angle. Dev Psychol 24:8–13.

Morrongiello BA, Clifton RK (1984) Effects of sound frequency on behavioral and cardiac orienting in newborn and five-month-old infants. J Exp Child Psychol 38:429–446.

Morrongiello BA, Robson RC (1984) Trading relations in the perception of speech by 5-year-old children. J Exp Child Psychol 37:231–250.

Morrongiello BA, Rocca PT (1987a) Infants' localization of sounds in the horizontal plane: effects of auditory and visual cues. Child Dev 58:918–927.

Morrongiello BA, Rocca PT (1987b) Infants' localization of sounds in the median sagittal plane: effects of signal frequency. J Acoust Soc Am 82:918–927.

Morrongiello BA, Rocca PT (1987c) Infants' localization of sounds in the median vertical plane: estimates of minimal audible angle. J Exp Child Psychol 43:181–193.

Morrongiello BA, Rocca PT (1990) Infants' localization of sounds within hemifields: estimates of minimum audible angle. Child Dev 61:1258–1270.

Morrongiello BA, Trehub SE (1987) Age related changes in auditory temporal perception. J Exp Child Psychol 44:413–426.

Morrongiello BA, Trehub SE, Thorpe LA, Capodilupo S (1985) Children's perception of melodies: the role of contour, frequency, and rate of presentation. J Exp Child Psychol 40:279–292.

Morrongiello BA, Fenwick K, Chance G (1990) Sound localization acuity in very young infants: an observer-based testing procedure. Dev Psychol 26:75–84.

Morrongiello BA, Hewitt KL, Gotowiec A (1991) Infants' discrimination of relative distance in the auditory modality: approaching versus receding sound sources. Infant Behav Dev 14:187–208.

Morse PA (1972) The discrimination of speech and nonspeech stimuli in early infancy. J Exp Child Psychol 14:447–492.

Movshon JA, Van Sluyters RC (1981) Visual neural development. In: Rosenzweig MR, Porter LW (eds) Annual Review of Psychology. Palo Alto, CA: Annual Reviews, pp. 477–522.

Muir D, Field T (1979) Newborn infants orient to sounds. Child Dev 50:431–436.

Muir D, Clifton RK, Clarkson MG (1989) The development of human auditory localization response: a U-shaped function. Can J Psychol 43:199–216.

Murphy WD, Shea SL, Aslin RN (1989) Identification of vowels in "vowel-less" syllables by 3-year-olds. Percept Psychophys 46:375–383.

Neff DL, Callaghan BP (1988) Effective properties of multicomponent simultaneous maskers under conditions of uncertainty. J Acoust Soc Am 83:1833–1838.

Nittrouer S, Studdert-Kennedy M (1987) The role of coarticulatory effects in the perception of fricatives by children and adults. J Speech Hear Res 30:319–329.

Nozza RJ (1987) The binaural masking level difference in infants and adults:

developmental change in binaural hearing. Infant Behav Dev 10:105–110.

Nozza RJ, Wilson WR (1984) Masked and unmasked pure tone thresholds of infants and adults: development of auditory frequency selectivity and sensitivity. J Speech Hear Res 27:613–622.

Nozza RJ, Wagner EF, Crandell MA (1988) Binaural release from masking for a speech sound in infants, preschoolers, and adults. J Speech Hear Res 31:212–218.

Nozza RJ, Miller SL, Rossman RNF, Bond LC (1991) Reliability and validity of infant speech-sound discrimination-in-noise thresholds. J Speech Hear Res 34:643–650.

O'Connor MJ (1980) A comparison of preterm and full-term infants on auditory discrimination at four months and on Bayley scales of infant development at eighteen months. Child Dev 51:81–88.

O'Connor MJ, Cohen S, Parmelee AH (1984) Infant auditory discrimination in preterm and full-term infants as a predictor of 5-year intelligence. Dev Psychol 20:159–165.

Okabe KS, Tanaka S, Hamada H, Miura T, Funai H (1988) Acoustic impedance measured on normal ears of children. J Acoust Soc Jpn 9:287–294.

Oller DK, Eilers RE (1983) Speech identification in Spanish- and English-learning 2-year-olds. J Speech Hear Res 26:50–53.

Olmstead CE, Villablanca JR (1980) Development of behavioral audition in the kitten. Physiol Behav 24:705–712.

Olsho LW (1984) Infant frequency discrimination. Infant Behav Dev 7:27–35.

Olsho LW (1985) Infant auditory perception: tonal masking. Infant Behav Dev 7:27–35.

Olsho LW, Schoon C, Sakai R, Turpin R, Sperduto V (1982a) Auditory frequency discrimination in infancy. Dev Psychol 18:721–726.

Olsho LW, Schoon C, Sakai R, Turpin R, Sperduto V (1982b) Preliminary data on frequency discrimination in infancy. J Acoust Soc Am 71:509–511.

Olsho LW, Koch EG, Halpin CF (1987) Level and age effects in infant frequency discrimination. J Acoust Soc Am 82:454–464.

Olsho LW, Koch EG, Halpin CF, Carter EA (1987) An observer-based psychoacoustic procedure for use with young infants. Dev Psychol 23:627–640.

Olsho LW, Koch EG, Carter EA (1988) Nonsensory factors in infant frequency discrimination. Infant Behav Dev 11:205–222.

Olsho LW, Koch EG, Carter EA, Halpin CF, Spetner NB (1988) Pure-tone sensitivity of human infants. J Acoust Soc Am 84:1316–1324.

Orchik DJ, Rintelman WF (1978) Comparison of pure-tone, warble-tone and narrow-band noise thresholds of young normal-hearing children. J Am Audiol Soc 3:214–220.

Panneton RK, DeCasper AJ (1984) Newborns prefer an intrauterine heartbeat sound to a male voice. Infant Behav Dev 7:281.

Patterson RD (1974) Auditory filter shape. J Acoust Soc Am 55:802–809.

Patterson RD, Nimmo-Smith I, Weber DL, Milroy R (1982) The deterioration of hearing with age: frequency selectivity, the critical ratio, the audiogram, and speech threshold. J Acoust Soc Am 72:1788–1803.

Pearson DA, Lane DM (1991) Auditory attention switching: a developmental study. J Exp Child Psychol 51:320–334.

Penner MJ (1978) A power law transformation resulting in a class of short-term integrators that produce time-intensity trades for noise bursts. J Acoust Soc Am 63:195–201.

Perris EE, Clifton RK (1988) Reaching in the dark toward sound as a measure of auditory localization in infants. Infant Behav Dev 11:473–491.

Peterzell DH, Werner, JS, Kaplan, PS (1995) Individual differences in contrast sensitivity functions: longitudinal study of 4-, 6- and 8-month-old human infants. Vision Res 35:961–979.

Philbin MK, Balweg DD, Gray L (1994) The effect of an intensive care unit sound environment on the development of habituation in healthy avian neonates. Dev Psychobiol 27:11–21.

Pick AD, Palmer CF, Hennessy BL, Unze MG, Jones RK, Richardson RM (1993) Children's perception of certain musical properties: scale and contour. J Exp Child Psychol 45:28–51.

Pillsbury HC, Grose JH, Hall JW III (1991) Otitis media with effusion in children. Arch Otolaryngol Head Neck Surg 117:718–723.

Pilz PKD, Schnitzler H-U, Menne D (1987) Acoustic startle threshold of the albino rat (*Rattus norvegicus*). J Comp Psychol 101:67–72.

Potash M, Kelly J (1980) Development of directional responses to sounds in the rat (*Rattus norvegicus*). J Comp Physiol Psychol 94:864–877.

Quinn PC, Eimas PD (1986) On categorization in early infancy. Merrill-Palmer Q 32:331–363.

Raab DH, Goldberg IA (1975) Auditory intensity discrimination with bursts of reproducible noise. J Acoust Soc Am 57:437–447.

Rebillard G, Rubel EW (1981) Electrophysiological study of the maturation of auditory responses from the inner ear of the chick. Brain Res 229:15–23.

Relkin EM, Saunders JC (1980) Displacement of the malleus in neonatal golden hamsters. Acta Otolaryngol 90:6–15.

Rewey HH (1973) Developmental change in infant heart rate response during sleeping and waking states. Dev Psychol 8:35–41.

Ritsma RJ (1967) Frequencies dominant in the perception of the pitch of complex sounds. J Acoust Soc Am 42:191–198.

Robertson D, Irvine DR (1989) Plasticity of frequency organization in auditory cortex of guinea pigs with partial unilateral deafness. J Comp Neurol 282:456–471.

Rosen S, Fourcin AJ (1986) Frequency selectivity and the perception of speech. In: Moore BCJ (ed) Frequency Selectivity in Hearing. New York: Academic Press, pp. 373–487.

Roush J, Tait CA (1984) Binaural fusion, masking level differences and auditory brain stem responses in children with language-learning disabilities. Ear Hear 5:37–41.

Rovee-Collier C (1987) Learning and memory in infancy. In: Osofsky J (ed) Handbook of Infant Development. New York: John Wiley and Sons, pp. 98–148.

Rubel EW (1978) Ontogeny of structure and function in the vertebrate auditory system. In: Jacobson M (ed) Handbook of Sensory Physiology. Vol. 9. Development of Sensory Systems. New York: Springer-Verlag, pp. 135–237.

Rübsamen R, Neuweiler G, Marimuthu G (1989) Ontogenesis of tonotopy in inferior colliculus of a hipposiderid bat reveals postnatal shift in frequency-place code. J Comp Physiol A Sens Neural Behav Physiol 165:755–769.

Ryals BM, Rubel EW (1985) Ontogenetic changes in the position of hair cell loss after acoustic overstimulation in avian basilar papilla. Hear Res 19:135–142.

Sanes DH, Merickel M, Rubel EW (1989) Evidence for an alteration of the tonotopic map in the gerbil cochlea during development. J Comp Neurol

279:436–444.

Saunders JC, Relkin EM, Rosowski JJ, Bahl C (1986) Changes in middle-ear admittance during postnatal auditory development in chicks. Hear Res 24:227–235.

Saunders SS, Salvi RJ (1993) Psychoacoustics of normal adult chickens: thresholds and temporal integration. J Acoust Soc Am 94:83–90.

Schlauch RS (1992) A cognitive influence on the loudness of tones that change continuously in level. J Acoust Soc Am 92:758–765.

Schlauch RS, Hafter ER (1991) Listening bandwidths and frequency uncertainty in pure-tone signal detection. J Acoust Soc Am 90:1332–1339.

Schlauch RS, Wier CC (1987) A method for relating loudness-matching and intensity-discrimination data. J Speech Hear Res 30:13–20.

Schneider BA, Trehub SE, Bull D (1980) High-frequency sensitivity in infants. Science 207:1003–1004.

Schneider BA, Trehub SE, Morrongiello BA, Thorpe LA (1986) Auditory sensitivity in preschool children. J Acoust Soc Am 79:447–452.

Schneider BA, Bull D, Trehub SE (1988) Binaural unmasking in infants. J Acoust Soc Am 83:1124–1132.

Schneider BA, Trehub SE, Morrongiello BA, Thorpe LA (1989) Developmental changes in masked thresholds. J Acoust Soc Am 86:1733–1742.

Schneider BA, Morrongiello BA, Trehub SE (1990) The size of the critical band in infants, children, and adults. J Exp Psychol Hum Percept Perform 16:642–652.

Schneider I, Gray L (1991) Rapid development of a sensory attribute in young chickens. Hear Res 52:281–287.

Schouten JF, Ritsma RJ, Cardozo BL (1962) Pitch of the residue. J Acoust Soc Am 34:1418–1424.

Segall ME (1972) Cardiac responsivity to auditory stimulation in premature infants. Nurs Res 21:15–19.

Sheets LP, Dean KF, Reiter LW (1988) Ontogeny of the acoustic startle response and sensitization to background noise in the rat. Behav Neurosci 102:706–713.

Shimojo S (1993) Development of interocular vision in infants. In: Simons K (ed) Early Visual Development, Normal and Abnormal. New York: Oxford University Press, pp. 201–223.

Shvachkin NK (1973) The development of phonemic speech perception in early childhood. In: Ferguson CA, Slobin DI (eds) Studies of Child Language Development. New York: Holt, Rinehart, and Winston, pp. 91–127.

Simon C, Fourcin AJ (1978) Cross-language study of speech-pattern learning. J Acoust Soc Am 63:925–935.

Sinnott JM, Aslin RN (1985) Frequency and intensity discrimination in human infants and adults. J Acoust Soc Am 78:1986–1992.

Sinnott JM, Pisoni DB, Aslin RM (1983) A comparison of pure tone auditory thresholds in human infants and adults. Infant Behav Dev 6:3–17.

Slater A, Morison V (1985) Selective adaptation cannot account for early infant habituation: a response to Dannemiller and Banks. Merrill-Palmer Q 31:99–103.

Spetner NB, Olsho LW (1990) Auditory frequency resolution in human infancy. Child Dev 61:632–652.

Spring DR, Dale PS (1977) Discrimination of linguistic stress in early infancy. J Speech Hear Res 20:224–232.

Srulovicz R, Goldstein JL (1983) A central spectrum model: a synthesis of auditory-nerve timing and place cues in monaural communication of frequency separation. J Acoust Soc Am 73:1266–1276.

Standley JM, Madsen CK (1990) Comparison of infant preferences and responses to auditory stimuli: music, mother, and other female voice. J Music Ther 27:54–97.

Stebbins WC (1993) The perceptual impasse in animal psychophysics. Abstr Assoc Res Otolaryngol 17:146.

Steinschneider A, Lipton EL, Richmond JB (1966) Auditory sensitivity in the infant: effect of intensity on cardiac and motor responsivity. Child Dev 37:233–252.

Stevens SS (1956) The direct estimation of sensory magnitudes – loudness. Am J Psychol 69:1–25.

Stevens SS (1975) Psychophysics: Introduction to Its Perceptual, Neural and Social Prospects. New York: John Wiley and Sons.

Stevens SS, Newman EB (1936) The localization of actual sources of sound. Am J Psychol 48:297–306.

Strange W, Broen PA (1980) Perception and production of approximant consonants by 3-year-olds: a first study. In: Yeni-Komshian GD, Kavanagh JF, Ferguson CA (eds) Child Phonology. Vol. 2. Perception. New York: Academic Press, pp. 117–154.

Stratton PM, Connolly K (1973) Discrimination by newborns of the intensity, frequency and temporal characteristics of auditory stimuli. Br J Psychol 64:219–232.

Streeter LA (1976) Language perception of 2-month-old infants shows effects of both innate mechanisms and experience. Nature 259:39–41.

Stuart A, Durieux-Smith A, Stenstrom R (1991) Probe tube microphone measures of loudness discomfort levels in children. Ear Hear 12:140–143.

Swoboda PJ, Morse PA, Leavitt LA (1976) Continuous vowel discrimination in normal and high risk infants. Child Dev 47:459–465.

Swoboda PJ, Kass J, Morse PA, Leavitt LA (1978) Memory factors in vowel discrimination of normal and at-risk infants. Child Dev 49:332–339.

Tarquinio N, Zelazo PR, Weiss MJ (1990) Recovery of neonatal head turning to decreased sound pressure level. Dev Psychol 26:752–758.

Tarquinio N, Zelazo PR, Gryspeerdt DH, Allen KM (1991) Generalization of neonatal habituation. Infant Behav Dev 14:69–81.

Terhardt E (1974) Pitch, consonance, and harmony. J Acoust Soc Am 55:1061–1069.

Thorpe LA, Schneider BA (1987) Temporal integration in infant audition. Abstr Soc Res Child Dev 6:273.

Thorpe LA, Trehub SE, Morrongiello BA, Bull D (1988) Perceptual grouping by infants and preschool children. Dev Psychol 24:484–491.

Trainor LJ, Trehub SE (1992) The development of referential meaning in music. Music Percept 9:455–470.

Trainor LJ, Trehub SE (1993) Musical context effects in infants and adults: Key distance. J Exp Psychol Hum Percept Perform 19:615–626.

Trehub SE (1973) Infants' sensitivity to vowel and tonal contrasts. Dev Psychol 9:91–96.

Trehub SE (1989) Infants' perception of musical sequences: implications for language acquisition. Speech Lang Pathol Audiol 13:3–11.

Trehub SE (1990) The perception of musical patterns by human infants: the provision of similar patterns by their parents. In: Berkley M, Stebbins WC (eds) Comparative Perception. Vol. 1. Basic Mechanisms. New York: John Wiley and

Sons, pp. 429–460.

Trehub SE, Rabinovitch MS (1972) Auditory-linguistic sensitivity in early infancy. Dev Psychol 6:74–77.

Trehub SE, Thorpe LA (1989) Infants' perception of rhythm. Categorization of auditory sequences by temporal structure. Can J Psychol 43:217–229.

Trehub SE, Unyk AM (1992) Music prototypes in developmental perspective. Psychomusicology 10:31–45.

Trehub SE, Schneider BA, Endman M (1980) Developmental changes in infants' sensitivity to octave-band noises. J Exp Child Psychol 29:282–293.

Trehub SE, Bull D, Thorpe LA (1984) Infants' perception of melodies: the role of melodic contour. Child Dev 55:821–830.

Trehub SE, Thorpe LA, Morrongiello BA (1985) Infants' perception of melodies: changes in a single tone. Infant Behav Dev 8:213–223.

Trehub SE, Cohen AJ, Thorpe LA, Morrongiello BA (1986) Development of the perception of musical relations: semitone and diatonic structure. J Exp Psychol Hum Percept Perform 12:295–301.

Trehub SE, Thorpe LA, Morrongiello BA (1987) Organizational processes in infants' perception of auditory patterns. Child Dev 58:741–749.

Trehub SE, Schneider BA, Morrongiello BA, Thorpe LA (1988) Auditory sensitivity in school-age children. J Exp Child Psychol 46:273–285.

Trehub SE, Endman MW, Thorpe LA (1990) Infants' perception of timbre: classification of complex tones by spectral structure. J Exp Child Psychol 49:300–313.

Trehub SE, Thorpe LA, Trainor LJ (1990) Infants' perception of *good* and *bad* melodies. Psychomusicology 9:5–19.

Trehub SE, Schneider BA, Henderson J (1995) Gap detection in infants, children, and adults. J Acoust Soc Am 98:2532–2541.

Turkewitz G, Birch HG, Cooper KK (1972) Responsiveness to simple and complex auditory stimuli in the human newborn. Dev Psychobiol 5:7–19.

Velleman SL (1988) The role of linguistic perception in later phonological development. Appl Psychol 9:221–236.

Veloso K, Hall JW III, Grose JH (1990) Frequency selectivity and comodulation masking release in adults and in 6-year-old children. J Speech Hear Res 33:96–102.

Viemeister NF (1988) Psychophysical aspects of auditory intensity coding. In: Edelman GM, Gall WE, Cowan WM (eds) Auditory Function: Neurobiological Bases of Hearing. New York: John Wiley and Sons, pp. 213–242.

Viemeister NF, Schlauch RS (1992) Issues in infant psychoacoustics. In: Werner LA, Rubel EW (eds) Developmental Psychoacoustics. Washington, DC: American Psychological Association, pp. 191–210.

Viemeister NF, Wakefield GH (1991) Temporal integration and multiple looks. J Acoust Soc Am 90:858–865.

Walley AC (1993) More developmental research is needed. J Phonet 21:171–176.

Walley AC, Carrell TD (1983) Onset spectra and formant transitions in the adult's and child's perception of place of articulation in stop consonants. J Acoust Soc Am 73:1011–1022.

Walley AC, Pisoni DB, Aslin RN (1984) Infant discrimination of two- and five-formant voiced stop consonants differing in place of articulation. J Acoust Soc Am 75:581–589.

Walsh EJ, McGee J (1986) The development of function in the auditory periphery.

In: Altschuler RA, Hoffman DW, Bobbin RP (eds) Neurobiology of Hearing: The Cochlea. New York: Raven Press, pp. 247–269.

Walton GE, Bower TG (1993) Newborns form "prototypes" in less than 1 minute. Psychol Sci 4:203–205.

Wang X, Sachs MB (1992) Coding of envelope modulation in the auditory nerve and anteroventral cochlear nucleus. In: Carlyon RP, Darwin CJ, Russell IJ (eds) Processing of Complex Sounds by the Auditory System. Oxford, UK: Clarendon Press, pp. 32–47.

Watrous BS, McConnell F, Sitton AB, Fleet WF (1975) Auditory responses of infants. J Speech Hear Disord 40:357–366.

Watson CS, Gengel RW (1969) Signal duration and signal frequency in relation to auditory sensitivity. J Acoust Soc Am 46:989–997.

Weir C (1976) Auditory frequency sensitivity in the neonate: a signal detection analysis. J Exp Child Psychol 21:219–225.

Weir C (1979) Auditory frequency sensitivity of human newborns: some data with improved acoustic and behavioral controls. Percept Psychophys 26:287–294.

Werker JF, Tees RC (1983) Developmental changes across childhood in the perception of non-native speech sounds. Can J Psychol 37:278–286.

Werker JF, Tees RC (1984) Cross-language speech perception: evidence for perceptual reorganization during the first year of life. Infant Behav Dev 7:49–63.

Werker JK, Polka L (1993) Developmental changes in speech perception: new challenges and new directions. J Phonet 21:83–101.

Werker JF, Gilbert JH, Humphrey K, Tees RC (1981) Developmental aspects of cross-language speech perception. Child Dev 52:349–355.

Werner LA (1992) Interpreting developmental psychoacoustics. In: Werner LA, Rubel EW (eds) Developmental Psychoacoustics. Washington, DC: American Psychological Association, pp. 47–88.

Werner LA (1995) Observer-based approaches to human infant psychoacoustics. In: Klump GM, Dooling RJ, Fay RR, Stebbins WC (eds) Methods in Comparative Psychoacoustics. Boston, MA: Birkhauser Verlag, pp. 135–146.

Werner LA (1996) The development of forward masking in human infants. J Acoust Soc Am 99:2562.

Werner LA, Bargones JY (1991) Sources of auditory masking in infants: distraction effects. Percept Psychophys 50:405–412.

Werner LA, Bargones JY (1992) Psychoacoustic development of human infants. In: Rovee-Collier C, Lipsitt L (eds) Advances in Infancy Research. Norwood, NJ: Ablex Publishing, pp. 103–146.

Werner LA, Gillenwater JM (1990) Pure-tone sensitivity of 2- to 5-week-old infants. Infant Behavior and Development 13:355–375.

Werner LA, Mancl LR (1993) Pure-tone thresholds of 1-month-old human infants. J Acoust Soc Am 93:2367.

Werner LA, Marean GC (1991) Methods for estimating infant thresholds. J Acoust Soc Am 90:1867–1875.

Werner LA, Marean GC, Halpin CF, Spetner NB, Gillenwater JM (1992) Infant auditory temporal acuity: gap detection. Child Dev 63:260–272.

Werner LA, Folsom RC, Mancl LR (1993) The relationship between auditory brainstem response and behavioral thresholds in normal hearing infants and adults. Hear Res 68:131–141.

Werner LA, Folsom RC, Mancl LR (1994) The relationship between auditory brainstem response latency and behavioral thresholds in normal hearing infants and adults. Hear Res 77:88–98.

Wier CC, Jesteadt W, Green DM (1977) Frequency discrimination as a function of frequency and sensation level. J Acoust Soc Am 61:178–184.

Wightman F, Allen P (1992) Individual differences in auditory capability in preschool children. In: Werner LA, Rubel EW (eds) Developmental Psychoacoustics. Washington, DC: American Psychological Association, pp. 47–88.

Wightman F, Allen P, Dolan T, Kistler D, Jamieson D (1989) Temporal resolution in children. Child Dev 60:611–624.

Wilmington D, Gray L, Jahrsdorfer R (1994) Binaural processing after corrected congenital unilateral conductive hearing loss. Hear Res 74:99–114.

Wilson WR, Thompson G (1984) Behavioral audiometry. In: Jerger J (ed) Recent Advances in Hearing Disorders. San Diego, CA: College-Hill Press, pp. 1–44.

Withington-Wray DJ, Binns KE, Keating MJ (1990) The developmental emergence of a map of auditory space in the superior colliculus of the guinea pig. Brain Res Dev Brain Res 51:225–236.

Wolf CG (1973) The perception of stop consonants by children. J Exp Child Psychol 16:318–331.

Wormith SJ, Moffitt AR, Pankhurst DB (1975) Frequency discrimination by young infants. Child Dev 46:272–275.

Yost WA (1991) Auditory image perception and analysis: the basis for hearing. Hear Res 56:8–18.

Yost WA, Sheft S (1989) Across-critical-band processing of amplitude-modulated tones. J Acoust Soc Am 85:848–857.

Yost WA, Watson CS (eds) (1987) Auditory Processing of Complex Sounds. Hillsdale, NJ: Lawrence Erlbaum Associates.

Zelazo PR, Komer MJ (1971) Infant smiling to nonsocial stimuli and the recognition hypothesis. Child Dev 42:1327–1339.

Zelazo PR, Brody LR, Chaika H (1984) Neonatal habituation and dishabituation of head turning to rattle sounds. Infant Behav Dev 7:311–321.

Zelazo PR, Weiss MJ, Papageorgiou AN, Laplante DP (1989) Recovery and dishabituation of sound localization in normal-, moderate-, and high-risk newborns: discriminant validity. Infant Behav Dev 12:321–340.

Zimmermann E (1993) Behavioral measures of auditory thresholds in developing tree shrews (*Tupaia belageri*). J Acoust Soc Am 94:3071–3075.

Zlatin MA, Koenigsnecht RA (1975) Development of the voicing contrast: Perception of stop consonants. J Speech Hear Res 18:530–540.

3

Early Embryology of the Vertebrate Ear

BERND FRITZSCH, KATE F. BARALD, AND MARGARET I. LOMAX

1. Introduction

Organogenesis of the vertebrate inner ear has been described as "one of the most remarkable displays of precision microengineering in the vertebrate body" (Swanson, Howard, and Lewis 1990). The initial morphological event in ear development in all vertebrates is the formation of the embryonic otic placode, a thickening of the head ectoderm in the region of the developing hindbrain. Through interaction with and incorporation of tissue from several other embryonic sources, the placode develops into the otocyst or otic vesicle, a differentiated structure with sharply defined borders (Noden and Van De Water 1986; Couly, Coltey, and Le Douarin 1993). The epithelium of the otic placode/vesicle also gives rise to the primary neurons of the statoacoustic ganglion, later in development called the cochleovestibular ganglion, the octaval, or the otic ganglion (probaby the most appropriate terminology), which contributes to cranial nerve VIII and to the specialized sensory structures known as hair cells (Fig. 3.1).

In this presentation we will, (1) discuss the commonalities and divergences found among species in the earliest events in the development of the verte-brate inner ear; (2) outline how the inner ear develops as an organ from a placode and the roles of the rhombencephalon (hindbrain) and neural crest in this development, as well as the genes that are thought to be involved in shaping the development of these transitory embryonic structures; (3) discuss the role of known genes and morphogenetic events in this process and their identification in deafness mutants; (4) discuss how the ear primordium diversifies; and (5) present data on the development of the ear innervation.

This chapter is intended to provide an overview of concepts and ideas with citations limited to major breakthrough papers and/or reviews that highlight a large body of past literature. We are aware that this chapter will be lagging behind the current literature in the vastly expanding field of molecular biology of ear development by the time it is in print. Nevertheless, we hope that the reader will find this chapter helpful in conceptualizing

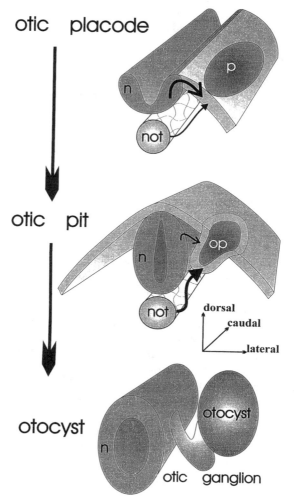

FIGURE 3.1. This drawing illustrates the presumed inductive interactions between the forming neural tube (n) and the mesoderm (represented by the notchord [not]) to induce an otic placode (p). Once induced and polarized through appropriate interactions with the neural tube and the mesoderm (*curved arrows*), the otic placode will invaginate into an otic pit (op). Subsequently, the otic pit will close and separate from the ectoderm (light gray) to become the otocyst. Experimental evidence indicates that there is variation in the relative importance of the neural and mesodermal inductive actions (*curved arrows*) probably over time (as shown here) and also between species. Several early genes that are important in the hindbrain or the otic placode have been characterized and to some extent tested for their alleged function. From both transplantation studies as well as from these gene-deletion experiments, it has become clear that the early interactions that lead to the formation of invagination and appropriate polarity formation of the otocyst are crucial for the developing ear.

both the broader issues of developmental mechanisms and the contribution molecular approaches have made to elucidating their basis.

1.1 Formation of the Vertebrate Ear: Commonalities of Gene Expression and Early Developmental Events in Placode Formation Are Shared by the Vertebrates

The commonalities of inner ear development in vertebrates are underscored by studies of gene expression in auditory anlagen, in the rhombomeres of the hindbrain, and in the cephalic neural crest that profoundly influence otocyst development. Progress has been made in recent years, and some regulatory genes have been identified (Holland et al. 1992; Marshall et al. 1992; Lumsden & Krumlauf 1996) that may play important roles in the formation of the auditory system.

A model of the cascades of gene expression that ultimately guide the transformation of the otic placode into an ear can be found in the developing limb (Tabin 1995). The earliest events in limb formation include a sequence of focal gene activations that lead to a polarization of the limb bud and a stable, topologically restricted expression of genes. It appears likely that genes and events comparable to those that are active in polarizing the limb bud are also involved in polarity determination in the inner ear. In the limb bud, these genes include wingless-int (*Wnt*), fibroblast growth factors (*Fgf*), homeobox (*Hox* genes), bone morphogenetic protein (*Bmp*), and Hedgehog. Genes of these families [with the exception of Hedgehog (*hh*), which has not yet been examined] have been shown to be expressed in the developing ear and have been implicated as well in patterning of the otic placode.

Genes postulated to be involved in ear development are largely those expressed in the rhombencephalon, the nonneuronal portion of the placode-derived otocyst, the neuronal precursors that arise from the placode, and the neuronal, glial and melanocyte precursors that come from the neural crest. Although there are still many open questions, it appears that examining gene expression in the hindbrain and cephalic neural crest and their subsequent effects on ear development will provide valuable new lines of research. The lessons learned from studies of epigenetic influences on gene expression and patterning in the limb will also provide new insights into the century-old problem of otic development and evolution.

1.2 Divergence in Patterning of the Vertebrate Ear: Functional and Evolutionary Considerations

Despite the conservation of gene expression and the similarities among vertebrates of the earliest events in inner ear development (formation of and invagination of the otic placode), the vertebrate ear shows a bewildering variation in form and formation of certain sensory epithelia (Retzius 1884;

Baird 1974; Henson 1974). Although the first vertebrate ear may have looked like the single toral structure found in the jawless hagfish, with two semicircular sensory epithelia, a single macula, and two distinct octaval ganglia and nerves (Loewenstein and Thornhill 1970; Fig. 3.2), the structure changed dramatically through the formation of the two distinct semicircular canals found in derived jawless vertebrates, the lampreys.

Jawed vertebrates subsequently evolved several distinct sensory epithelia (e.g., lagena, basilar papilla, amphibian papilla), one additional canal (horizontal canal), and three recesses (lagena, amphibian papilla, basilar papilla; Lewis, Leverenz, and Bialek, 1985). In addition, there is loss of some recesses (Fig. 3.3) and transformation of one epithelium (papilla neglecta/amphibiorum; Fritzsch and Wake 1988; Fritzsch 1992). These losses and gains involve several distinct structures that may be developmentally coupled. All of the vestibular organs of the ear are positioned within the skull to detect head orientation and movement. In addition, the ear is able to perceive sound either through specialized organs positioned near the sound-conducting perilymphatic pathways (De Burlet 1934; Werner 1960; Van Bergijk 1967; Fritzsch 1992) or, in aquatic vertebrates, through direct impact of sound on the otolithic organs (Schellart and Popper 1992).

Evolution has transformed the periotic tissue around the ear into important sound-conducting structures, including the vestibular and tympanic scalae of the mammalian cochlea (Fig. 3.2). In birds, this periotic

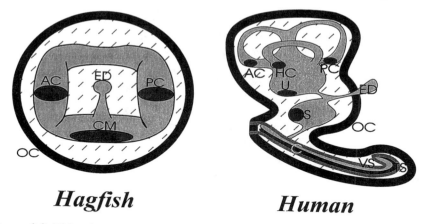

Hagfish   Human

FIGURE 3.2. This scheme compares the ear of a jawless vertebrate (hagfish) with a mammal (human) to emphasize the degree of morphological changes between these taxa. Please note that hagfish and other jawless vertebrates have only two canals, a single macula, and a uniform periotic mesenchyme (*hatch*). In contrast, humans and other tetrapods have three canals, at least two vestibular maculae, and the cochlea. Note also the formation of the periotic labyrinth around the cochlea (light gray). All sensory epithelia are dark gray. AC, anterior canal; C, cochlea; CM, common macula; ED, endolymphatic duct; HC, horizontal canal; OC, otic capsule; PC, posterior canal; S, saccule; TS, tympanic scala; U, utricle; VS, vestibular scala.

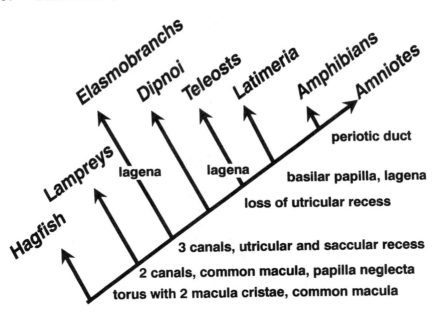

FIGURE 3.3. The most parsimonious arrangement of otic character evolution is shown. Please note that this may not necessarily reflect the evolutionary history of vertebrates. The ancestors of vertebrates had no ear and it is concluded that a single torus with three sensory epithelia is primitive for craniates. Note that the formation of a basilar papilla (cochlea) is a late event in vertebrate evolution and the formation of the periotic sound-conducting system evolved only in terrestrial vertebrates but independently also in some bony fish.

mesenchyme develops from three embryonic sources: neural crest, somitic mesoderm, and paraxial cephalic mesoderm (Couly, Coltey, and Le Douarin 1993) and forms the otic capsule and the connective, perilymphatic tissue between the otic capsule and the inner ear. The functionally essential evolutionary change that transformed the periotic tissue into perilymphatic, sound-conducting pathways (De Burlet 1934; Werner 1960) for sound-pressure perception (Van Bergijk 1967) happened independently several times (Fritzsch 1992; Schellart and Popper 1992). In addition, there are modifications of the otic capsule to form windows through which the sound pressure-induced particle motion can enter and leave the inner ear without affecting the vestibular organs.

These changed morphologies of the ear are the outcome of altered ontogenies (Noden 1991; Northcutt 1992) that modify a number of closely spaced tissues of four different embryonic origins. Only in their concerted modification do they provide the morphological substrate for sound-pressure reception. Therefore, to understand the evolution of this system, we need to understand how mutations of structural and regulatory genes have modified the tissue interactions in this most complex area of the head

(Couly, Coltey, and Le Douarin 1993) to induce and stabilize the development of the different structures seen in various vertebrates (Fig. 3.2). Unfortunately, this is still as unclear for the ear as it is for most other major parts of the vertebrate body, with the possible exception of the limb.

2. From Placodes to Cochlea: an Outline of Major Embryological Stages and Influences in the Development of the Vertebrate Ear

2.1 Induction of Placodes

Placodes arose in vertebrates (Northcutt and Gans 1983; Couly, Coltey, and Le Douarin 1993; Fritzsch and Northcutt 1993; Webb and Noden 1993) and are a major embryonic source of contributions to the sensory organs of the vertebrate head (Von Kupffer 1895). The nose derives from the olfactory placode, the lens forms from a placodal thickening, the ophthalmic placode forms part of the trigeminal ganglion, the epibranchial placodes form the gustatory ganglia (Farbman and Mbiene 1991; Barlow and Northcutt 1995; Stone et al. 1995), and the dorsolateral placodes form the ampullary electroreceptors, the mechanosensory neuromasts, and the inner ear (Northcutt 1992; Northcutt, Brändle, and Fritzsch 1995). The origin of the ear is tightly coupled to the evolution of these placodes. Although apparently novel, placodes may have evolved through transformation of primordial primary sensory cells (with an axon to the brain) into primordia, which form both sensory cells and ganglia that connect these axonless sensory cells to the brain (Von Kupffer 1895; Fritzsch 1993; 1996a).

The placode of the ear (the otic placode) was one of the first placodes to be recognized (Huschke 1831; Von Kupffer 1895). It is not unique in its capacity to form sensory cells or ganglion cells but shares with other placodes a number of characteristics, such as invagination, skeletogenic interactions with mesoderm (Webb and Noden 1992), and even migration under certain conditions (Gutknecht and Fritzsch 1990). What makes the otic placode different from other placodes and what induces the differentiation of the placode into the different ears recognized in vertebrates? This question will be considered in detail in Section 4.1 of this review, which deals with comparative development of the otic placode.

2.2 The Rhombomeres

The vertebrate hindbrain (rhombencephalon) is formed as part of the neural tube, which will eventually comprise the brain and spinal cord, and thus is part of the central nervous system. The hindbrain has a complex series of bulges called rhombomeres (reviewed in Keynes and Krumlauf 1994), which appear transiently after neurulation. The rhombomeres are separated by prominent ridges that delineate morphological compartments in which differential gene expression allows segment-specific differentiation of the developing hindbrain, the cranial ganglia, and the branchial arches (Keynes and Krumlauf 1994). In the mouse and the chick, there are eight rhombencephalic rhombomeres. Studies in the chick and other vertebrates (Gilland and Baker 1993) demonstrated that the rhombomeres form the basis for the patterning of nerves in the hindbrain. Several genes have been identified that have unique early rhombomeric expression patterns, suggesting that they may contribute to the formation or differentiation of the rhombomeres (Lumsden and Krumlauf 1996). Whereas rhombomeres and gene expression patterns specific to each rhombomere are rather constant building blocks of the hindbrain (Guthrie 1995), their cellular differentiation may vary to some extent among vertebrates for a given rhombomere (Fritzsch 1995, 1996b).

2.2.1 Gene Expression in the Rhombencephalon and Developing Otocyst: Homeobox Genes and Segmentation Genes

Some of the earliest known molecular events in ear development involve expression of genes required for axis specification, segmentation, and body plan patterning (Keynes and Krumlauf 1994). In invertebrate embryos, the genes determining these events have been elucidated through a combination of genetic and molecular biological approaches (Nüsslein-Vollhard 1994). These include a large number of genes that either establish or coincide with the establishment of segmentation patterns in the vertebrate head and nervous system, particularly in the rhombomeres (Keynes and Krumlauf 1994). The isolation of vertebrate homologues of these genes has shed light on their role in patterning both the hindbrain and forebrain regions of vertebrate embryos and the associated neuronal structures. Each major class of genes encodes transcription factors and thus represents a genetic cascade.

Combinations of functionally active *Hox* genes, or *Hox* codes, are thought to play a determinant role in the development of the vertebrate hindbrain, the branchial arches, the vertebrae, and the limbs. Both the pattern of expression of *Hox* genes and the phenotype of mice carrying targeted disruptions of these genes suggest that they are involved in patterning branchial regions of the head and nervous system (Keynes and Krumlauf 1994). *Hox* genes were originally isolated and studied in *Drosophila*, where they affect patterning of the embryo; they encode transcription factors containing a highly conserved 60 amino acid DNA-binding domain, the homeodomain (Keynes and Krumlauf 1994). Vertebrate *Hox* genes comprise four separate *Hox* gene clusters (*Hoxa–d*, formerly designated *Hox1–4*), each of which is located on a different chromosome. The temporal and spatial (anterior to posterior) expression of these genes is colinear with their order within the gene cluster and reflects a functional hierarchy among genes of the *Hox* complex. The arrangement of the vertebrate *Hox* complex apparently evolved in the chordate ancestors (Garcia-Fernàndez and Holland 1994) before ears evolved (Fritzsch 1993).

The regional expression of *Hox* genes has been shown to define the segmental pattern of the vertebrate hindbrain (Keynes and Krumlauf 1994). Expression of each *Hox* gene terminates at defined segmental borders, i.e., the rhombomeres of the hindbrain, and combinations of *Hox* genes, the *Hox* codes, determine segment identity. Thus loss of a given *Hox* gene function in a specific rhombomere leads to "posteriorization" of a given segment, whereas ectopic expression of a *Hox* gene leads to a more anterior segment identity. Those with a more anterior boundary of expression, such as *Hoxa1*, which is discussed in detail in Section 2.5.1, specifically affect the rhombencephalon, the neural crest and inner ear development.

A number of *Hox* genes, as well as other gene types, are expressed in specific patterns in the rhombomeres and appear to be essential for correct rhombomere identity. *Krox20*, which encodes a zinc-finger protein that regulates *Hox* gene expression, appears long before the hindbrain is segmented morphologically and is expressed in two alternating rhombomeres, rhombomeres 3 and 5, in the mouse, chick, and *Xenopus* hindbrain (Fig. 3.4). *Krox20* has been proposed as one of the genes essential for rhombomere formation (Keynes and Krumlauf 1994). *Hoxb3* (*Hox2.7*) and *Hoxb4* (*Hox2.9*) are expressed at the r4/5 and r6/7 boundaries, respectively, in all of these vertebrates. *Hoxb1* is normally expressed in rhombomere 4. Engrailed (*En*) genes are expressed in rhombomeres 4–7 in the chick (Gardner and Barald 1992) but may be expressed in single rhombomeres in specific planes of section (from dorsal to ventral). *Pax2* is expressed in the entire rhombencephalon in the mouse (Nornes et al. 1990) and in the chick (Lindberg et al. 1995). Both *Pax2* message and gene product are expressed uniformly in the placode and early otocyst of in the chick (Lindberg et al. 1995) at stage 12 (2 embryonic days) and in the 9.5 days post coitus (dpc) mouse embryo otocyst (Nornes et al. 1990). However, by stage 20 (3 embryonic days) in the chick, *Pax2* protein product is limited to the side of the otocyst closest to the rhombencephalon (Lindberg et al. 1995).

The *dlx3* gene is expressed in the preotic mesenchyme and in the developing otocyst (Ekker and colleagues 1992a,b; Akimenko et al. 1994) and appears to be crucial for the formation of the otic and other placodes (Fritz and Westerfield 1996). The gene for bone morphogenetic protein-4 (*Bmp4*) is expressed in two distinct regions of the developing otocyst in the frog (Hemmati-Brivanlou and Thomsen 1995). In the chick otocyst, *Bmp4* expression is restricted to the future ampullae of the semicircular canals, whereas *Bmp5* mRNA was found only transiently in the future ampula of the posterior semicircular canal (Wu and Oh 1996; Oh et al. 1996). Contact between the otic anlage and the rhombencephalon coincides with the period of invagination of the otic vesicle (Meier 1978; Hilfer, Esteves, and Sanzo 1989). The fibroblast growth factor (FGF-3), the int-2 gene product, is expressed in and secreted by the rhombencephalon (rhombomeres 4, 5, and,

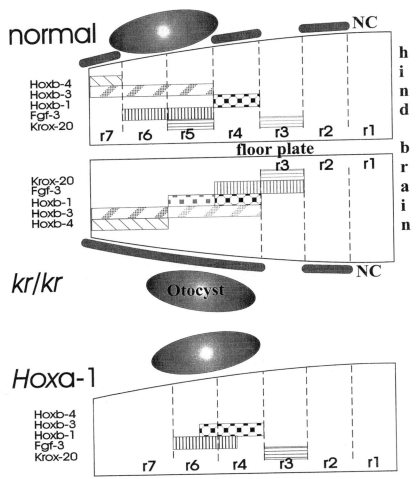

FIGURE 3.4. The expression of some homeobox genes with respect to hindbrain rhombomeres (r1–r7), the otocyst, and the cranial neural crest (NC; light gray) in mice of embryonic day 9–9.5 is shown. The gene expression is restricted to certain rhombomeres, and the fibroblast growth factor (FGF)-3 gene is expressed adjacent to the ear in rhombomeres 5 and 6. Note that in the kreisler mutant (*kr/kr;* lower half hindbrain) there is a shift of the otocyst (more distant from and more rostral to the hindbrain), more extensive neural crest formation (continuous between hindbrain and otocyst), loss of boundaries between rhombomeres 4 and 7, and a changed expression of homeobox genes, some of which have additional, less extensive expression domains (lighter gray). A nonfunctional mutant of a single homeobox gene, *Hoxa1*, leads also to an anterior shift of the otocyst, loss of a single rhombomere (r5), and overlapping expression domains of two genes normally restricted to a single rhombomere. The two mutants have abnormal ears, but the causality of the anterior shift of the otocyst, changes in neural crest, and changes in gene expression in these otic malformations is unclear. Combined according to Frohman et al. (1993) and Carpenter et al. (1993).

later, 6) in the vicinity of the developing otocyst (Tannahill et al. 1992; Mahmood et al. 1995). *GATA3* is expressed in subsets of neuronal cells and in otocysts during *Xenopus laevis* (unpublished observations), chick (Kornhauser et al. 1994), and mouse (George et al. 1994) development. *GH6*, a homeobox gene, is expressed in the dorsolateral otocyst at stage 23 in the chick (Stadler and Solursh 1994). These genes and what is known about their regulation in the auditory system are discussed in more detail in Sections 2.4 and 2.5.

2.2.2 Role of the Rhombencephalon in the Earliest Events in Ear Development

The close proximity of the developing otocyst to the rhombencephalon (Model, Jarret, and Bonazzoli 1981) is essential during the early stage of otic development in the mouse, chick, and amphibian (Li et al. 1978; reviewed in Van De Water 1983) and has been postulated as a source of molecular cues in these animals and in the zebra fish as well (Ekker and colleagues 1992a,b). The interaction between the otocyst and the rhombencephalon in chicks and amphibians may include contact- or gap junction-mediated interactions between the two structures, although gap junctions have not been detected in the mouse (Van De Water 1983). Interactions could also be mediated through extracellular matrix components and/or soluble factors released by both the rhombencephalon and the otocyst.

Epithelial (placodal)-mesenchyme interactions are also critical for morphogenesis of the labyrinth, cochlear sensory structures, vestibular sensory structures, and differentiation of sensory hair cells. The mesenchyme may also be a source of factors that modulate the genetic influence of the hindbrain on the otocyst or that act independently. The mesenchyme, in turn, is dependent on the presence of the otocyst for formation of an otic capsule (Van De Water 1983).

The role of the hindbrain in patterning the ear has largely been demonstrated through two kinds of studies: (1) mouse mutants that were identified because of ear defects (deafness and/or impaired vestibular system models) in which the hindbrain patterning is grossly affected (Deol 1964; Ruben et al 1991; Frohman et al. 1993; Cordes and Barsh, 1994) and (2) targeted disruptions of specific genes in mouse such as *Hoxa1* (*Hox1.6*) that produce strikingly similar defects in the inner and/or middle ears through primary effects on the hindbrain (Fig. 3.4). Alterations in hindbrain patterning in these mutants and knockouts are traced through effects on the elaborate rhombomere-specific expression of *Hox* genes described above. In some cases, ectopic application of agents such as retinoic acid (RA) has had similar effects on hindbrain organization and secondary effects on the inner ear (Papalopulu et al. 1991; Marshall et al. 1992; Morriss-Kay 1993).

In the most severely affected inner ears, for example, in homozygotes of the *kreisler* mutation *kr/kr* (Frohman et al. 1993), *Hoxa1* (*Hox 1.6*) knockouts (Lufkin et al. 1991; Chisaka, Musci, and Capecchi 1992), or embryos treated with high concentrations of RA (Papalopulu et al. 1991; Manns and Fritzsch 1992; Marshall et al. 1992; Morriss-Kay 1993), the inner ear is found only as an epithelial sac, the development of which is truncated at early stages. Even higher concentrations of retinoic acid at early stages may lead to complete block of placodal invagination (Neary and Fritzsch 1992; Fig. 3.5). These mutations and targeted disruptions of genes and gene expression appear not to act on the otocyst directly but through the otocyst's interactions with the hindbrain rhombomeres and probably also the neural crest. Likewise, some of the effects of retinoic acid may be through inductive effects on *Hox* genes.

2.3 Neural Crest

The neural crest is a transitory embryonic stem cell population that arises at the crests of the neural folds and migrates away from the neural tube along

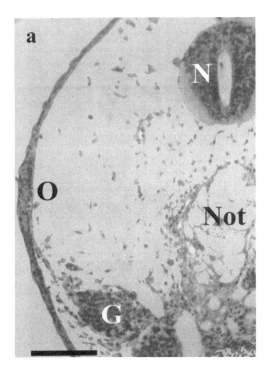
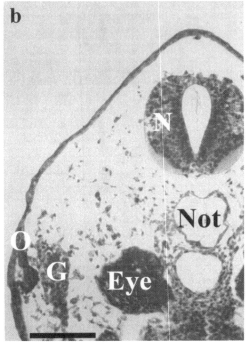

FIGURE 3.5. These pictures show the effect of 5×10^{-6} M retinoic acid treatment at two different stages in the clawed toad *Xenopus* (a; stage 12; b; stage 13). Note that the invagination of the otic placode (O) but not the formation of otic ganglion cells (G) is completely (a) or partially (b) blocked with retinoic acid. Not, notochord; N, neural tube. Bar, 100 μm.

nearly the entire length of the embryo. The neural crest contributes to a myriad of structures in the body, including all of the peripheral nervous system neurons and glia not made by the placodes, as well as melanocytes, glial cells, and many other specialized cell types (Ayer-Le Liver and Le Douarin 1982). Vital to the function of the inner ear, the neural crest contributes all of the glial (satellite) and Schwann sheath cells in the statoacoustic (otic) ganglion (Ayer-Le Liver and Le Douarin 1982; D'Amico-Martel 1982; D'Amico-Martel and Noden 1983).

A melanocyte-like cell population of neural crest origin is also proposed to have an important function in the stria vascularis. Steel and Barkway (1989) postulate that abnormalities in melanocyte lineages, including their gene expression or migration, could lead to absent or abnormal melanocyte-like cells, the role of which in establishing critical endocochlear potentials in the stria is hypothesized but still undocumented. Further support for this hypothesis comes from studies in which abnormal development of auditory receptors is seen in mutations in animals that have defects in the neural crest (Noden and Van De Water 1992; Fleischman 1993). Absence or alteration of these melanocytes contributes to the absence of the endolymphatic potential and thus to deafness in mice carrying mutations at the W (c-kit) locus, such as viable dominant spotting and other W alleles, including, W^X, W^{sh}, and W^{41} (reviewed in Cable et al. 1994; see Sections 2.7 and 3.0).

Mutations leading to pigment disorders in both humans and mice also demonstrate that neural crest-derived melanocytes are important in sensorineural deafness. Mutations in the $Pax3$ gene, which is expressed in neural crest cells, lead to sensorineural deafness in some patients with Waardenburg syndrome type I (Steel and Kimberling 1996). The role of Pax genes in development, and in inner ear development in particular, is described in detail in Section 2.7. Waardenburg syndrome type 2, another typical auditory-pigmentary defect, is caused by mutations in the human microphthalmia ($MITF$) gene (Tassabehji, Newton, and Read et al. 1994), which encodes a transcription factor (microphthalmia transcription factor, [$MITF$]) that affects melanocyte differentiation. Thus mutations in the $MITF$ gene lead to hypopigmentation, sensorineural deafness, and underdeveloped eyes (Jackson and Raymond, 1994; Tachibana et al. 1994).

The majority of the neural crest cells that play a role in ear development emerge above rhombomere 1 and even-numbered rhombomeres in the hindbrain. A study by Graham, Heyman, and Lumsden (1993) reports that in normal development cell death of the neural crests above rhombomeres 3 and 5 prevents crest cells from emerging over these rhombomeres; however, no cell death of neural crest cells occurs in the other rhombomeres, resulting normally in neural crest migration only from rhombomeres (1 + 2), 4, and 6, at chick developmental stages 10–12. Rhombomeres 3 and 5 express members of the msx family of homeobox genes, with $msx2$ expression preceding apoptosis in a precisely colocalized pattern.

Rhombomeres 3 and 5 are depleted of cephalic neural crest cells by an interaction within the neighboring neural epithelium. For example, both rhombomeres 3 and 5 produce migrating neural crest cells if separated from their even-numbered neighbors. Furthermore, when rhombomeres 3 or 5 are transplanted to ectopic locations or are placed in culture, neural crest cells emerge from these rhombomeres and *msx2* expression is concomitantly downregulated (Graham, Heyman, and Lunsden 1993). The signalling molecule BMP4 is expressed in both of these odd-numbered rhombomeres and is dependent on interactions with the even-numbered rhombomeres (Graham et al. 1994). Addition of recombinant BMP4 to explant cultures of rhombomere 3 or 5 upregulates *msx2* and reinstates neural crest apoptosis in these rhombomeres (Graham et al. 1994). Most critical for vertebrate inner ear development is the correct expression of specific genes in the hindbrain rhombomeres in the vicinity of the otocyst because this gene expression also appears to pattern the neural crest cells that contribute to this development (Graham, Heyman, and Lumsden 1993). The inner ear does not develop correctly in mice if, among other things, neural crest cells develop between the invaginating otocyst and the brain (Deol 1983).

2.4 Mutations That Affect Rhombomeric Development and/or Organization

Nature has provided a rich repertoire of naturally occurring mutants with various defects of the ear (Deol 1983; Ruben, Van De Water, and Steel 1991; Noden and Van De Water 1992; Steel and Harvey 1992; Steel and Brown, 1994; Steel and Kimberling 1996). Several mouse mutants are known in which the defect is believed to affect early morphogenesis of the ear (Deol 1983). One of these mutants is the kreisler (*kr*) mouse. Homozygous mutant kreisler mice (*kr/kr*) are both deaf and have balance defects (Rubin 1973; Frohman et al. 1993). The defects seen in these animals are found in the rhombencephalon and the inner and middle ears. Although the invagination and closure of the otic pit are normal, rhombomeres 4–7 are replaced by a morphologically featureless and unsegmented neural tube (Frohman et al. 1993; McKay et al. 1994; Cordes and Barsh 1994; Fig. 3.4).

The *kr/kr* mice were found to have altered expression patterns of hindbrain markers, including *Hoxb3* (*2.7*), *Hoxb4*, *Krox20*, and *Fgf3*. For example, the rostral expression boundaries of *Hoxb3* and *Hoxb4* are displaced from their normal positions at rhombomeres 4/5 and 6/7 to the approximate positions of rhombomeres 3/4 and 4/5, respectively. Expression domains of *Krox20* and *Fgf3* are displaced rostrally and the intensity of *Fgf3* hybridization is greatly reduced (Frohman et al. 1993). The expression domain of *Hoxb1* is affected differently in these embryos: its rostral boundary at rhombomeres 3/4 is intact, but the caudal boundary is displaced from its normal location at rhombomeres 4/5 to the approximate position of rhombomeres 5/6. Because boundaries of gene expression for *Hoxb1* and *Hoxb4* are found in a region of the *kr/kr* hindbrain that lacks visible rhombomeres, the authors conclude that establishment of regional identity, as reflected by differential gene expression, does not require morphological segmentation. The altered rhombomeres are critical in that r5 and r6 affect otocyst development and rhombomere 4 plays a role in providing vital neural crest populations that are involved in the process of inner ear innervation by providing supporting cells and a subpopulation of neurons.

In *kr/kr* mice, the neural crest that normally emanates from rhombomeres (1 + 2), 4, and 6, instead appears as a continuous sheet extending caudally from the junction of rhombomeres 3 and 4. In addition, the position of the developing otic vesicle is shifted from its normal location near the junction of rhombomeres 5/6 to a more lateral location and is separated from the neural tube by preganglionic neural crest. This separation may have severe consequences for subsequent development of the otic vesicle if either soluble factors, extracellular matrix molecules produced by one or both sets of cells, or cell-cell contact with the rhombencephalon are critical for correct inner ear formation. The position of the placode is also anomalous, and the otocyst invaginates away from the brain (Fig. 3.4). Clearly, the hindbrain is abnormal and lacks formation of rhombomere boundaries caudal to rhombomere 3. On the basis of the defects created by experimental manipulation of the otocyst in amphibians, it has been suggested that the primary defect is in the hindbrain and the ear defects, which are often asymmetric, are secondary and caused by altered induction (Deol 1983; Frohman et al. 1993).

The kreisler gene encodes a novel basic domain, leucine zipper DNA binding protein that is required for correct cell identity in rhombomeres 4 and 5 (Cordes and Barsh 1994). *kr* mRNA is highest at the rhombomeres 4/5 border before the establishment of rhombomere boundaries. In addition, the hindbrain shows differences in the expression patterns of some homeobox containing genes and *Fgf3* (Frohman et al. 1993; McKay, Lewis and Lumsden 1996). The recent identification of the *kreisler* gene contrasts with the results of earlier experiments in which an attempt was made to correct the *kr* defect by culturing it with normal hindbrain (Van De Water and Represa 1991). In this study, the authors cocultured *kr/kr* otocyst and/or neural tube with normal (CBA/5) otocyst and/or neural tube. Normal brain stem did not rescue the *kr/kr* ear phenotype, nor could a *kr/kr* phenotype be induced by the *kr/kr* brain stem in normal ears. The authors therefore concluded that the action of the kreisler gene was not on the neural tube but was at the level of notochordal mesoderm (Van De Water and Represa 1991). However, the Frohman et al. (1993) study suggests that the target genes of this transcription factor encode proteins that function within the rhombomeres and are not secreted. Recently, the zebra fish homologue of kreisler has been identified (Moens et al. 1996).

2.5 Gene Inactivation by Targeted Disruptions Through Homologous Recombination in Mice

Molecules that influence ear development have been identified by gene-inactivation experiments through homologous recombination in mice. These include transcription factors, growth factors, and growth factor receptors (Table 3.1).

2.5.1 *Hoxa1* (*Hox1.6*)

Two groups (Lufkin et al. 1991; Chisaka, Musci, and Capecchi 1992) produced targeted disruptions of the homeobox containing *Hoxa1* (*Hox1.6*) gene through homologous recombination. Mouse mutants carrying targeted disruptions of the *Hoxa1* gene showed developmental defects in the auditory system, thus implicating this gene in early ear development. In the independent experiments performed by Chambon's group (Lufkin et al. 1991; Dolle et al. 1992; Mark et al. 1992) and Capecchi's group (Chisaka, Musci, and Capecchi 1992; Carpenter et al. 1993; Mansour, Goddard, and Capecchi 1993), disruption of *Hoxa1* expression results in defects of the inner ear and also of moto- and sensory neurons in cranial nerves VII, VIII, IX, and X. Chisaka, Musci, and Capecchi (1992) further demonstrated that in their *Hoxa1$^{-/-}$* mice the otic vesicle is reduced and displaced rostrally and laterally and the cochlear duct is reduced in size. Furthermore, sensory receptors are abnormally located, and the vestibular and auditory ganglia are reduced in both neuronal and glial cell numbers. Ganglia VII and V are frequently contiguous. The embryonic rhombomeres are indistinct and rhombomere 5 is missing (Dolle et al. 1992; Mark et al. 1992; Carpenter et al. 1993; Fig. 3.4). Homozygous *Hoxa1$^{-/-}$* mouse mutants display somewhat different phenotypes, however, that may be related to which *Hoxa1* proteins are disrupted, i.e., both the 133- and 331-amino acid *Hoxa1* proteins, or only the 331-amino acid-long protein.

The normal expression pattern of the *Hoxa1* gene is in the caudal and middle regions of the hindbrain before the onset of otic placode formation. Although the gene reportedly is not expressed either in the neural crest or its derivatives or in the otic placode, the current studies (Lufkin et al. 1991; Chisaka, Musci, and Capecchi 1992) have not examined *Hoxa1* expression in very early embryos of either mouse or birds, and their in situ hybridization studies may be insufficiently sensitive to detect expression in just a few cells, as pointed out by Hogan and Wright (1992). The *Hoxa1* gene appears to be associated with the neurogenic neural crest because structures expected to contain crest-derived neurons are also deficient or absent in the transgenic animals.

Noden and Van De Water (1992) point out that the tissues affected in these knockout experiments do not come from a single embryonic source. In the inner ear, the affected tissues are mostly derived not from the neural crest or the hindbrain but from the otic placode. Lesions to any one of the three affected tissues alone would not result in the defects observed in the transgenic animals. However, the three affected tissues arise from progenitors that are normally located close together in the middle hindbrain region (beside and beneath rhombomeres 4–7), and there is much interaction among the neural crest, mesenchyme, and placode on which their correct development depends. Noden and Van De Water (1992) also conclude that the missing middle ear ossicles and outer ear hypoplasias reported in the study by Chisaka, Musci, and Capecchi (1992) cannot be attributed to a primary lesion in the neural crest unless only subpopulations of neural crest cells are affected. That identifiable neural crest subpopulations are affected has been postulated as evidence of very early diversification of some neural crest subpopulations (Barald 1989).

2.5.2 *Fgf3*

In the clawed toad *Xenopus*, *Fgf3* (formerly known as *int2*), which codes for the receptor tyrosine kinase ligand FGF-3, is expressed from just before the onset of gastrulation to prelarval stages. It is expressed around the blastopore lip (Tannahill et al. 1992). Later in tail-bud to prelarval embryos, it is expressed in the stomodeal mesenchyme, the endoderm of the pharyngeal pouches, and the cranial ganglia flanking the otocyst. Later sites of expression include rhombomeres 5 and 6 adjacent to the developing otocyst, the sensory region of the inner ear, and the tooth mesenchyme (Wilkinson, Bhatt, and McMahon 1989; McKay, Lewis & Lumsden 1996; Fig. 4).

In the chick embryo (Mahmood et al. 1995), *Fgf3* transcripts are observed in rhombomeres 4 and 5 and later in rhombomere 6. Subsequently, expression becomes restricted to rhombomere boundaries. *Fgf3* expression is induced in reforming boundaries when even-numbered rhombomere tissue is grafted next to odd but not when even-numbered or odd-numbered rhombomeres are juxtaposed. FGF-3 disappears from the rhombomeres just before to the loss of morphological boundaries (Mahmood et al. 1995).

Otic vesicle formation is inhibited through the application of antisense oligonucleotides or antibodies against *Fgf-3* (Represa et al. 1991). Blocking expression of *Fgf3* inhibits formation of the otic vesicle, whereas application of exogenous FGF-3 protein induces vesicle formation. The *Fgf3* gene is transiently expressed in the hindbrain (Wilkinson, Bhatt, and McMahon 1989), and FGF-3 is thought to be a diffusable molecule that can act on the adjacent placodal cells. Expression of the *Fgf3* gene may be regulated by early genes like *Krox20* or *Hox* that are expressed in the hindbrain. It is thought that FGF-3 is responsible in some capacity for proper morphogenesis of the otocyst (Wilkinson, Bhatt, and McMahon 1989; Mahmood et al. 1995).

The otocyst of mice homozygous for a targeted disruption of *Fgf3*

TABLE 3.1. Genes and phenotypes.

Gene	Species	Mutation/Knockout/ Overexpression	Localization/ Manipulation	References
Bmp4 growth factor downstream of hedgehog; controls *msx*	Chick, frog (*Xenopus*)	– / – / –	Chick. Neural crest in rhombomeres 3 and 5. Chick/Frog. Small patches at anterior and posterior regions of otocyst.	Graham et al. 1994 Hemmati-Brivanlou and Thomsen 1995 Wu and Oh 1996 Oh et al. 1996
dlx3, dlx4 (fish *dlx3* is *DLX-5* in chick and *dlx-3* in mouse	Zebrafish, chick, mouse	– / – / –	Zebrafish. Cells in the ectoderm where placode will form express *dlx3* before induction of the otic vesicle; *presumptive* precursor cells of the olfactory placodes express *dlx3* and *dlx4* but not *dlx2d*. Cells aligned with the future axes of the semicircular canals specifically express either *dlx3* or *msh-D. dlx3* expression is "sided," covering one-half of the vesicle on the dorsal side. *dlx-3* in mouse is expressed in the developing inner ear and in semicircular canals.	Ekker et al. 1992a, 1992b Akimenko et al. 1994 Robinson and Mahon 1994

| GH6; transcription factor | Chick | −/−/− | Expressed in dorsolateral otocyst; At stage 23, high levels in the second branchial arch, neural retina, lens epithelium, optic nerve and infundibulum, ventricular myocardium, and sensory spinal and cranial ganglia. | Stadler and Solursh 1994 |
| W locus (c-Kit) (mouse); KIT (human) c-Kit protooncogene, a transmembrane tyrosine protein kinase receptor | Mouse, human | M. Mouse. W^x, W^{sh}, W^{41}, Viable dominant spotting. In control animals, melanocytes in the vestibular part of the ear were found in the utricle, crus commune, and ampullae, but in many mutants only one or two of the regions were pigmented (affects also hematopoiesis and proliferation and or migration or primordial germ cells). KO. − OE. − M. Human. piebaldism (human equivalent of mouse viable dominant spotting). The only human presumed homozygote for piebaldism was deaf. KO. − OE. − | Affects melanocytes in the stria vascularis thought to contribute to the endocochlear potential. Melanocytes thought to come from the neural crest that arises over rhombomere 4. | Manova and Bachvarova 1991 Cable et al. 1994 |

(Continued)

TABLE 3.1. (*Continued*)

Gene	Species	Mutation/Knockout/ Overexpression	Localization/ Manipulation	References
Hoxa1(*Hox1.6*) *Hox* gene	Mouse	M. – KO. Mouse. Rhombomere 4 greatly reduced; rhombomere 5 missing. No development of inner ear. Remnants of rhombomeres 4 and 5 appear to be fused caudally with rhombomere 6 to fom a single fourth rhombomeric structure. Migration of neural crest cells contributing to the glossopharyngeal and vagus nerves occurs in a more rostral position, resulting in abnormalities of these cranial nerves. OE. –	In *wt* embryos, *Hoxa1* is expressed between days 7.5 and 8 in the neural epithelium and mesoderm of the rhombencephalic region caudal to the level of the presumptive rhombomere 4; the rostral boundary is the preotic sulcus.	Lufkin et al.1991 Chisaka, Musci, and Capecchi 1992 Carpenter et al. 1993 Mark et al.1993

Kreisler *kr/kr* Valentino	Mouse, zebrafish	M. Mouse Kreisler is a recessive mutation in which gross malformations of the inner ear are found in homozygous mice. These defects are related to abnormalities in the hindbrain of the embryo, basically a loss of rhombomeres 5 and 6. The zebrafish mutation Valentino is the fish homologue. Embryonic lethal, same hindbrain pattern as mouse. KO.– OE.–	No localization studies in mouse.	Doel 1964 Cordes and Barsh 1994 McKay et al. 1994 Moens et al. 1996
msh-C *msh-D*: belong to a family of vertebrate genes related to *Drosophila* muscle segment homeobox gene *msh*	Zebrafish	–/–/–	The *msh-C* and *msh-D* genes are expressed in distinct regions of the otic vesicle during its early development in zebrafish embryos. The expression of *msh-D* is "sided," covering one-half of the vesicle on the dorsal side.	Ekker et al. 1992a

(Continued)

TABLE 3.1. (*Continued*)

Gene	Species	Mutation/Knockout/ Overexpression	Localization/ Manipulation	References
Fgf3/int2 growth factor	*Xenopus*, chick, mouse	M. – KO. Otocysts fail to form an endolymphatic duct. Normal invagination of the otocyst but reported additional ear defects that were highly variable and sometimes different on the left and the right side. OE. –	In *Xenopus*, *Fgf3* is responsible in some capacity for proper morphogenesis of the otocyst. Expressed from just before the onset of gastrulation to prelarval stages and around the blastopore lip. Later in tail bud to prelarval embryos and in the cranial ganglia flanking the otocyst. Later sites of expression include rhombomeres 5 and 6, adjacent to the developing octocyst, the sensory region of the inner ear, and the tooth mesenchyme. In chick, expressed in and secreted by the rhombencephalon	Wilkinson Bhatt, and McMahon 1989, 1990 Represa et al. 1991 Tannahill et al. 1992 Mansour, Goddard, and Capecchi 1993 McKay, Lewis and Lumsden 1996

Gene	Species	Phenotype	References
		in the vicinity of the developing otocyst. Otic vesicle closure is inhibited through the application of antisense oligonucleotides or antibodies against *Fgf3/int2*. Blocking expression of *Fgf3/int2* inhibits formation of the otic vesicle while application of exogenous FGF-3 protein induces vesicle formation. Not determined.	Tassabehji, Newton, and Read 1994 Jackson and Raymond 1994
Microphtalmia (Mi) transcription factor	Mouse, human	M. Mutations in the *MITF* gene lead to hypopigmentation, sensorineural deafness, and underdeveloped eyes. Waardenburg syndrome type 2, another typical auditory-pigmentary syndrome, is caused by mutations in the human microphthalmia (*MITF*) gene. KO. – OE. –	

(Continued)

TABLE 3.1. (*Continued*)

Gene	Species	Mutation/Knockout/Overexpression	Localization/Manipulation	References
Pax2 Paired-domain transcription factor	Chick, mouse, human	M. Mouse.*Krd*, kidney and retinal defects, a transgene insertion that deletes the *Pax2* locus. Human. Syndrome with kidney and retinal defects, possible hearing loss. KO. Recently done for *Pax2* alone; larger region knockout including *Pax2* (Kellor et al. 1994) is early embryonic lethal in homozygotes. Pax-2 KO (Torres et al. 1996) has abnormal or missing cochlea OE. Delays kidney differentiation; maintains embryonic structures; prevents differentiation of neuronal structures in cochlea.	Neural tube, otic placode (chick), otocyst.	Mouse: Nornes et al. 1990 Dressler et al. 1990 Kellor et al. 1994 Chick: Lindberg et al. 1995
Pax3 Paired-domain transcription factor	Chick, mouse, human	M. Splotch mouse; Waardenburg syndrome type I in humans. KO. – OE. –	Neural crest?	Review: Stracham and Read 1994

M, mutation; KO, knockout; OE, overexpression; BMP, bone morphogenetic protein; FGF, fibroblast growth factor.

invaginate but fail to form an endolymphatic duct (Mansour, Goddard, and Capecchi 1993). The authors found a normal invagination of the otocyst but reported additional ear defects that were highly variable and sometimes different on the left and the right sides. The initial size of the otocyst (which is seen to be reduced in *Hoxa1* mutant embryos) is not affected in *Fgf3* null embryos. This implies that the *Hoxa1* embryos' alterations in the FGF-3-dependent pathway for inner ear development cannot account for all the inner ear defects in the *Hoxa1* mutant embryos. In the otocyst, endogenous FGF-3 proteins might specify the areas that will form sensory epithelium. These defects might be a direct consequence of the abnormal relationship between the rhombencephalon and the otic pit. The effect of FGF-3 on cell proliferation may also be crucial to the formation of the otocyst from the otic placode in the chick embryo (Represa et al. 1991).

Represa et al. (1991) have provided experimental evidence that this gene may be involved in otocyst formation in the chicken. To explain their data, Represa et al. (1991) invoked the known mitogenic effect of the gene product (FGF-3) but also other mechanisms. Although intuitively convincing, it appears that the crucial experiment, block of proliferation alone in chickens by, e.g., hydroxyurea (Harris and Hartenstein 1991), is needed to show whether enhanced proliferation is indeed the crucial step to transform the otic placode into an otocyst. In addition, the reagents involved were mammalian, but the embryological material was avian, a system in which *Fgf3* has only been characterized recently (Mahmood et al. 1995); nor was the specificity and sensitivity of the antibody tested. Nevertheless, *Fgf3* appears to be expressed at the right time and the right place in mice (Wilkinson, Bhatt, and McMahon 1989), frogs (Tannahill et al. 1992), and chickens (Mahmood et al. 1995) to play a role in the early development of the auditory system.

Although all the evidence obtained in *Fgf3* null mutants (Mansour, Goddard, and Capecchi 1993) argues against a direct role played by *Fgf3* in otocyst formation, it clearly underscores the potential role of *Fgf3* for otocyst differentiation. Most of the conclusions obtained thus far with targeted disruption add phenotypic examples of ear deformations but require closer analysis of correlated expression of genes suggested to be involved in pattern formation of the ear. Moreover, although theoretically simple, recent data show a vast array of reorganizations in homeobox expression patterns in these cases (Carpenter et al. 1993; Frohman et al. 1993; Fig. 3.4), thus complicating the interpretation of these experiments.

2.6 Distal-less (dlx), the Earliest Gene Expressed in the Auditory Anlagen

A number of additional *Hox* genes related to the *distal-less* gene of *Drosophila* have been found to be expressed in a spatiotemporally restricted pattern in the developing ear in zebrafish (Ekker et al. 1992a; Akimenko et al. 1994) and in frogs and chickens (Ramirez and Solursh 1993). Ekker and colleagues (1992a,b) have begun to delineate the expression of genes involved in development of the inner ear in zebra fish; the *msh-C* gene and two new homeobox genes, *msh-D* and a gene related to *distal-less*, *dlx-3*, are each expressed in distinct regions of the otic vesicle during its early development in zebrafish embryos.

Ekker et al. (1992a) speculate that there are three steps in inner ear formation. In the first, *dlx-3* may specify the ectodermal field, which will form the otic placode. Further evidence for this suggestion is provided by a zebrafish mutation that lacks *dlx-3* and does not form placodes (Fritz and Westerfield 1996). The *dlx-3* gene encodes a product that presumably acts as a transcription factor because it is localized within the nucleus, making it more likely that it plays a regulatory role rather than a direct role in the induction. The initial induction of the otic placode is thought to occur in response to signals from the chordal mesoderm. Later, additional signals from the hindbrain neuroectoderm are required. By the stage when *dlx-3* transcripts are localized in the developing otic placode in the zebrafish, the placode is not yet evident morphologically. Induction of the vesicle by the hindbrain neuroectoderm has probably not yet begun; however, the *dlx-3* gene product is present and could be playing an inducing role. Thus *dlx-3* expression could be permissive, allowing cells to respond to an as yet undefined inductive signal.

In their proposed second phase of ear development (Ekker et al. 1992a), FGF-3 is postulated to be the molecular inductive signal produced by the hindbrain ectoderm to induce the otic vesicle (see also Represa et al. 1991). Finally, expression of *msh-D* and *dlx-3* in specific regions of the otic vesicle might specify the morphogenesis and subsequent orientation of the semi-circular canals (Ekker et al. 1992a). The expression patterns of these mRNAs are aligned along the same axes that the extending semicircular canals later follow; therefore, their expression may help specify position within the epithelium.

Akimenko et al. (1994) comment that the restricted expression patterns of *msh-D* and *dlx-3* are seen after the otic vesicle has been induced and may be a consequence of the induction by factors like retinoic acid or FGF-3 (Represa et al. 1990). However, there is no experimental evidence showing that either *dlx-3* or its cognate in other vertebrates is directly involved in otic morphogenesis or changes its expression pattern with retinoic acid treatment or experimental manipulations known to affect otic differentiation.

Although causality has not yet been established between these genes and identified morphogenetic steps, their distribution makes them good candidates to specify dorsoventral, anteroposterior, and mediolateral axes. Whether their gene product is responsible for axis determination or whether these genes are turned on as a consequence of a process that specifies polarity needs to be examined with appropriate transplantation experiments and correlated with in situ hybridization experiments.

2.7 Role of Pax Genes in Organogenesis During Early Embryogenesis

Pax genes (pair-rule genes) were originally identified in *Drosophila* as genes specifying segment polarity (Noll 1993). Pax proteins are developmentally regulated transcription factors (Gruss and Walther 1992) with a highly conserved, 128 to 129-amino acid DNA-binding domain, the paired domain. Whereas mutations in *Hox* genes are recessive and their phenotype can only be analyzed in mice homozygous for targeted disruptions of these genes, mutations in *Pax* genes are either semidominant or show gene dosage effects and produce phenotypes in heterozygotes. Pax proteins are both activators and repressors of transcription (Chalepakis et al. 1993). Their role as critical master regulatory proteins was demonstrated by the remarkable experiment by Halder, Calaerts, and Gehring (1995) in which ectopic expression of the mouse *Pax6* (or the *Drosophila* homologue *eyeless*) in the fly produces a nearly complete eye on legs and antennae.

In addition to the paired domain, all Pax proteins contain a conserved octapeptide of unknown function, and many contain either a complete, a truncated, or no paired-type homeodomain (Gruss and Walther 1992). Of the nine vertebrate *Pax* genes, *Pax1* to *Pax9*, four (*Pax1, -2, -3* and *-6*) have been associated with mutations in mouse mutants or human syndromes (Erickson 1990; Hill and von Heyningen 1992; Morell et al. 1992; Hoth et al. 1993, Pierpoint and Erickson 1993; Tassabehji, Newton, and Read 1994).

Pax genes have recently been implicated in vertebrate development (Dressler et al. 1990). With the exception of *Pax1*, *Pax* genes appear to be expressed in and important for specification of the developing nervous system and brain (Stoykova and Gruss 1994). *Pax* genes have recently been shown to be upstream of engrailed-2 genes and critical for engrailed-2 expression in the developing brain (Song et al. 1996). Two *Pax* genes, *Pax2* and *Pax3*, have also been implicated in ear development. Their role in ear development is described below.

2.7.1 Role of *Pax3* Genes in Deafness

Mutations in the paired domain of *HuP2*, the human equivalent of *Pax3*, are associated with Waardenburg's syndrome type I (Morell et al. 1992; Tassabehji, Newton, and Read 1994) and Waardenburg's syndrome type III (Hoth et al. 1993) in humans and the splotch mutant in mice (Erickson 1990; Hill and von Heyningen 1992; Pierpoint and Erickson 1993). Individuals with Waardenburg's syndrome type I have decreased pigmentation of eyes and skin and craniofacial abnormalities, and many have sensorineural deafness. Splotch mutants (Steel and Smith 1992; Steel and Kimberling 1996), so-called because of the white spotting, show, in addition to numerous neural crest-based deformations, anatomic deficits of the inner ear. These defects appear to be related to localized loss of specific migratory neural crest precursors. Thus the membranous labyrinth of the ear does not form normally and auditory sensory receptors are also affected (Moase and Trasler 1990). *Pax3* is normally expressed in the dorsal spinal cord and hindbrain before and during otocyst formation, although not in the otocyst itself (Goulding, Lumsden, and Gruss 1993).

2.7.2 *Pax2* in Development

Pax2 shows transient, high-level expression in the developing eye, ear, and kidney (Dressler et al. 1990; Nornes et al. 1990). Zebrafish *pax(b)*, the homologue of vertebrate *Pax2* (Krauss and colleagues 1991, 1992), shows a similar pattern of expression as in mammals. *Pax2* is overexpressed in Wilms' tumor, a kidney tumor of embryonic origin; conversely, overexpression of *Pax2* in transgenic mice prevents normal differentiation of kidney structure and leads to tumors (Dressler and Douglas 1992). Keller et al. (1994) generated a transgenic mouse with a 5-kb deletion that removed the entire *Pax-2* gene. The heterozygotes showed retinal and kidney abnormalities, but no obvious inner ear abnormalities. The effect of this deletion could not be studied in homozygotes, which are preimplantation embryonic lethals. A mutation of the *Pax2* gene in humans, which is due to a single nucleotide deletion in exon 5, causes a frame shift of the *Pax2* coding region in the octapeptide domain. The phenotype in a family with variable optic nerve colobomas, renal anomalies, and some evidence of sensorineural hearing loss (Sanyanusin et al. 1995) is similar to abnormalities described in *Krd* mutant mice (Keller et al. 1994) in which the entire *Pax2* gene is deleted. The role of the *Pax2* gene in ear development should be investigated further.

2.8 Role of GATA Genes in Lineage Specification

In contrast to the Hox and Pax proteins, which are required for correct segmentation and pattern formation, the *GATA* genes appear to encode lineage-specific transcription factors, which may be downstream targets of either *Hox* or *Pax* transcription factors. *GATA1* was identified first in globin genes and is erythroid specific; it is required for the generation and/or maintenance of the erythroid lineage. *GATA3* is expressed in subsets of neuronal cells and in otocysts during embryogenesis in *Xenopus laevis* (L. Zon, personal communication), the chick (Kornhauser et al. 1994), and the mouse (George et al. 1994). Chick *GATA2* and *GATA3* are expressed in the developing chick optic tectum (Kornhauser et al. 1994), and *cGATA3* may play a unique regulatory role in developing brain and in T cells (Yamamoto et al. 1990). The mouse *GATA3* gene (*mGATA3*) is expressed in restricted cells in the developing central nervous system, parasympathetic nervous system, embryonic kidney, and thymic rudiment and throughout T-cell differentiation. A 3-kb region of the *mGATA3* gene containing the promoter and 5′-flanking region was fused to the *lacZ* gene and used to generate transgenic mice containing this construct. An 10.5-day embryo displayed *lacZ* expression patterns that paralleled the expression of the endogenous *mGATA3* gene in trigeminal and facioacoustic ganglia and the otic vesicle (George et al. 1994). To our knowledge, this is the first report of correct temporal and spatial expression in the otic vesicle of a reporter gene driven by the promoter of a vertebrate gene.

2.9 GH6, Another Gene Expressed in the Otocyst

GH6 is a novel chicken homeobox gene that appears to be related to the human homeobox gene *H6* (93% homology at the amino acid level; Stadler and Solursh 1994). Expression of *GH6* is detectable as early as stage 13 in the chick embryo by in situ hybridization. At stage 23, *GH6* mRNA is found at high levels in the second branchial arch, neural retina, lens epithelium, optic nerve and infundibulum, ventricular myocardium, and sensory spinal and cranial ganglia. It is also found in the dorsolateral otocyst. *GH6* is not detected in premigratory or migrating neural crest cells. *GH6* is also expressed in a temporal and spatial pattern that correlates well with the regions and developmental stages most susceptible to exogenous retinoic acid treatment, including derivatives of the first and second branchial arches.

2.10 Role of Thyroid and Retinoic Acid Receptors

Alterations in endogenous retinoic acid or thyroid hormone levels are known to cause inner ear defects. The role of these factors in ear development has been extensively reviewed recently (Corey and Breakefield 1994). The transcription factors are members of the steroid hormone-receptor superfamily that binds ligand (i.e., retinoic acid or thyroid hormone) in the cytoplasm, then diffuses to the nucleus, where the receptor-ligand complex binds to enhancer or repressor sequences in the target genes either to activate or repress transcription. Both compounds are known teratogens, and retinoic acid has been implicated as a morphogen in the ear (Kessel 1992; Morriss-Kay 1993).

Retinoic acid, a known teratogen and potential morphogen (Manns and Fritzsch 1992; Pijnappel et al. 1993) can not only block differentiation of the otocyst (Represa et al. 1990) but can completely block formation of an otocyst resulting in a placode instead (Fig. 3.5). Although these data may be taken in support of the theory by Represa et al. (1991) concerning the importance of cell proliferation in otocyst formation, it is clear that retinoic acid can drastically change the expression pattern of different homeobox-containing genes (Sive and Cheng 1991; Marshall et al. 1992). Therefore, other mechanisms for the action of retinoic acid cannot be excluded at the moment. For example, upregulation (Sive and Cheng 1991) or a shift of homeobox domains to more anterior rhombomeres (Marshall et al. 1992) could have disrupted hindbrain/otocyst interactions (Fig. 3.4). Whether retinoic acid changes the expression pattern of *Fgf3* or other genes, e.g., *hox* genes, expressed in the otocyst (see below) is not yet known.

3. Genetics of Ear Development

No review of early embryogenesis of the ear would be complete without a discussion of recent insights gained from the genetics of deafness in both mice and humans (Steel and Kimberling 1996). Mouse deafness mutants have played an important role in understanding the molecular genetic basis of deafness, particularly with the advent of positional-cloning techniques, which have resulted in the isolation of several genes first identified by their mutated state.

The association of pigmentary disturbances with hearing loss suggested that melanocytes in the stria vascularis play an important but as yet undefined role in hearing. Three genes have been identified that cause white spotting and deafness in the mouse: *W* (dominant white spotting), Steel, and *Mi* (Microphthalmia). The *W* gene encodes *c-kit*, a growth factor receptor, and the *Steel* locus encodes the ligand for the *c-kit* receptor, referred to as stem cell or steel factor. These two genes represent a receptor-ligand complex that must be important for melanocyte differentiation. A number of mutations have been identified in the human *KIT* gene, the homologue of the mouse *W* gene (Spritz et al. 1992). The mouse *Mi* gene encodes a transcription factor. The human homologue of *MITF* was recently mapped (Tassabehji, Newton, and Read 1994) and cloned (Tachibana et al. 1992). Mutations in the *MITF* gene were shown to cause Waardenburg syndrome type II. Our understanding of the function of this transcription factor, however, comes from extensive analysis of *Mi* mutations in mice. Mutations may be dominant or recessive, and all affect both neural crest-derived melanocytes and the retinal pigmented epithelium. Mice homozygous for the extreme alleles are completely unpigmented and are deaf due to a lack of inner ear melanocytes. A model for the *Mi* phenotype is the failure to activate specific genes, which leads to a failure of melanocyte differentiation.

A large number of syndromes associated with hearing loss have been mapped genetically in humans (Duyk, Gastier, and Mueller 1992) and should lead to the identification of the gene product through positional cloning. Many of these syndromes involve collagen isoform genes. Other syndromes involve head, ear and kidney defects (branchio-oto-renal syndrome), which suggest that a factor common to the development of each of these organs is defective. Individuals with Usher syndrome have both eye and ear defects, implicating factors involved in development of these two organs. At least seven genes for Usher syndrome have been mapped (Steel and Kimberling 1996), and there remain Usher families that do not show linkage to any of these seven loci. Positional cloning of the mouse shaker-1 (*shl*) mutation (Gibson et al. 1995) led to identification of the mutation in the gene for Usher syndrome type I (*USH1B;* Weil et al. 1995). The *shl* homozygotes show hyperactivity, head tossing, and circling due to vestibular dysfunction, together with typical neuroepithelial-type cochlear defects involving dysfunction and progressive degeneration of the organ of Corti. Both the *shl* and the *USH1B* genes have been identified as the gene for an unconventional myosin, myosin VIIA. The defective myosin VIIA could affect microtubular structures, the microvilli of the retinal pigmented epithelium of the eye, the inner ear supporting cells, or the stereocilia of hair cells.

In summary, as of this writing, many genes thought to be involved in ear development are largely expressed in the rhombencephalon and the neural crest, rather than in the otocyst or otic placode. Mutations of, or knockouts of, these genes have primary effects on the development of the rhombencephalon and neural crest; effects on the inner ear are secondary to these effects. The genes identified to date include *Hox* genes and *Pax* genes, particularly *Hoxal* (1.6), which is expressed in the developing hindbrain, and *Pax3*, which is expressed in the neural crest and hind brain, as well as the kreisler gene, mutations of which affect the rhombencephalon and secondarily, the otocyst. *Hoxal* (1.6) knockout mice have defective inner ears as well as defective neural crest-derived structures. Known mutations of *Pax3* result in deafness in recognized syndromes such as Waardenburg syndrome type 1. Growth factors such as FGF-3, which is a product of the *Fgf3* gene, also have effects on the otocyst. *Fgf3* is also expressed in the rhombencephalon, but FGF-3, the soluble product of the gene, is found in the vicinity of and is known to have effects on the development of the otocyst. Therefore, any studies that seek to define the molecular cues involved in the development of the inner ear also need to include the hindbrain and the neural crest as primary foci of the study. The *dlx3* gene is a candidate for the earliest gene expressed in the vicinity of the developing placode (Ekker et al. 1992a), and *Pax2* is one of the first genes to show regionalization in the developing otocyst in the mouse (Nornes et al. 1990) and the chick (Lindberg et al. 1995) and is expressed in the placode (Lindberg et al. 1995).

4. Comparative Development of the Ear

4.1 What Is Unique About the Otic Placode?

There is no detailed space map of the otic placode available for most vertebrates. However, through a comparative embryological approach, there is now a chance to study the spatial configuration of the otic placode with respect to the notochordal mesoderm, paraxial mesoderm, neural crest, and rhombencephalon. All of these elements have either been implicated or shown to play a role in otic placode induction (Yntema 1955; Van De Water 1983; Jacobson and Sater 1988). Recently, the existing data on placodal distribution were reviewed for the lateral line (Northcutt 1992). From these data it appears that there is a rather stable topographic relationship of the otic placode to the various lateral-line placodes and the eye (Northcutt 1992). However, it should be stressed that in the earliest stages of placode formation there may be problems in delimiting the boundaries between the otic placode and the lateral-line placodes despite the fact that lateral-line placodes tend to appear somewhat later in development.

   In this context, it would be important to find a molecular marker that allows unequivocal identification of the two sets of dorsolateral placodes at the earliest possible stages of their formation. A good candidate is *dlx-3* (Akimenko et al. 1994; Section 2.6), which appears to be expressed in zebrafish in the region of the future otic placode before the latter is histologically distinct. Perhaps in situ hybridization of this homeobox gene or immunocytochemistry of its protein(s) may help to identify the otic placode unequivocally in various vertebrates. Such data could allow us to describe the boundaries of the otic placode in all anamniotic vertebrates that develop the lateral-line susbset of dorsolateral placodes (Fritzsch 1992; Northcutt 1992). Equally important, such a marker would allow us to analyze the earliest stage of placodal formation in direct-developing amphibians and in amniotes, all of which lack formation of lateral-line organs (Fritzsch 1992). Such an analysis could help in the determination of whether any rudiments of lateral-line placodes are present in these species or whether these systems are, in fact, selectively suppressed at the earliest possible stage of their development, as previously suggested (Fritzsch 1992). For our analysis of ontogenetic changes that may be related to the evolution of different ears, it is important to note here that, even at the earliest stages of their formation, otic placodes may vary somewhat with respect to size, relative position, and the expression of the *dlx-3* transcripts (Akimenko et al. 1994). However, otic placodes apparently show a rather stable relationship to certain areas of the hindbrain.

4.2 The Position of Otic Placodes: Conservation of a Space Map

Of apparent importance for induction and differentiation of the otic placode is the position with respect to the hindbrain, which, as already discussed in Section 2.2.2, is suggested to play a role in the induction and development of the ear (Harrison 1936; Yntema 1955; Van De Water 1983;

Van De Water and Represa 1991). This topology can only be partly inferred from existing developmental data on a limited sample of vertebrate embryos. Induction and invagination of the otic placodes in hagfish occurs rather far from the hindbrain (Von Kupffer 1900), a feature that sets hagfish apart from other vertebrates except for mutant mice such as the kreisler mutant (Hertwig 1942; Deol 1983; Frohman et al. 1993). It is noteworthy that only hagfish and sometimes kreisler mutants have two octaval nerve roots.

In lampreys, Balfour (1885), Von Kupffer (1895), Neal (1918), and Damas (1944) have described the position of the otic placode. Damas (1944) found continuity with other placodes, whereas Von Kupffer (1895) did not. Moreover, lampreys have a somewhat different arrangement and fate of somites in the area of the developing ear than jawed vertebrates (Von Kupffer 1895; Damas 1944). These developmental data on hagfish and lampreys are not detailed enough to be conclusive. Nevertheless, they indicate a different interaction among the hindbrain, mesoderm, and otic placode than in jawed vertebrates.

The most detailed analysis of spatial relationships of mesoderm, neuroectoderm, and placodal tissue in jawed vertebrates is the work of Meier (1978; Jacobson 1988 for a review). This work shows that the ear appears to be stable with respect to certain landmarks in the surrounding tissue. Nevertheless, the hindbrain position may differ with respect to the otic placode, at least in some vertebrates (Neal 1918). Moreover, boundaries in somitomeres are different in amphibia compared to amniotes and bony fish (Jacobson 1988). As a consequence, the ear forms adjacent to the anterior half of somitomere 4 of salamanders and adjacent to somitomeres 6/7 in amniotes (Jacobson 1988) but at the same hindbrain level. It may be possible that this difference in the otic placode relates to the formation of the amphibian papilla of all amphibians (Fritzsch and Wake 1988). Alternatively, it may be a curious developmental difference without any apparent consequence, comparable to the differences in the formation of the notochord in salamanders and frogs (Hall 1987, 1991) or the absence (mouse) or presence (salamanders) of cellular contact between the invaginating otic cup and the hindbrain (Van De Water 1983). Undoubtedly, the relative position of the otic placode is important because the ear will not develop fully in chickens if the placode is shifted into more anterior or posterior positions (Noden and Van De Water 1986) or in mice if the neural crest develops between the invaginating otocyst and the brain (Deol 1983).

In summary, the only large scale topological differences in otic placode formation may be between hagfish and lampreys on one hand and jawed vertebrates on the other. It is tempting to speculate that this difference is causally related to the different organization of the ear in jawless vertebrates that all lack a horizontal canal. Xenoplastic transplantations between, for example, lampreys and amphibians might be one way of evaluating the conservation of the morphogenetic field of otic induction and ear formation, provided host vs. graft and/or graft vs. host reactions do not interfere with the interpretation of the results of these experiments.

4.3 Heterochronic Shifts in the Induction of the Otic Placode

As outlined above, there are a number of unresolved issues in the formation of the otic placode, including the relative importance of the adjacent tissue and the temporal plan of the inductive events. There are even more unresolved issues with respect to an experimental understanding of inductive interactions of these tissues with both mesoderm and neuroectoderm (Harrison 1945; Yntema 1955; Van De Water et al. 1992).

Extirpation of the otic placode in salamanders, frogs, and birds will result in induction of a new ear in grafted foreign ectoderm, suggesting that both the hindbrain and adjacent mesoderm are involved in establishing the morphogenetic field for otic induction. Consistent with this interpretation, in some amphibians, hindbrain transplanted to other areas of the body can induce small, imperfect ear vesicles (Harrison 1945). Recent experiments on grafts of rhombomere 4 indicate that this may be the area that exerts this capability (Sechrist, Scherson, and Bronner-Fraser 1994). However, if the hindbrain is removed and foreign ectoderm is grafted instead, the underlying mesoderm is able to induce otic placodes and an imperfect ear (Harrison 1945). Thus either the periotic mesoderm alone or the hindbrain alone is able to induce ear formation. Yntema (1950) showed a temporal effect of induction through a series of transplantations of variously aged donor ectoderm on hosts. Harrison (1945), Yntema (1955), and later Jacobson (1966) provided further evidence that these inductive actions of mesoderm and neural tissue are overlapping and, to some extent, redundant because either tissue is able to create an otocyst on its own.

Jacobson (1963) provided clear evidence that in a salamander the proximity to the hindbrain is crucial for invagination. He rotated either the neural plate alone, the neural plate with the adjacent neural fold, and/or this combination with the placodal epidermis. He obtained ear formation along the rostrocaudal axis of the graft in rotations involving all ectoderm, no ears in rotations of the neural plate with the neural fold alone, and ear formation in the appropriate position in rotations of the neural plate without the neural fold. These data demonstrate that positional information is present in the underlying mesoderm/endoderm and that this information can be overruled only if the entire placodal/neural crest/neural plate tissue is rotated. Whatever the nature of these inductive signals may be, xenoplastic transplantations between salamanders and frogs can clearly lead to donor-specific ears under the inductive influence of the host (Andres 1949), thus indicating a high degree of conservation, at least among amphibians.

Nevertheless, the sequence of induction apparently differs among vertebrates. For example, in some ranid frogs, the otic placode is able to undergo morphogenesis when transplanted into the posteroventral flank at slit blastopore/early neural plate stage (Zwilling 1941). In contrast, the otic placode of at least one salamander tested for comparable epithelial autonomy is unable for self-differentiation before the late neurula (Yntema 1939). Data from studies of birds suggest that the autonomy of the ear is not established before the otic pit stage, several hours after the primary brain vesicles have formed (Waddington 1937). Autonomy of the otic vesicle is established as late as a day after its formation in mice (Van De Water and Represa 1991). Although these data are not easily comparable, they indicate a substantial shift in the timing of otic independence (i.e., a heterochronic shift) for further differentiation in these species. Moreover, the relative importance of the mesoderm and neuroectoderm may vary among species because otic placodes are apparently autonomous at the neural plate stage (i.e., before formation of neural folds begins) in some amphibians (Harrison, 1924).

In summary, the molecular mechanisms involved in mesodermal and neuronal induction in otic placode and otocyst formation are not yet clear. However, it appears that different factors are acting in succession to create the otic placode and, subsequently, the otocyst. Obviously, as long as these individual factors are not known, there is no easy way to relate the minute topological differences in various species causally to differences in the adult ear. Moreover, there is a need to study these early inductive events in more vertebrates than in amphibians and birds in greater detail to show the relative importance of mesoderm and neuroectoderm in this induction as well as the timing of induction. It would not be surprising to find various degrees of epithelial autonomy and even changes in the inductive competence of mesoderm and neuroectoderm because this is already well established for other placodally derived tissue, e.g., the lens (Grainger, Henry, and Henderson 1988).

4.4 Invagination of the Otocyst: a Novel Achievement or Opportunistic Use of Existing Programs?

At the center of the discussion about the ancestry of lateral-line and inner ear placodes is the fact that the otic placode always invaginates to form the otocyst. Based, among other things, on this developmental pattern, the lateral line was considered to be ancestral to the ear (Van Bergeijk 1967). However, Wever (1974) clearly pointed out the weakness of this argument in suggesting that the sequence of developmental appearance of the ear and lateral-line placodes may as well be taken as evidence that the ear is ancestral to the lateral line. Nevertheless, although invagination can occur in lateral-line primordia as well, it is usually after the placode is induced and has invaginated (Gutknecht and Fritzsch 1990; Northcutt 1992).

Invagination as a morphogenetic process existed in chordates before the evolution of craniates. It was associated with gastrulation and neurulation and occurs in other placodes in addition to the otic placode (Webb and Noden 1993). Although superficially similar, it is unclear whether otocyst invagination follows the same principles, such as forceful intercalation to produce convergent extention, known for gastrulation (Keller, Shish, and Sater 1992) and neurulation (Jacobson 1991). In addition to these known factors, it has been suggested that proliferation may play a role (Represa et al. 1991), but experimental data do not support this hypothesis for gastrulation and neurulation (Jacobson 1991). In fact, it has been shown that neurulation and lens, nose, and ear placode invagination take place even in the complete absence of cell division (Harris and Hartenstein 1991).

The work of Hilfer suggests that, at least in birds, otic invagination is different from other placodal invaginations because it does not rely on intracellular movement but presumably on changes in the surrounding tissue (Hilfer, Esteves, and Sanzo 1989). Hilfer and Randolph (1993) went on to show that it is likely that extracellular matrix attachment between the ear and the hindbrain may play a role in otocyst invagination. Invagination is rather different in teleosts and other vertebrates, and in some amphibians, invagination occurs even if the placode is transplanted away from the hindbrain. Thus there is clearly a need for further experimental manipulations to consolidate these suggestions. Moreover, similar mechanisms may be responsible for invagination of the neural plate and the ear in species such as bony fish because similar evolutionary changes in invagination modes of the neural plate and the ear compared with other vertebrates are found in these animals. There are some molecular candidates that might be involved in the invagination of the otocyst: these include *Fgf3 (int2)* (Represa et al. 1991; Mansour, Goddard, and Capecchi 1993; Mahmood et al. 1995), retinoic acid (Represa et al. 1990; Manns and Fritzsch 1992; Pijnappel et al. 1993), and possibly such genes as *dlx-3* (Akimenko et al. 1994), all of which have been discussed previously.

In summary, the detailed mechanisms of otocyst invagination and its molecular governance are not understood. Despite our reservations as outlined above, good candidates appear to include trophic factors such as FGF-3. A major difference between the otic placode and the adjacent lateral-line placodes could be potential access to these factors based on spatiotemporal differences in their location and release. In this context, the demonstration of Yntema (1950) that grafted ectoderm may still be induced to form lateral-line organs after the ability to form ear structures has ceased needs to be reexamined. The outcome of such experiments could strengthen the idea outlined above that only a restricted area of placodal tissue will invaginate owing to the spatiotemporally restricted influence of certain rhombomeres of the hindbrain, access to growth factors, and/or specific areas of cephalic mesoderm.

The old discussion about ancestral relationships of the ear and lateral line may be resolved by assuming that a specific combination of factors may have caused invagination of otic placodal tissue as soon as it evolved in the region of the ear, i.e., without ever forming lateral-line organs. It is possible that the developmental sequence of otic and lateral-line placode formation (in *Xenopus* at stage 21 and stage 31, respectively; Nieuwkoop and Faber 1967; Winkelbauer and Hausen 1983) may indeed reflect the sequence of evolutionary events and indicate that the lateral-line placode is nothing more than an otic placode that has failed to invaginate because it arose too late and too far away form the otic region to be induced to invaginate. Retinoic acid treatment may disrupt these signals and therefore prevent subsequent development beyond placodal formation of the ear (Fig. 3.5).

4.5 Fixation of Polarity in the Otocyst: Experimental and Genetic Evidence

Numerous experiments in amphibians have demonstrated that the polarity of axes is fixed at early stages of otic pit formation (Harrison 1945). This fixation of polarity provides topologic specificity and precedes differentiation of the otocyst in amphibians, the only vertebrates mapped in that respect in detail. It would be important for the comparative embryological analysis attempted here to know at which stage this fixation of axis is established in other vertebrates and how much difference in timing of this event exists among different species. This knowledge about the time of fixation of polarity is critical for the interpretation of all transplantation experiments and may help to resolve conflicting data obtained in different species with respect to epithelial autonomy.

In contrast to the otocyst, the placode is still an equipotential system and can regulate to form a whole ear from parts or from two amalgamated placodes in salamanders (Harrison 1945). Transplantation of rotated placodes produces enantiomorphic twins in many instances, i.e., mirror duplicates across a transverse plane that may consist of two anterior or two posterior halves (Fig. 3.6; Harrison 1945; Yntema 1955), that somewhat resemble the symmetric ear of the hagfish (Fig. 3.2). This reversal in polarity will happen along the anteroposterior axis alone until the dorsoventral axis becomes fixed somewhat later during otic cup formation. Yntema (1955) suggested that the fixation of polarity in the ear rudiment may be a local expression of a general body polarity (Fig. 3.6). The formation of enantiomorphic twins in transplanted otic placodes demonstrates that placode polarity largely follows the pattern of comparable duplication phenomena in developing and regenerating limbs (Harrison 1945; Bryant, French, and Bryant 1981; Tickle et al. 1982).

As in the developing eye and the developing limb, the anteroposterior axis is specified before the dorsoventral axis. More detailed transplantations are necessary to see whether an area equivalent to the zone of polarizing activity

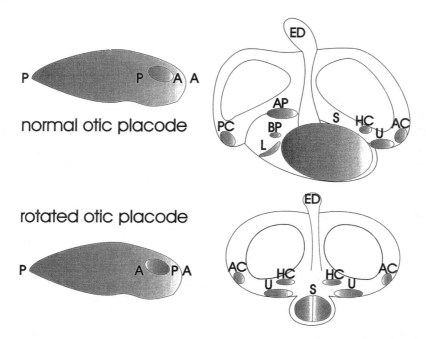

FIGURE 3.6. During development, it is at the level of otic pit formation that the polarity of the ear becomes fixed and leads to the normal differentiation of an asymmetric ear (*top*). Rotation of the otic placode with a reversal of the antero-posterior axis with respect to the anteroposterior body axis leads to the formation of an enantiomorphic ear in *Ambystoma*. This ear can consist of either two anterior halves (as shown) or two posterior halves. More detailed analysis of the sensory epithelia in these ears are necessary to learn about the hair cell polarity in the sensory epithelia. A, anterior; AC, anterior canal; BP, basilar papilla; ED, endolymphatic duct; HC, horizontal canal; L, lagena; P, posterior; PC, posterior canal; S, saccule; U, utricle. Modified according to Harrison (1945) and Yntema (1955).

(Gilbert 1991) exists at the posterior margin of the otic placode that will form a gradient of a putative morphogen. Whether retinoic acid and FGF play any role in this specification, as suggested for the developing limb (Thaller, Hofmann, and Eichele 1993; Tabin 1995), remains to be shown. At any rate, the invaginating otic cup is no longer a harmonic and equipotent system in amphibians. It is at the level of the otic placode that the isotropy is lost and polarization becomes fixed along the anteroposterior axis and soon afterward in the dorsoventral axis.

Quite recently, a number of genes have been found to be expressed in a spatiotemporal restricted pattern in the developing ear of zebra fish (Akimenko et al. 1994), frogs (Ramirez and Solursh 1993), and chickens (Ramirez and Solursh 1993; Lindberg et al. 1995). Although a causality has not yet been established among these genes and morphogenetic steps identified, their distribution makes them good candidates to specify dorsoventral, anteroposterior, and mediolateral axes. Whether their gene products are responsible for axis determination or whether these genes are turned on as a consequence of a process that specifies the polarity needs to be examined with appropriate transplantation and correlated with in situ hybridization and protein localization experiments.

In this context, it should be pointed out that the expression of *Fgf3* in the differentiating hindbrain implies a possible role in axis determination and the formation of the endolymphatic duct (Fig. 3.4). In fact, recent data with a targeted disruption of the *Fgf3 (int2)* gene in mice show reduction or absence of the endolymphatic duct and the development of larger than usual otic chambers. Lack of the endolymphatic duct causes excessive pressure of endolymph that normally may, in part, be drained through this duct (Mansour, Goddard, and Capecchi 1993).

Alternatively, lack of the endolymphatic duct and the anomalous swelling may both be defects resulting from disruption of the normal dorsoventral polarity because similar swellings have been reported in dorsoventrally inverted (Harrison 1945) or transplanted ears (Detwiler and van Dyke 1950) without loss of the endolymphatic duct. Moreover, the absence of more posterior parts of the ear, such as the posterior vertical canal (Mansour, Goddard, and Capecchi 1993), suggests an incomplete fixation of the early axis in these embryos. Therefore, *Fgf3* may play a role in fixing the embryonic axes of the ear as well as establishing axes during embryogenesis (Tannahill et al. 1992). In this context, it is notable that FGF-1, FGF-2, or FGF-4 can induce limb formation (Cohn et al. 1995). Given that FGF-3 is expressed in the ear (Tannahill et al. 1992) and that a grafted ear is able to induce ectopic limbs (Balinsky 1925), it may be possible that patterns of interactions recently discussed for limb bud formation (Tabin 1995) may also be important for ear polarity formation. Together, these data underscore the conclusion that both the retinoic acid data (Represa et al. 1990) and FGF data (Represa et al. 1991) need to be evaluated in the context of disruption of polarity, which is known to jeopardize differentiation of the otocyst (Harrison 1945; Yntema 1955) and can alter differentiation of limbs (Tabin 1995).

In summary, experimental manipulations described in the classic embryonic literature and modern studies of gene expression are compatible with the idea of an early fixation of polarity at the otic cup stage. Unfortunately, the data were obtained in different models, and it would be useful to combine both in the same animal to see whether embryonic manipulation leads to a changed expression of genes, supposedly involved in axis specification. Two possible factors involved in polarity formation (retinoic acid and *FGF-3*) could be tested in paradigms similar to those established for limb pattern formation (Riley et al. 1993; Tabin 1995). In addition, embryological manipulations of the otocyst in combination with gene expression studies could further our understanding of the mechanism(s) of polarity fixation.

5. From Otocyst to Inner Ear: Formation and Segregation of Sensory Epithelia

In the previous discussion, the early inductive influences that establish an otocyst with polarized anteroposterior, dorsoventral, and mediolateral axis are outlined. In the following section, we will describe in vivo and in vitro experiments that demonstrate the morphogenesis of the ear and some of the factors necessary to transform an otocyst into a differentiated ear. Obviously, on the basis of Harrison's, Yntema's and Jacobson's experiments, it is clear that although both the hindbrain and the mesoderm can induce an otocyst even in ventral ectoderm, this otocyst will not differentiate appreciably in most amphibians unless the hindbrain is adjacent to it at (and for) a critical time. Furthermore, this organogenesis is apparently orchestrated into a sequence of inductive events acting in concert to produce complete differentiation. Thus, for normal development of the ear, all factors appear to be important, and absence of a single one can compromise further differentiation.

Xenoplastic transplantations of an otic placode have shown that differentiation can proceed in grafted tissue without major disruptions in distantly related hosts (Andres 1949). The latter approach may be used to test the interaction in vastly different tissues such as lamprey otocyst in amphibian hosts or mouse otocyst in chicken host. To our knowledge, these xenoplastic transplantations have not been tried but may potentially help to dissect the spatiotemporal pattern of species-specific gene activation in the induction and differentiation of the otocyst (Andres 1949).

In particular, grafting of an otocyst between apparently similar developmental stages of vertebrates (e.g., neural plate stage, otic pit stage) may help to determine the degree of polarity fixation and independence of further development that results in the host environment. These experiments may provide insight into possible heterochronic shifts among species (Arnold et al. 1989). Heterochronic and heterotypic transplants of quail otocysts into chick embryos, known to be possible at these early developmental stages, may provide a more feasible model to test some of these hypotheses. However, these experiments must be done with the caveat that graft vs. host or host vs. graft rejection, even in short-term transplants, might at least complicate the interpretation of these studies involving transplants of distantly related species.

5.1 Segregation of Sensory Epithelia

The ontogeny of the sensory epithelia apparently closely follows the known and inferred evolutionary history (Werner 1960). In a frog, for example, the utricle/anterior vertical/horizontal canal epithelia segregate together from the saccular/lagenar/papilla amphibiorum/papilla basilaris epithelia to form the anterior (dorsal, superior) compared with the posterior (ventral, inferior) subdivision of the ear in most vertebrates (Fig. 3.7; Norris 1892). In mammals, this early subdivision is modified by the addition of a further subdivision that forms the cochlea. Apparently this differentiation and segregation of sensory epithelia can take place if otocysts are transplanted to various locations or placed in vitro, even when the otocyst is reopened (Harrison 1945; Van De Water 1983; Swanson, Howard, and Lewis 1990; Sokolowski, Stahl, and Fuchs 1993). However, what is unclear in most of the studies dealing with transplantation is the degree of axis fixation in the otocyst at the time of transplantation. Other unknown elements include the time during which the hindbrain exerts its influence and the timing and factors (cell bound to or secreted by mesoderm) that are necessary for complete differentiation. Given that there is some apparent heterochronic shift among species, discrepancies among different studies may be explained on this basis. Because of these uncertainties, we will deal with transplantations first and in vitro experiments second. Moreover, we will deal with amphibians, birds, and mammals separately.

5.2 Amphibians

Transplantation and grafting experiments have been performed mostly in these species, but only one attempt has been made to cultivate the otocyst. Numerous transplantation studies have clearly shown that the further differentiation of the otocyst requires the presence of cephalic mesenchyme (which is of neural crest and mesodermal origin in the otic region) and the hindbrain (Kaan 1930; Detwiler and van Dyke 1950). Transplanted ears will develop normally in most species only when the hindbrain is adjacent. Sensory cell differentiation in the ear, even though apparently dependent on the otic mesenchyme, can proceed even if the mesenchyme is reduced. This underscores the necessity of evaluating the apparent independence of sensory epithelial differentiation and morphogenesis of the ear. Whereas the latter requires both the hindbrain and the otic mesenchyme, the former can proceed to some extent. The degree of development is presumably correlated with the degree of autonomy already established in the grafted otocyst.

In this context, Fritzsch (unpublished observations) has performed late and early transplantations of the otocyst in *Xenopus* (stages 28 and 38, Nieuwkoop and Faber 1967) from the otic region into the dorsal fin of the same animal. In none of 46 cases was there morphogenesis of the ear into the known subdivisions, i.e., the ear remained cystic. However, the late-transplanted ears showed a complete segregation of up to seven discrete sensory epithelia, which probably represent all sensory epithelia known in frogs (Fritzsch and Wake 1988). Given that all of these transplantations were performed before the morphological segregation of sensory epithelia had occurred, it would indicate that at this stage the anlagen of all sensory

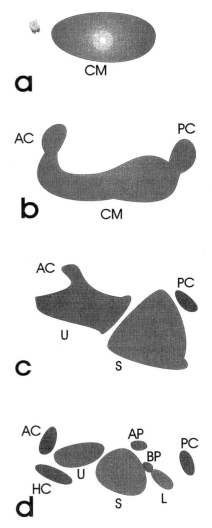

FIGURE 3.7. The segregation of sensory epithelia in a salamander, *Ambystoma*, is shown. Note that the common macula (CM; a) first gives rise to the anterior and posterior canal cristae (AC and PC, respectively; b). At later stages, the utricle and saccule (U and S, respectively; c) and even later the horizontal canal (HC), the amphibian papilla (AP), the lagena (L), and the basilar papilla (BP) segregate, thus recapitulating the known history of otic evolution (Figs. 3.2, 3.3). Before segregation, there may already be a commitment to form certain epithelia from specific regions of the otocyst, as indicated by selective gene expression and the formation of enantiomorphic twins in otic placode rotation (Fig. 3.6). Modified according to Norris (1892).

epithelia are induced and will segregate even in the complete absence of any morphogenesis of the ear as they do during normal development (Norris 1892; Fig. 3.7).

The only in vitro study of amphibian ears thus far (Jacobson 1963) showed that normal differentiation of the ear requires the synergistic action of mesoderm and neural plate at least until neurulation is complete. Unfortunately, this study did not illustrate the details of morphological differentiation that can be obtained in vitro nor did it dissect the chronology of inductive events from mesoderm and neural plate further. New experiments testing for windows in inductive competence for mesoderm and neuroectoderm in otocyst formation, polarization, and differentiation are needed. Combinations of tissues analogous to experiments that have been performed in mammals (Van De Water et al. 1992) would provide information on these points.

5.3 Birds

In the chick, there is an extensive literature on the normal differentiation, transplantation, and cultivation of chick otocysts. Induction of the otic placode appears to happen between 24 and 48 hours, and it is rhombomere 4 that seems to be able to induce otic vesicle formation during this period in the ectoderm (Sechrist, Scherson, and Bronner-Fraser 1994). Hilfer, Esteves, and Sanzo (1989) have studied the invagination in detail, and Hilfer and Randolph (1993) and Knowlton (1967) have described the transformation of the invaginating otic placode into a differentiated ear: The otocyst is formed at 2.5 days and soon after is transformed into a pars superior, a pars inferior, and the endolymphatic duct. At 4 days, the future sensory epithelia consist of three segregated patches: (1) an anteroventral one forming the utricular macula and the anterior and lateral (horizontal) cristae, (2) a ventromedial one with the basilar papilla and the maculae sacculi and lagena, and (3) a posteroventral one with the anlage of the posterior crista and papilla neglecta. Further differentiation of the basilar papilla consists of elongation, constriction of the ductus reuniens to the saccular lumen (Cohen and Cotanche 1992), and proliferation of hair cells (Katayama and Corwin 1989). Differentiation of hair cells partly overlaps with the proliferative phase (5–10 days; Katayama and Corwin 1989; Cohen and Cotanche 1992) and is largely complete when the apical stereocilia have almost achieved their adult configuration (17 days; Tilney, Cotanche, and Tilney 1992) and orientation (Cotanche and Corwin 1991). Transformation of the periotic mesenchyme into fluid-filled scalae next to the basilar papilla is completed at hatching (Knowlton 1967). A recent excellent summary is presented in Fekete (1996) and Wu and Oh (1996).

An extensive literature exists describing the development of transplanted ears. Waddington (1937) was the first to show the autonomy of ear differentiation in transplanted ears. This finding was later confirmed and extended by Corwin and Cotanche (1989) and Swanson, Howard, and Lewis (1990). From these data, it can be concluded that at the late otic cup/early otocyst stage the sensory epithelia of the ear are able to form even when the ear is transplanted to the chicken wing and cut open so that the inside of the otocyst is exposed to the amniotic fluid. Similarly, otocysts transplanted to the chorioalantoic membrane develop a basilar papilla even in the absence of nerve fibers (Corwin and Cotanche 1989). Neuroepithelial autonomy, at least at the otocyst level, may perhaps be due to the early expression of genes and/or the distribution of cell adhesion molecules (Richardson et al. 1987). Despite this early autonomy, the present data do not show that the proliferation and segregation of all sensory epithelia in the ectopically transplanted otocysts is normal. In fact, Swanson, Howard, and Lewis (1990) stress the limited differentiation of morphologically identifiable sensory epithelia that seem to require interaction with the appropriate mesenchyme to undergo separation from the single anlage. Corwin and Cotanche (1989) report dramatically reduced numbers of hair cells and a changed morphology of the basilar papilla in their transplants.

Similarly, cultivating otocysts has long been known to result in variably differentiated sensory epithelia, possibly related to the amount of cotransplanted otic mesenchyme (Fell 1928; Friedmann 1956; Orr 1981, 1986; Ard, Morest, and Hauger 1985; Sokolowski, Stahl, and Fuchs 1993). In the study, it was shown that despite some differentiation of sensory epithelia, they formed predominantly type II vestibular hair cells with rather normal electrophysiological characteristics. A number of factors may be responsible for the different success rates of differentiation of cultivated otocysts. These include slight differences in actual age of the otocyst at the time of explantation or differences in the amount of otic mesenchyme of either of the three cellular sources (Couly, Coltey, and Le Douarin 1993) attached to the transplanted otocyst. Most recently the possible role played by neuronal factors or interactions such as efferent innervation was suggested (Sokolowski, Stahl, and Fuchs 1993). Cocultivation of otocysts with the hindbrain is needed to determine the role of the hindbrain in otocyst development at very early stages. Culture of otocysts with the brain stem should show how much (if any) effect the brain stem may have on the maturation of cochlear hair cells, some of which are exclusively innervated by efferent fibers (Fischer 1992). Transplantation of the cochlear duct is possible as late as day 10, and the maturation of these explants will continue without major deviations (Stone and Cotanche 1991).

5.4 Mammals

An extensive descriptive literature exists on the normal development of the otocyst in humans and some other mammals (Anniko 1983; Lim and Rueda 1992; Sulik and Cotanche 1994). The most detailed knowledge of the differentiation of the ear in any mammal exists in mice in which the timing of terminal mitosis (Ruben 1967), morphological differentiation (Tello 1931; Sulik and Cotanche 1994), and ingrowth of fibers (Sher 1971; Sobkowicz and Rose 1983) have been studied. The morphogenesis and maturation of the cochlea will be dealt with in Chapter 4. In mice, the cochlea is derived from the ventral half of an 11-day otocyst (Li et al. 1978; Fig. 3.8), which represents a mosaic of cells with disparate future fates at this stage of development (Van De Water 1983).

Although there is an extensive literature on the in vitro cultivation of the mouse otocyst, much less is known with respect to transplantation or early inductive events. However, Waterman (1925) described rather normal development of transplanted otocysts into the bursa omentalis of rabbits.

cultivation of ventral half
produces: Cochlea, Utricle
and Saccule

cultivation of dorsal half
produces: Canals, Utricle
and Saccule

FIGURE 3.8. The experimental evidence for anisotropy in the mouse otocyst as revealed in tissue culture experiments is shown. In the mouse, in vitro cultivation only of the ventral half of the otocyst leads to formation of the cochlea. In contrast, cultivation of the dorsal half of the otocyst leads to the formation of canals, utricle, and saccule. Modified according to Li et al. (1978).

To obtain more detailed information, either this approach or xenoplastic grafts to chickens could be done to establish the early inductive events in mammals. However, again some reservations must be entertained concerning the results of xenoplastic grafting experiments, particularly between birds and mammals, between which immune responses may clearly complicate the interpretation of results.

By the time the otocyst closes (9.5 days of gestation), it has acquired some degree of autonomy but will not undergo complete differentiation if transplanted into tissue culture. Moreover, from closure of the otocyst until ~1 day later, the explanted otocyst requires the presence of the hindbrain for its differentiation (Van De Water 1983). Only after embryonic day 10.5 is the otocyst apparently autonomous in its differentiation except for the cochlea and related structures (Van De Water and Represa 1991). However, it will undergo normal differentiation and nonvestibular hair cell differentiation only if coexplanted with surrounding mesenchyme (Sobkowicz, Bereman, and Rose 1975; Sobkowicz and Rose 1983; Van De Water 1983). It is only after embryonic day 11.5 that the otocyst can develop to a large extent without the mesenchyme around it.

These data strongly suggest that there is a sequence of inductive events leading to developmental competence, which requires the hindbrain (until embryonic day 10.5) and the periotic mesenchyme (until embryonic day 11). From studies that exclusively eliminate the ventral periotic mesenchyme, it is apparent that the periotic mesenchyme is essential for the formation of the cochlea (Li and McPhee 1979; Van De Water 1983) and coiling of the cochlea (Fig. 3.9).

Cochlear morphogenesis therefore may reflect an epitheliomesenchymal interaction. It would be important to know whether this interaction is possible between a mouse cochlea and a chicken periotic mesenchyme; i.e., are the primary differences between these two types of auditory sensors within the sensory epithelium or within the surrounding mesenchyme? Appropriate transplantations or in vitro cultivations may allow us (caveats observed) to answer these questions, which would be important for an understanding of the evolution of the mammalian cochlea out of a therapsid basilar papilla. Similarly, further characterization of the trophic interactions between these two tissues (Van De Water et al. 1992) in mice and chicken may help to identify unique molecular interactions present in either developing system.

In summary, the descriptive and experimental evidence on the formation of subdivisions and segregation of sensory epithelia in amphibians, birds, and mammals has shown that the basilar papilla/cochlea is an outgrowth of the ventral part of the otocyst that may become developmentally committed soon after the formation of the otocyst is complete. The segregation of the sensory anlage and its morphogenesis and histogenesis are critically dependent on the interaction between the otocyst and the surrounding periotic

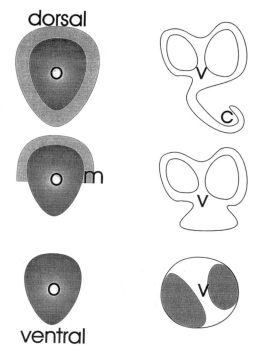

FIGURE 3.9. This diagram shows the effect of coculturing the otocyst of an embryonic day 11 mouse with the surrounding mesenchyme (m; *top*), mesenchyme only around the dorsal half (*middle*), or without mesenchyme (*bottom*). The resulting degree of otic differentiation shows that the mesenchyme is necessary for the formation of the cochlea (c) and to a lesser degree the vestibular part of the ear (v). Without any mesenchyme, the ear will remain cystic (*bottom*) but may form isolated patches of sensory epithelia. Modified according to Van De Water (1983).

mesenchyme. Although aspects of this process can be replicated in vitro, more electrophysiological studies are needed to show the degree of auditory hair cell maturation in these preparations.

6. Shaping the Ear: the Periotic Tissue and the Formation of the Perilymphatic Ducts

After otocyst formation is completed, the subdivision of the epithelia and the morphogenesis of the ear may proceed along the previously induced pathways. Although formation of the epithelia may have reached a large degree of autonomy at this stage, the morphogenesis of the ear requires further interaction with the periotic mesenchyme to form the cochlea and the semicircular canals (Van De Water 1983).

The reciprocal influence of periotic mesenchyme on otocyst morphogenesis and of the otocyst on cartilage and bone formation in the surrounding mesenchyme has been studied in a variety of transplantations and in vitro in mice, chickens, and frogs (Yntema 1955; Van De Water 1983). These data show that morphogenesis of the ear is largely eliminated in the absence of periotic mesenchyme, and cytodifferentiation is reduced to the vestibular type of hair cells in mice (Van De Water 1983). More specifically, the coiling of the cochlea was absent in grafts that lacked periotic mesenchyme around the ventral aspect of the otocyst (Li and McPhee 1979; Fig. 3.9). However, the contrary is not true, and cartilage may develop in the absence of an otocyst in some species but will also be induced by the otocyst in foreign mesoderm, thus providing a double security for ear capsule formation (Yntema 1955). This chondrogenic ability of the otocyst is apparently stage dependent in mice (Van De Water et al. 1992) and may be fixed at different stages in different species, thus explaining the apparent formation of cartilage in the absence of an otocyst in bony fish (Yntema 1955; Hall 1991; for review). Moreover, the otocyst may act as a suppressor of chondrogenesis at later stages allowing the sculpting of the ear. It appears that certain growth factors like the transforming growth factors (Frenz and Van De Water 1991, Van De Water et al. 1992) may be the mediators of this chondrogenic signal and will also act as a suppressor at later stages (Van De Water et al. 1992).

Evidence was also presented suggesting that the extracellular matrix molecule hyaluronan may act as a propellant for the formation of the horizontal canal in *Xenopus* (Haddon and Lewis 1991) and also in a bony fish (Chapman and Fraser 1993). If hyaluronidase is injected, the formation of the horizontal canal is blocked. Nevertheless, the sensory cells of the horizontal canal segregate, thus showing that both events are clearly distinct as already pointed out in a number of transplantation and in vitro experiments outlined above. Moreover, hyaluronidase converts these ears into cyclostome-like ears, which lack a horizontal canal (Fig. 3.1). It would be important to know whether similar defects can be achieved in the growing vertical canals or whether the hyaluronidase effect is specific to the horizontal canal.

More specific effects of certain extracellular matrix proteins in the differentiation of the cochlea of mice were recently suggested. For example, loss of type I collagen can result in fusion of the stapes footplate with the otic capsule (Altschuler et al. 1991). Other mutants with defects in chondrogenesis may show selective absence of parts of sensory epithelia in the cochlea (Cho, Yamada, and Yoo 1991; Yoo, Cho, and Yamada 1991). Thus far the molecular mechanism(s) of the influence of the extracellular matrix on cochlea differentiation and maturation is not fully understood, but it clearly emphasizes the necessity for proper tissue interaction between otocyst and periotic mesenchyme for complete morphogenesis and differentiation.

In summary, the placodally derived otocyst and the mesodermally and neural crest-derived periotic mesenchyme need to interact to allow complete morphogenesis of the ear. It is unclear at which time in development complete autonomy of either tissue for its specific differentiation is achieved. There appears to be a need for continuous, probably reciprocal, interaction of these tissues to ensure complete differentiation, at least for the morphogenesis of the semicircular canals and for the cytodifferentiation of the cochlea.

7. Innervating the Ear: Origin of Otic Ganglion Cells

Formation of the otic ganglion (statoacoustic or VIII nerve ganglion) and the role of innervation for trophic support of hair cells and otic differentiation is a controversial issue. The data can be divided into two sets: descriptive data and experimental data. Whereas the former, by and large, tend to support the idea of a neural crest contribution to otic ganglia and see correlation between ingrowing fibers and hair cell differentiation, the latter suggest an almost exclusive formation of otic ganglia from placodal tissue and a minimal, if any, influence of the ganglia on ear morphogenesis and differentiation.

Understanding and resolving the controversy require an understanding of how the ear and its ganglion are related to each other. A number of authors have claimed that otic ganglia are derived from neural crest (Adelman 1925) and even that neural crest may induce the ear (e.g., Szepsenwol 1933). Later descriptive analysis has provided different evidence (Rubel 1978), and experimental studies have demonstrated the origin of otic ganglia exclusively from the otic placode in amphibians (Yntema 1955) and almost exclusively in the chicken (D'Amico-Martel 1982; D'Amico-Martell and Noden 1983). The otic ganglia seem to be no different from other placodally derived ganglia and therefore fit into the general scheme of other placodes that form ganglia as a first step in development (Von Kupffer 1895; Noden and Van De Water 1986).

The migration of otic ganglia away from the invaginating placode can be as readily observed as the delamination of the neural crest (Von Kupffer 1895; Hemond and Morest 1991; Webb and Noden 1993; Haddon and Lewis 1996). This delamination of ganglia from the "otic crest" begins in chickens before otocyst formation is completed (Hemond and Morest 1991), and it apparently occurs even when invagination is suppressed with retinoic acid in frogs (Fig. 3.5), much like lateral-line and epibranchial placodes (Webb and Noden 1993). Extirpation of various parts of the otocyst in mice provides experimental evidence that the statoacoustic ganglion is predominantly derived from the anteroventral aspect of the otocyst (Li et al. 1978). Comparable extirpations need to be performed in other vertebrates to elucidate how early and at which position of the otocyst the ganglion cell formation takes place. The apparent early formation of ganglion cells at a specific site also offers some insight into the timing of polarity fixation of the otocyst of which the formation of ganglion cells is an integral part. It would be important to establish what clonal relationship, if any, exists between the sensory epithelia and the ganglion cells. Even though it is now clear that hair cells and supporting cells are clonally related (Muthukumar and Fekete 1994), it is not yet clear what sets the clones of sensory/supporting cells apart from the other areas of the developing otocyst: is it clonal restriction or induction via differential gene expression? The distribution of BMP-3 in the developing ear of the chick (Wu and Oh 1996) and BMP-4 in the developing ear of *Xenopus* (Hemmati-Brivanlou and Thomsen 1995) suggests the latter. In situ hybridization of these genes demonstrates an early distribution in two patches at the placode, which remains stable in position throughout the process of invagination and is found ultimately in the sensory epithelia of the anterior and posterior vertical canal.

In conclusion, both comparative and experimental data overwhelmingly demonstrate the placodal origin of the otic ganglion cells, whereas the majority of glial cells and a small subpopulation of neurons in the vestibular portion of the ganglion are contributed by the neural crest. It is unclear whether the ganglion cells in various species derive from the same area of the otocyst and show a comparable time course of delamination. Is otic induction conserved with respect to the earliest mosaic of otocyst subdivisions? In particular, hagfish should be examined in that respect because they are the only known vertebrate with two distinct vestibular nerves (Amemiya et al. 1985).

7.1. The Ear and Neurotrophic Substances: Who Is Supporting Whom?

There is a large body of evidence demonstrating arrival of afferent fibers in the developing ear before the onset of hair cell maturation. For example, most of the spiral ganglion neurons are postmitotic before cochlear hair cell proliferation peaks (Ruben 1967), and Sher (1971) described afferents in the otocyst wall before differentiation of hair cells in mice. Similarly, it appears that in chickens afferent fibers penetrate the future sensory epithelium before all hair cells are postmitotic (Whitehead and Morest 1985; Cohen and Cotanche 1992). Thus the arrival of afferents in the developing cochlea would be consistent with suggestions that they play a role in hair cell maturation (Pirvola, Lehtonen, and Ylikoski 1991). However, numerous in vitro studies in mice (Van De Water 1983; Sobkowicz et al. 1986) and in chickens (Sokolowski, Stahl, and Fuchs 1992) as well as grafting procedures with or without adjacent statoacoustic ganglia (Corwin and Cotanche 1989) indicate that innervation is not necessary for histogenesis of the cochlea.

More recent experiments in other placodally derived sensory systems strongly support this conclusion for the lateral line (Kelley and Corwin 1992) and electroreceptors (Fritzsch, Zakon and Sanchez 1990). However, it is clear that hair cells require some kind of innervation for long-term maintenance, and the detailed electrophysiological maturation of all types of hair cells may depend on some neurotrophic interaction (Sokolowski, Stahl, and Fuchs 1992).

In contrast to the rather limited effect of ganglion cells on the differentiating ear, the developing cochlear ganglion cells have better survival rates in vivo when cocultured with the otocyst or the rhombencephalon (Ard, Morest, and Hauger 1985; Orr 1986; Van De Water et al. 1992). Possible mediators of this neurotrophic influence may be nerve growth factor and/or brain-derived neurotrophic factor and neurotrophin 3, which are known to be present in the developing otocyst of rats and chickens around the critical time (Von Bartheld et al. 1991; Pirvola et al. 1992; Hallböök et al. 1993). In addition, the octaval ganglion cells express the appropriate receptors during development (Pirvola et al. 1994). These factors are known to minimize cell death in other systems with naturally occurring cell death (Oppenheim 1991; Oppenheim et al. 1992), which eliminates ~25% of cochlear ganglion cells in rat (Rueda et al. 1987) and chicken embryos (Ard and Morest 1984). However, the effect of nerve growth factor on statoacoustic ganglion cell survival in vitro appears to be rather limited (Represa and Bernd 1989), but anti-nerve growth factor antibodies can disturb the pattern of innervation in vitro, thus implying a role for nerve growth factor in the establishment and maintenance of innervation (Van De Water et al. 1992).

Direct proof for the suggested role of brain-derived neurotrophic factor (Ernfors, Lee, and Jaenisch 1994) and neurotrophin 3 (Farinas et al. 1994) and their respective receptors (TrkB and TrkC) was recently gained in mice, with a targeted disruption of either the neurotrophin or the neurotrophin-receptor genes (Fig. 3.10). Mice doubly homozygous for both disrupted neurotrophins or neurotrophin receptors have no innervation at all at birth (Ernfors et al. 1995; Fritzsch et al. 1995). Although there were minor differences in the details of the single neurotrophin or neurotrophin-receptor lesions, these data clearly demonstrate that the development of the ear proceeds normally in the complete absence of any innervation (Fritzsch et al. 1997). It also shows that these two neurotrophins and their respective receptors are both necessary for the maintenance of inner ear innervation.

7.2. Timing and Arrival of Afferents

Tissue culture experiments were the first to indicate that distal processes of ganglia can be attracted to innervate sensory epithelia of a second otocyst freed from its own ganglion (Van De Water 1983; Van De Water et al. 1992). Delaminating otic ganglion cells lose contact with the otocyst. Their

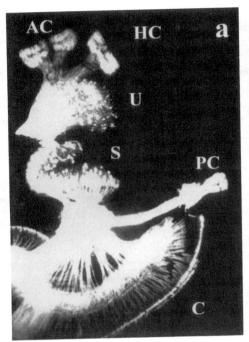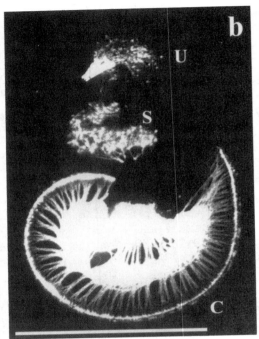

FIGURE 3.10. The appearance of the afferent innervation of a postnatal day 1 mouse as revealed by transganglionic diffusion of DiI is shown. a: normal ear. b: ear of a *trkB*[-/-] littermate. *Right*, posterior; *top*, dorsal. Note the prominent innervation of all sensory epithelia in the control ear but complete absence of innervation to all sensory epithelia of the three semicircular canals. The innervation to the utricle and saccule, while qualitatively unaffected, is less dense in the transgenic mouse (b). AC, anterior canal; C, cochlea; HC, horizontal canal; PC, posterior canal; S, saccule; U, utricle. Bar, 100 μm.

distal processes need to regrow towards the otocyst and enter at specific sites (Hemond and Morest 1992). In vitro studies suggest that the developing otocyst releases a tropic factor that may attract the growing ganglionic processes toward the differentiating sensory epithelia (Hemond and Morest 1992; Bianchi and Cohan 1993). The nature of this molecule(s) is currently unknown. However, the early appearance of innervation to all sensory epithelia in *trkB/trkC* knockout mice (Fritzsch et al. 1995) supports the conclusions derived from in vitro studies that this factor must be different from brain-derived neurotrophic factor and neurotrophin 3.

7.3. The Efferent System to the Ear: Timing of Arrival in Mice and Chicken

Traditionally, the efferent system to the ear (Roberts and Meredith 1992; Warr 1992) was thought to arrive after the 10th embryonic day in chickens

(Cohen and Cotanche 1992) and predominantly after birth in mammals (Shnerson, Devigne, and Pujol 1982, Pujol 1986). However, it was pointed out recently that early stages of afferent and efferent synaptogenesis can not be distinguished. Thus efferents may be present earlier (Emmerling et al. 1990; Sobkowicz 1992). Reexamination of arrival of efferent fibers based on acetylcholinesterase histochemistry, the catabolizing enzyme of the putative efferent transmitter, suggested a prenatal arrival in mice (Sobkowicz 1992). This conclusion is also supported by studies employing the lipophilic tracer DiI (Cole and Robertson 1992). Most recently, it was shown that efferent fibers may enter the developing otocyst of mice (Fritzsch and Nichols 1993) and chickens (Fritzsch, Christensen, and Nichols 1993; Fritzsch, 1996b) much earlier then previously suggested and at about the same time as afferents. At the moment, it is unclear whether efferent fibers are important at all for the differentiation of the ear or of hair cells (Sokolowski, Stahl, and Fuchs 1992). However, there is the intriguing possibility that these fibers, being derived from a subpopulation of facial motoneurons (Fritzsch and Northcutt 1993), may respond to the brain-derived neurotrophic factor present at that time in the developing ear (Hallböök et al. 1993) and known to be important for motoneuron survival (Oppenheim et al. 1992). Recent data on brain-derived neutrophic factor and neurotrophin 3 mutant mice suggest that efferents reach sensory epithelia devoid of afferent fibers (Ernfors et al. 1995). In contrast, a detailed analysis of *trkB* and *trkC* knockout mice strongly suggests that efferents predominantly navigate to their target along the few remaining intact afferents and respond only indirectly to brain-derived neurotrophic factor or TrkB defects (Fritzsch et al. 1995, 1997). In summary, the neurotrophin(s) that supports the efferent system is not yet known. All effects on efferents published in neurotrophin-deficient mice could represent indirect effects achieved through the action of these neurotrophins and their receptors on the afferents.

8. Summary

This paper summarizes the role of specific genes in the early development of the vertebrate inner ear, specifically in relation to the interaction of the otic placode/otocyst with the hindbrain and neural crest, and then outlines the evolution of ontogenetic changes in early ear formation in an attempt to elucidate the comparative embryology of the otocyst. The comparative data show strong similarities in early spatial configurations as well as the early inductive interactions but also significant differences, in particular in the timing of certain degrees of autonomy. Even though induction of an otic placode can be achieved by either the hindbrain or the periotic mesenchyme, both are necessary for complete morphogenesis. This conclusion of the classic experimental embryology is reinforced by more recent in vitro experiments and by gene expression patterns in normal vertebrates and mutant mice.

Transformation of the otic placode to the otocyst fixes the pattern induced earlier and allows a variable degree of autonomous differentiation, at least of the sensory epithelia. However, complete morphogenesis of the ear, in particular the cochlea, requires a longer interaction with the periotic mesenchyme. This interaction is likely to be mediated by certain growth factors. Although the otocyst can differentiate to a large degree without innervation, the otocyst-derived otic ganglion cells require some support from the otocyst provided by two otocyst-derived trophic factors, brain-derived neurotrophic factor and neutrotrophin 3.

8.1 Epilogue: Evolution as a Continuous Shaping of Developmental Processes

This chapter has provided an overview of the structural and molecular events involved in the transformation of undifferentiated epidermal cells and the underlying mesenchyme into a highly complex, three-dimensional ear. Although tremendous progress has been achieved in a few model systems like the chicken, mouse, and some amphibians, we do not yet know enough about the issues of comparative embryology of ear development to explain the adult structural differences found in vertebrate ears. We can only speculate that changes in selector and/or structural genes (particularly transcription factors) and changes in the spatiotemporal induction of structural gene activation and growth factors, as well as possible changes in the interactions between the various embryonic sources that contribute to the ear, will soon be understood. The most promising new avenue in this context appears to combine studies involving classic transplantation experiments, modern gene expression studies, and in vitro assays of the role of putative morphogens or trophic factors. Another promising approach is the analysis of the various mutations of the ear recently described in zebrafish (Malichi et al. 1996; Whitfield et al. 1996).

It is relevant to emphasize that we do not as yet understand the inductive and molecular basis of the developmental differences leading to the basilar papilla of amphibians, birds, and mammals. In addition, we do not understand what is missing in the developmental program of vertebrates that do not develop a basilar papilla or a lagena. Direct comparison of gene expression patterns in salamanders of comparable stages, which either do or do not develop a basilar papilla, may help to understand the molecular events that have led to this major evolutionary novelty of the developing ear.

Acknowledgments. This work was supported in part by National Institute on Deafness and Other Communication Disorders Grant #P50-DC-00215-

09 to Bernd Fritzch and DC02492 to Margaret I. Lomax, by National Institute of Neurological Disorders and Stroke Grant #RO1-NS-3164-01 and National Science Foundation Grant #IBN9219666 to Kate F. Barald, and by the Deafness Research Foundation to Margaret I. Lomax. Bernd Fritzsch expresses his sincere gratitude to his wife Dr. M.-D. Crapon de Caprona for helping him through the gestation of this work and to J. Schneider for his excellent assistance with the computer graphics. Kate F. Barald thanks Drs. Drew M. Noden of Cornell University, Monte Westerfield of the University of Oregon, Steve Wilson of Kings College, London, Ali Hemmati-Brivanlou of Rockefeller University, Peter Hitchcock of the University of Michigan, Julian Lewis of Oxford University, and Marc Ekker of the University of Ottawa for very helpful discussions, permission to quote unpublished material and/or comments on the manuscript and Kendra Lindberg for editorial comments in preparation of the manuscript.

References

Adelman HB (1925) The development of the neural folds and cranial ganglion of the rat. J Comp Neurol 39:19–123.

Akimenko M-A, Ekker M, Wegner J, Lin W, Westerfield M (1994) Combinatorial expression of three zebra fish genes related to distal-less: part of a homeobox gene code for the head. J Neurosci 14:3475–3486.

Altschuler RA, Dolan DF, Ptok M, Gholizadeh G, Bonadio J, Hawkins JE (1991) An evaluation of otopathology in the MOV-13 transgenic mutant mouse. Ann NY Acad Sci 630:249–255.

Amemiya F, Kishida R, Goris RC, Onishi H, Kustinoki T (1985) Primary vestibular projections in the hagfish, *Eptatretus burgeri*. Brain Res 337:73–79.

Andres G (1949) Untersuchungen an Chimären von Triton und Bombinator. Teil I. Entwicklung xenoplastischer Labyrinthe und Kopfganglien. Genetica 24:387–534.

Anniko M (1983) Embryonic development of vestibular sense organs and their innervation. In: Romand R (ed) Development of Auditory and Vestibular Systems. New York: Academic Press, pp 375–423.

Ard MD, Morest DK (1984) Cell death during development of the cochlear and vestibular ganglia of the chick. Neuroscience 2:535–547.

Ard MD, Morest DK, Hauger SH (1985) Trophic interactions between the cochleovestibular ganglion of the chick embryo and its synaptic targets in culture. Neuroscience 16:151–170.

Arnold SJ, Alberch P, Csanyi V, Dawkins RC (1989) How do complex organisms evolve? In: Wake DB, Roth G (eds) Complex Organismal Functions: Integration and Evolution in Vertebrates. Chichester, UK: John Wiley and Sons, pp. 403–434.

Ayer-Le Liver CS, Le Douarin NM (1982) The early development of cranial sensory ganglia and the potentialities of their component cells studied in quail-chick chimeras. Dev Biol 94:291–310.

Baird IL (1974) Anatomical features of the inner ear in submammalian vertebrates. In: Keidel WD, Neff WD (eds) Handbook of Sensory Physiology. Vol. V/1. Auditory System. Berlin: Springer-Verlag, pp. 159–212.

Balfour FM (1885) Comparative Embryology. London: Macmillan, Vol. 1, pp. 492.

Balinsky BI (1925) Transplantation des Ohrbläschens bei Triton. Roux's Arch Dev Biol 143:718–731.

Barald KF (1989) Culture conditions affect the cholinergic development of an isolated subpopulation of chick mesencephalic neural crest cells. Dev Biol 135:349–366.

Barlow LA, Northcutt RG (1995) Embryonic origin of amphibian taste buds. Dev Biol 169:273–285.

Bianchi LM, Cohan CS (1993) Effects of the neurotrophins and CNTF on developing statoacoustic neurons: comparison with an otocyst-derived factor. Dev Biol 159:353–365.

Bryant SV, French V, Bryant PJ (1981) Distal regeneration and symmetry. Science 212:993–1002.

Carpenter EM, Goddard JM, Chisaka O, Manley NR, Capecchi MR (1993) Loss of Hox-A1 (Hox-1.6) functions results in the reorganization of the murine hindbrain. Development 118:1063–1075.

Chalepakis G, Stoykova A, Wijnholds J, Tremblay P, Gruss P (1993) Pax: gene regulators in the developing nervous system. J Neurobiol 24:1367–1384.

Chapman B, Fraser SE (1993) Locations of vestibular hair cells in developing zebrafish embryos visualized with a fluorescent vital dye. Soc Neurosci Abstr 19:1580.

Chisaka O, Musci TS, Capecchi MR (1992) Developmental defects of the ear, cranial nerves, and hindbrain resulting from targeted disruption of the mouse homeobox gene Hox-1.6. Nature 355:516–520.

Cho H, Yamada Y, Yoo TJ (1991) Ultrastructural changes of cochlea in mice with hereditary chondrodysplasia (cho/cho). Ann NY Acad Sci 630:259–261

Cohen GM, Cotanche DA (1992) Development of the sensory receptors and their innervation in the chick cochlea. In: Romand R (ed) Development of Auditory and Vestibular Systems 2. Amsterdam: Elsevier, pp. 101–138.

Cohn MJ, Izipisua-Belmonte JC, Abud H, Heath JK, Tickle C (1995) Fibroblast growth factors induce additional limb development from the flank of chick embryos. Cell 80:739–746.

Cole KS, Robertson D (1992) Early efferent innervation of the developing rat cochlea studied with a carbocyanine dye. Brain Res 575:223–230.

Cordes SP, Barsh GS (1994) The mouse segmentation gene kr encodes a novel basic domain-leucine zipper transcription factor. Cell 79:1025–1034.

Corey DP, Breakefield XO (1994) Transcription factors in inner ear development. Proc Natl Acad Sci USA 91:433–436.

Corwin, JT, Cotanche DA (1989) Development of location-specific hair cell stereocilia in denervated embryonic ears. J Comp Neurol 288:529–537.

Cotanche DA, Corwin JT (1991) Stereociliary bundles reorient during hair cell development and regeneration in the chick cochlea. Hear Res 52:379–402.

Couly GF, Coltey PM, Le Douarin NM (1993) The triple origin of skull in higher vertebrates: a study in quail-chick chimeras. Development 117:409–429.

Damas H (1944) Research on the development of the lamprey (Lampetra fluviatilis L.) Arch Biol 55:1–284.

D'Amico-Martel A (1982) Temporal patterns of neurogenesis in avian cranial sensory and autonomic ganglia Am J Anat 163:351–372.

D'Amico-Martel A, Noden DM (1983) Contribution of placode and neural crest cells to avian cranial peripheral ganglia. Am J Anat 166:445–468.

De Burlet HM (1934) Vergleichende Anatomie des statoakustischen Organs. a) Die innere Ohrsphäre; b) Die mittlere Ohrsphäre. In: Bolk L, Göppert E, Kallius E, Lubosch W (eds). Handbuch der Vergleichenden Anatomie der Wirbeltiere. Vol. 2. Berlin: Urban and Schwarzenberg, pp. 1293–1432.

Deol MS (1964) The abnormalities of the inner ear in kreisler mice. J Embryol Exp Morphoal 12:475–490.

Deol MS (1983) Development of auditory and vestibular systems in mutant mice. In: Romand R (ed) Development of Auditory and Vestibular Systems. New York: Academic Press, pp. 309–333.

Detwiler SR, Van Dyke RH (1950) The role of the medulla in the differentiation of the otic vesicle. J Exp Zool 113:179–199.

Dolle P, Lufkin T, Krumlauf R, Mark M, Dubolle D, Chambon P (1992) Local alterations of Krox-20 and Hox gene expression in the hindbrain suggest lack of rhombomeres 4 and 5 in homozygote null *Hoxa-1* (Hox-1.6) mutant embryos. Proc Nat Acad Sci USA 90:7666–7670.

Dressler GR, Douglas EC (1992) Pax-2 is a DNA-binding protein expressed in embryonic kidney and Wilms tumor. Proc Nat Acad Sci USA. 89:1179–1183.

Dressler GR, Deutsch U, Chowdhury K, Nornes HO, Gruss, P. (1990) Pax2, a new murine paired-box-containing gene and its expression in the developing excretory system. Development 109:787–795.

Duyk G, Gastier JM, Mueller RF (1992) Traces of her workings. Nature Genet 2:5–8.

Ekker M, Akimenko MA, Bremiller R, Westerfield M (1992a) Regional expression of three homeobox transcripts in the inner ear of zebra fish embryos. Neuron 9:27–35.

Ekker M, Wegner J, Akimenko M-A, Westerfield M (1992b) Coordinate embryonic expression of three zebrafish engrailed genes. Development 116:1001–1010.

Emmerling MR, Sobkowicz HM, Levenick CV, Scott GL, Slapnick SM, Rose JE (1990) Biochemical and morphological differentiation of acetylcholineesterase-positive efferent fibers in the mouse cochlea. J Electron Microsc Technol 15:123–143.

Erickson RP (1990) Mapping dysmorphic syndromes with the aid of human/mouse homology map. Am J Hum Genet 46:1013–1016.

Ernfors P, Lee K-F, Jaenisch R (1994) Mice lacking brain-derived neurotrophic factor develop with sensory deficits. Nature 368:147–150.

Ernfors P, Van De Water T, Loring J, Jaenisch R (1995) Complementary roles of BDNF and NT-3 in vestibular and auditory development. Neuron 14:1153–1164.

Farbman AI, Mbiene J-P (1991) Early development and innevation of taste bud-bearing papillae on the rat tongue. J Comp Neurol 304:172–186.

Farinas I, Jones, KR, Backus C, Wang X-Y, Reichardt GF (1994) Severe sensory and sympathetic deficits in mice lacking neurotrophin-3. Nature 369:658–661.

Fekete DM (1996) Cell fate specification in the inner ear. Curr Op Neurobiol 6:533–541.

Fell HB (1928) The development *in vitro* of the isolated otocyst of the embryonic fowl. Arch Exp Zellforsch Besonders Gewebezuech 7:69–81.

Fischer FP (1992) Quantitative analysis of the innervation of the chicken basilar papilla. Hear Res 61:167–178.

Fleischman RA (1993) From white spots to stem cells: the role of the kit receptor in mammalian development. Trends Genet 9:285–290.

Frenz DA, Van De Water TR (1991) Epithelial control of periotic mesenchyme chondrogenesis. Dev Biol 144:38–46.

Friedmann I (1956) In vitro culture of the isolated otocyst of the embryonic fowl. Ann Otol 65:98–107.

Fritz A, Westerfield M (1996) Analysis of two mutants affecting neuroectodermal patterning in zebrafish. CSH, Zebrafish Development & Genetics, p. 216.

Fritzsch B (1992) The water-to-land transition: evolution of the tetrapod basilar papilla, middle ear, and auditory nuclei. In Webster DB, Popper AN, Fay RR (eds) The Evolutionary Biology of Hearing. New York: Springer-Verlag, pp. 351–375.

Fritzsch B (1993) Evolutionary gain and loss of non-teleostean electroreceptors. J Comp Physiol 173:710–712.

Fritzsch B (1995) Evolution of the ancestral vertebrate brain. In: Arbib MA (ed) The Handbook of Brain Theory and Neural Networks. Cambridge, MA: MIT Press, pp. 373–377.

Fritzsch B (1996a) Similarities and differences in lancelet and craniate nervous systems. Israel J Zool, 42:147–160.

Fritzsch B (1996b) Development of the labyrinthine efferent system. Ann NY Acad Sci 781:21–33.

Fritzsch, B, Silos-Santiago I, Bianchi L, Farinas I (1997) The role of neurotrophic factors in regulating inner ear innervation. TINS 20:159–164.

Fritzsch B, Nichols DH (1993) DiI reveals a prenatal arrival of efferents at developing ears of mice. Hear Res 65:51–60.

Fritzsch B, Northcutt RG (1993) Cranial and spinal nerve organization in amphioxus and lampreys. Acta Anat 148:96–110.

Fritzsch B, Wake MH (1988) The inner ear of gymnophione amphibians and its nerve supply: a comparative study of regressive events in a complex sensory system. Zoomorphology 108 210–217.

Fritzsch B, Zakon HH, Sanchez DY (1990) Time course of structural changes in regenerating electroreceptors of a weakly electric fish. J Comp Neurol 300:386–404.

Fritzsch B, Christensen MA, Nichols DH (1993) Fiber pathways and positional changes in efferent perikarya of 2.5 to 7 day chick embryos as revealed with DiI and dextran amines. J Neurobiol 24:1481–1499.

Fritzsch B, Silos-Santiago I, Smeyne D, Fagan A, Barbacid M (1995) Reduction and loss of inner ear innervation in trkB and trkC receptor knock out mice: a whole mount DiI and SEM analysis. Aud Neurosci 1:401–417.

Frohman MA, Martin GR, Cordes SP, Halamek LP (1993) Altered rhombomere-specific gene expression and hyoid bone differentiation in the mouse segmentation mutant, kreisler (kr). Development 117:925–936.

Garcia-Fernàndez J, Holland PWH (1994) Archetypal organization of the lancelet Hox gene cluster. Nature, 370:563–566.

Gardner CA, Barald KF (1992) Expression patterns of engrailed-like proteins in the chick embryo Dev Dyn 193:370–388.

George KM, Leonard MW, Roth MW, Lieuw KH, Kloussis D, Grosveld F, Engel JD (1994) Embryonic expression and cloning of the murine GATA-3 gene. Development 120:2673–2686.

Gibson F, Walsh H, Mburu P, Varea A, Brown KA, Autonio M, Beisel KW, Steel KP, Brown, SDA (1995) A type VII myosin is encoded by the mouse deafness gene shaker-1. Nature 374:62–64.

Gilbert SF (1991) Developmental Biology. Sunderland: (USA) Sinauer, pp. 891.

Gilland E, Baker R (1993) Conservation of neuroepithelial and mesodermal segments in the embryonic vertebrate head. Acta Anat 148:110-123.

Goulding MD, Lumsden A, Gruss P (1993) Signals from the notochord, floor plate regulate the region-specific expression of two Pax genes in the spinal cord. Development 117:1001-1016.

Graham A, Heyman I, Lumsden A (1993) Even-numbered rhombomeres control the apoptotic elimination of neural crest cells from odd-numbered rhombomeres in the chick hind brain. Development 119:233-245.

Graham A, Francis-West P, Brickell P, Lumsden A (1994). The signalling molecule BMP4 mediates apoptosis in the rhombencephalic neural crest. Nature 372:684-686.

Grainger RM, Henry JJ, Henderson RA (1988) Reinvestigation of the role of the optic vesicle in embryonic lens induction. Development 102:517-526.

Gruss P, Walther C (1992) Pax in development. Cell 69:719-722.

Guthrie S (1995) The status of the neural segment. Trends Neurosci 200:74-79.

Gutknecht D, Fritzsch B (1990) Lithium induces multiple ear vesicles in *Xenopus laevis* embryos. Naturwissenschaften, 77:235-237.

Haddon CM, Lewis JH (1991) Hyaluronan as a propellant for epithelial movement: the development of semicircular canals in the inner ear of *Xenopus*. Development 112:541-550.

Haddon C, Lewis J (1996) Early ear development in the embryo of the zebrafish, Danio rerio. J. Comp Neurol 365:113-128.

Halder G, Calaerts P, Gehring WJ (1995) Induction of ectopic eyes by targeted expression of the eyeless gene in Drosophila. Science 267:1788-1792.

Hall BK (1987) Tissue interactions in the development and evolution of the vertebrate head. In: Maderson PFA (ed) Developmental and Evolutionary Aspects of the Neural Crest. New York: John Wiley and Sons, pp. 215-259.

Hall BK (1991) Cellular interactions during cartilage and bone development. J Craniofac Genet Dev Biol 11:238-250.

Hallböök F, Ibanez CF, Ebendal T, Persson H (1993). Cellular localization of brain-derived neurotrophic factor and neurotrophin-3 mRNA expression in early chick embryo. Eur J Neurosci 5:1-14.

Harris WA, Hartenstein V (1991) Neuronal determination without cell division in *Xenopus* embryos. Neuron 6:499-515.

Harrison RG (1936) Relations of symmetry in the developing ear of *Amblystoma punctatum*. Proc Natl Acad Sci USA 22:238-247.

Harrison RG (1945) Relations of symmetry in the developing embryo. Trans Conn Acad Arts Sci 36:277-330.

Hemond SG, Morest DK (1991) Ganglion formation from the otic placode and the otic crest in the chick embryo: mitosis, migration, and the basal lamina. Anat Embryol 184:1-13.

Hemond SG, Morest DK (1992). Trophic effects of otic epithelium on cochleo-vestibular ganglion fiber growth in vitro. Anat Rec 232:273-284.

Hemmati-Brivanlou A, Thomsen G (1995) Ventral mesoderm patterning in *Xenopus* embryos: the expression patterns and activities of BMP-2 and BMP-4. Dev Genet 17:78-89.

Henson OW (1974) Comparative anatomy of the middle ear. In: Keidel WD, Neff WD (eds) Handbook of Sensory Physiology. Vol. V/1. Auditory System. Berlin: Springer-Verlag, pp. 40-110.

Hertwig P (1942) Neue Mutationen und Kopplungsgruppen bei der Hausmaus. Z

Indukt Abstammungs Vererbungsl 80:220–247.

Hilfer SR, Randolph GJ (1993) Immunolocalization of basal lamina components during development of chick otic and optic primordia. Anat Rec 235:443–452.

Hilfer SR, Esteves RA, Sanzo JF (1989) Invagination of the otic placode: normal development and experimental manipulation. J Exp Zool 251:253–264.

Hill R, Von Heyningen V (1992) Mouse mutations, human disorders are paired. Trends Genet 8:119–120.

Hogan B, Wright C (1992) The making of the ear. Nature 355:494–495.

Holland PWH, Holland LZ, Williams NA, Holland ND (1992) An amphioxus homeobox gene: sequence conservation, spatial expression during development and insights into vertebrate evolution. Development 116:653–661.

Hoth CF, Milunsky A, Lipsky N, Sheffer R, Clarren SK, Baldwin CT (1993) Mutations in the paired domain of the human Pax3 gene cause Klein-Waardenburg syndrome (WS-III) as well as Waardenburg syndrome type I (WS-I). Am J Hum Genet 52:455–462.

Huschke E (1831) Erste Bildungsgeschichte des Auges und des Ohres. (Versammlung Naturforscher und Ärzte zu Hamburg). Isis von Oken 1831. (As cited in Rubel [1978].)

Jackson IJ, Raymond S (1994) Manifestations of microphthalmia. Nature Genet 8:209–210.

Jacobson AG (1963) The determination and positioning of the nose, lens and ear. I. Interactions within the ectoderm, and between ectoderm and underlying tissue. J Exp Zool 154:273–284.

Jacobson AG (1966) Inductive processes in embryonic development. Science 152:25–34.

Jacobson AG (1988) Somitomeres: mesodermal segments of vertebrate embryos. Development 104:209–220.

Jacobson AG (1991) Experimental analysis of the shaping of the neural plate and tube. Am Zool 311:628–643.

Jacobson AG, Sater AK (1988) Features of embryonic induction. Development 104:341–359.

Kaan H (1930) The relation of the developing auditory vesicle to the formation of the cartilage capsule in Amblystoma punctatum. J Exp Zool 55:263–291.

Katayama A, Corwin JT (1989) Cell production in the chicken cochlea. J Comp Neurol 281:129–135.

Keller R, Shish J, Sater A (1992) The cellular basis of the convergence and extension of the *Xenopus* neural plate. Dev Dyn 193:199–217.

Keller SA, Jones JM, Boyle A, Barrow LL, Killen PD, Green DG, Kapousta NV, Hitchcock PF, Swank RT, Meisler MH (1994) Kidney and retinal defects (*Krd*), a transgene-induced mutation with a deletion of mouse chromosome 19 that includes the *Pax2* locus. Genomics 23:309–320.

Kelley MW, Corwin JT (1992) Development of hair cell structure and function in fish and amphibians. In: Romand R (ed) Development of Auditory and Vestibular Systems 2. Amsterdam: Elsevier, pp. 139–159.

Kessel M (1992) Respecification of vertebral identities by retinoic acid. Development 115:487–501.

Keynes R, Krumlauf R (1994) Hox genes and regionalization of the nervous system. Ann Rev Neurosci 17:109–132.

Knowlton VY (1967) Correlation of the development of membraneous and bony labyrinths, acoustic ganglia, nerves, and brain centers of the chick embryo. J

Morphol 121:179–208.

Kornhauser JM, Leonard MW, Yamamoto M, LaVail JH, Mayo KE, Engel JD (1994) Temporal, spatial changes in GATA transcription factor expression are coincident with development of the chicken optic tectum. Mol Brain Res 23:100–110.

Krauss S, Johansen T, Korzh V, Fjose A (1991) Expression pattern of zebrafish pax genes suggest a role in early brain regionalization. Nature 353:267–270.

Krauss S, Maden M, Holder N, Wilson S (1992) Zebrafish pax[b] is involved in the formation of the midbrain-hind brain boundary. Nature 360:87–89.

Lewis ER, Leverenz EL, Bialek W (1985) The verebrate inner ear. Boca Raton, FL: CRC Press, pp. 256.

Li CW, McPhee J (1979) Influences on the coiling of the cochlea. Ann Otol Rhinol Laryngol 88:280–287.

Li CW, Van De Water TR, Ruben RJ, Shea CA (1978) The fate mapping of the eleventh and twelfth day mouse otocyst: an *"in vitro"* study of the sites of origin of the embryonic inner ear sensory structures. J Morphol 157:249–268.

Lim DJ, Rueda J (1992) Structural development of the cochlea. In: Romand R (ed) Development of Auditory and Vestibular Systems 2. Amsterdam: Elsevier, pp. 33–58.

Lindberg KH, Lomax MI, Hegeman AD, Barald KF (1995) *Pax-2*: An early marker of otic placode induction that defines otocyst development in the chick. Midwinter Meet Assoc Otolaryngol, pp.

Lowenstein O, Thornhill RA (1970) The labyrinth of *Myxine*: anatomy, ultrastructure and electrophysiology. Proc R Soc Lond B Biol Sci 176:21–42.

Lufkin T, Dierich A, LeMeur M, Mark M, Chambon P (1991) Disruption of the Hox-1.6 homeobox gene results in defects in a region corresponding to its rostral domain of expression. Cell 66:1105–1119.

Lumsden A, Krumlauf R (1996) Patterning the vertebrate axis. Science 274: 1109–1115.

Mahmood R, Kiefer P, Guthrie S, Dickson C, Mason J (1995) Multiple roles for FGF-3 during cranial neural development in the chicken. Development 121:1399–1410.

Malicki J, Schier AF, Solnica-Krezel L, Stemple DL, Neuhaus SCF, Stainier DYR, Abdelilah S, Rangini Z, Zwartkruis F, Driever W (1996) Mutations affecting development of the zebrafish ear. Development 123:275–283.

Manns M, Fritzsch B (1992) Retinoic acid affects the organization of reticulospinal neurons in developing *Xenopus*. Neurosci Lett 139:253–256.

Mansour SL, Goddard JM, Capecchi MR (1993) Mice homozygous for a targeted disruption of the proto-oncogene *int-2* have developmental defects in the tail and inner ear. Development 117:13–28.

Mark M, Lufkin T, Vonesch J-L, Ruberte E, Olivo J-C, Dolle P, Gorry P, Lumsden A, Chambon P (1992) Two rhombomeres are altered in *Hoxa-1* mutant mice. Development 119:319–338.

Marshall H, Nonchev S, Sham MH, Muchamore I (1992) Retinoic acid alters hindbrain Hox code and induces transformation of rhombomeres 2/3 into 4/5 identity. Nature 360:737–741.

McKay IJ, Muchamore I, Krumlauf R, Maden M, Lumsden A, Lewis J (1994) The kreisler mouse: a hindbrain segmentation mutant that lacks two rhombomeres. Development 120:2199–2211.

McKay IJ, Lewis J, Lumsden A (1996) The role of FGF-3 in early inner ear development: an analysis in normal and kreisler mutant mice. Dev Viol 174:370–378.

Meier S (1978) Development of the embryonic chick otic placode. II. Electron microscopic analysis. Anat Rec 191:459–465.

Moase CE, Trasler DG (1990) Splotch locus mouse mutants: model for neural tube defects, Waardenburg syndrome type I in humans. Teratology 42:171–182.

Model PG, Jarret LS, Bonazzoli R (1981) Cellular contacts between hindbrain and prospective ear during inductive interaction in the axolotl embryo. J Embryol Exp Morphol 66:27–41.

Moens CB, Kimmel CB (1995) Hindbrain patterning in the zebrafish embryo. Soc Neurosci Abstr 21:277.

Moens CB, Yan Y-L, Appel B, Force A, Kimmel CB (1996) valentino: a zebrafish gene required for normal hindbrain segmentation. Development 122:3981–3990.

Morell R, Friedman TB, Moeljopawiro S, Soewito H, Hartono, Soewilo, JH Jr. (1992) A frameshift mutation in the HuP2 paired domain of the probable human homolog of murine Pax-3 is responsible for Waardenburg syndrome type I in an Indonesian family. Hum Mol Genet 1:43–59.

Morriss-Kay GM (1993) Retinoic acid and craniofacial development: molecules and morphogenesis. BioEssays 15:9–15.

Muthukumar S, Fekete DM (1994) Hair cells and supporting cells share a common progenitor in the developing chicken inner ear. Soc Neurosci Abstr 20:1079.

Neal HV (1918) The history of the eye muscles. J Morphol 30:433–453.

Neary TJ, Fritzsch B (1992) Stage and concentration specific effects of retinoic acid on the differentiation of Xenopus hindbrain and ear. Soc Neurosci Abstr 18:328.

Nieuwkoop PD, Faber J (1967) Normal table of Xenopus laevis (Daudin). Amsterdam: North-Holland Press, pp. 252.

Noden DM (1991) Vertebrate craniofacial development: the relation between ontogenetic process and morphological outcome. Brain Beh Evol 38:190–225.

Noden DM, Van De Water TR (1986) The developing ear: tissue orgins and interactions. In: Ruben RJ, Van De Water TR, Rubel EW (eds) The Biology of Change in Otolaryngology. Amsterdam: Elsevier, pp. 15–46.

Noden DM, Van De Water TR (1992) Genetic analyses of mammalian ear development. Trends Neurosci 15:235–237.

Noll M (1993) Evolution and role of Pax genes. Curr Opin Genet Dev 3:595–605.

Nornes HO, Dressler GR, Knapik EW, Deutsch U, Gruss P (1990) Spatially and temporally restricted expression of Pax2 during murine neurogenesis. Development 109:797–809.

Norris HW (1892) Studies on the development of the ear in Amblystoma. I. Development of the auditory vesicle. J Morphol 7:23–34.

Northcutt RG (1992) The phylogeny of octavolateralis ontogenies: a reaffirmation of Garstang's phylogenetic hypothesis. In: Webster DB, Fay RR, Popper AN (eds) The Evolutionary Biology of Hearing. New York: Springer-Verlag, pp. 21–47.

Northcutt RG, Gans C (1983) The genesis of neural crest and epidermal placodes: a reinterpretation of vertebrate origin. Q Rev Biol 58:1–28.

Northcutt RG, Brändle K, Fritzsch B (1995) Electroreceptors and mechanosensory lateral line organs arise from single placodes in axolotls. Dev Biol 168:358–373.

Nüsslein-Vollhard C (1994) Of flies and fishes. Science 266:572–574.

Oh, SH, Johnson R, Wu DK (1996) Differential expression of bone morphogenetic proteins in the developing vestibular and auditory sensory organs. J Neurosci

15:6463–6475.

Oppenheim RW (1991) Cell death during development of the nervous system. Ann Rev Neurosci 14:453–503.

Oppenheim RW, Qin-Wei Y, Prevette D, Yan Q (1992). Brain-derived neurotrophic factor rescues developing avian motoneurons from cell death. Nature 360: 755–757.

Orr MF (1981) Anatomical development of the embryonic chick otocyst in organ culture. Anat Rec 199:188A.

Orr MF (1986) Development of acoustic ganglia in tissue cultures of embryonic chick otocysts. Exp Cell Res 40:68–77.

Papalopulu N, Clarke JD, Bradley L, Wilkinson D, Kramlauf R, Holder N (1991) Retinoic acid causes abnormal development and segmental patterning of the anterior hind brain in *Xenopus* embryos. Development 113:1145–1158.

Pierpoint JW, Erickson RP (1993) Facts on Pax. Am J Hum Genet 52:451–454.

Pijnappel WWM, Hendriks HFK, Folkers GE, Van Den Brink CE, Durston T (1993) The retinoid ligand 4-oxo-retinoic acid is a highly active modulator of positional specification. Nature 366:340–344.

Pirvola U, Lehtonen E, Ylikoski J (1991) Spatiotemporal development of cochlear innervation and hair cell differentiation in the rat. Hear Res 52:345–355.

Pirvola U, Ylikoski J, Palgi J, Lehtonen E, Arumae U, Saarma M (1992) Brain-derived neurotrophic factor and neurotrophin 3 mRNAs in the peripheral target fields of developing inner ear ganglia. Proc Natl Acad Sci USA 89:9915–9919.

Pirvola U, Arumae U, Moshnyakov M, Palgi J, Saarma M, Ylikoski J (1994) Coordinated expression and function of neurotrophins and their receptors in the rat inner ear during target innervation. Hear Res 75:131–144.

Pujol R (1986) Synaptic plasticity in the developing cochlea. In: Ruben RJ, Van De Water TR, Rubel EW (eds) The Biology of Change in Otolaryngology. Amsterdam: Elsevier, pp. 47–54.

Ramirez F, Solursh M (1993) Expression of vertebrate homologs of the Drosophila msh gene during early craniofacial development. Am Zool 33:457–461.

Represa J, Bernd P (1989) Nerve growth factor and serum differentially regulate development of embryonic otic vesicle and vestibular ganglion *in vitro*. Dev Biol 134:21–29.

Represa J, Sanchez A, Miner C, Lewis J, Giraldez F (1990) Retinoic acid modulation of the early development of the inner ear is associated with the control of c-fos expression. Development 110:1081–1090.

Represa J, Leon Y, Miner C, Giraldez F (1991) The *int-2* proto-oncogene is responsible for induction of the inner ear. Nature 353:561–563.

Retzius, G. (1884) Das Gehörorgan der Wirbeltiere: II. Das Gehörorgan der Amnioten. Stockholm: Samson und Wallin, pp. 345.

Richardson GP, Crossin KL, Chuong, CM, Edelman GM (1987) Expression of cell adhesion molecules during embryonic induction. III. Development of the otic placode. Dev Biol 119:217–230.

Riley BB, Savage MP, Simandl BK, Olwin BB, Fallon JF (1993) Retroviral expression of FGF-2 (bFGF) affects patterning in chick limb bud. Development 118:95–104.

Roberts BL, Meredith GE (1992). The efferent innervation of the ear: variations on an enigma. In: Webster DB, Fay RR, Popper AN (eds) The Evolutionary Biology of Hearing. New York: Springer-Verlag, pp. 182–210.

Robinson A, Mahon KA (1994) Differential and overlapping expression domains of Dlx-2 and Dlx-3 suggest distinct roles for Distal-less homeobox genes in craniofacial development. Mech Dev 48:199–215.

Rubel EW (1978) Ontogeny of structure and function in the vertebrate auditory system. In: Jacobson M (ed) Development of Sensory Systems. Berlin: Springer-Verlag, pp. 135–237.

Ruben RJ (1967) Development of the inner ear of the mouse: a radioautographic study of terminal mitosis. Acta Otolaryngol 220:1–44.

Ruben RJ, Van De Water TR, Steel KP (1991) Genetics of hearing impairment. Ann NY Acad Sci 630:329.

Rueda J, De La Sen D, Juiz JM, Merchan JA (1987) Neuronal loss in the spiral ganglion of young rats. Acta Otolaryngol 104:417–421.

Sanyanusin P, Schimmenti LA, McNoe LA, Ward TA, Pierpont ME, Sullivan MJ, Dobyns WB, Eccles MR (1995) Mutation of the Pax2 gene in a family with optic nerve colobomas, renal anomalies and vesicoureteral reflux. Nature Genetics 9:358–364.

Schellart NAM, Popper AN (1992) Functional aspects of the evolution of the auditory system of actinopterygian fish. In: Webster DB, Fay RR, Popper AN (eds) The Evolutionary Biology of Hearing. New York: Springer-Verlag, pp. 295–321.

Sechrist J, Scherson T, Bronner-Fraser M (1994) Rhombomere rotation reveals that multiple mechanisms contribute to the segmental pattern of hind brain neural crest migration. Development 120:1777–1790.

Sher AE (1971) The embryonic and postnatal development of the inner ear of the mouse. Acta Otolaryngol 285:1–77.

Shnerson A, Devigne C, Pujol R (1982) Age-related changes in the C57BL/6J mouse cochlea. II. Ultrastructural findings. Dev Brain Res 2:77–88.

Sive HL, Cheng PF (1991) Retinoic acid perturbs the expression of Xhox.lab genes and alters mesodermal determination in *Xenopus laevis*. Genes Dev 5:1321–1332.

Sobkowicz HM (1992) The development of innervation in the organ of Corti. In: Romand R (ed) Development of Auditory and Vestibular Systems 2. Amsterdam: Elsevier, pp. 59–100.

Sobkowicz HM, Rose JE (1983) Innervation of the organ of Corti of the fetal mouse in culture. In: Romand R (ed) Development of Auditory and Vestibular Systems. New York: Academic Press, pp. 27–45.

Sobkowicz HM, Bereman B, Rose JE (1975) Organotypic development of the organ of Corti in tissue culture. J Neurocytol 4:543–572.

Sobkowicz HM, Rose JE, Scott GL, Levenick CV (1986) Distribution of synaptic ribbons in the developing organ of Corti. J Neurocytol 15:693–714.

Sokolowski BHA, Stahl LM, Fuchs PA (1993) Morphological and physiological development of vestibular hair cells in the organ-cultured otocyst of the chick. Dev Biol 155:134–146.

Song D-L, Chalepakis G, Gruss P, Joyner AL (1996) Two Pax binding sites are required for early embryonic brain expression of engrailed-2 transgene. Development 122:627–635.

Spritz RA, Holmes SA, Ramesar R, Greenberg J, Curtis D, Beighton P (1992) Mutations of the KIT (mast/stem cell growth factor receptor) proto-oncogene account for a continuous range in phenotypes in human piebaldism. Am J Hum Genet 51:1058–1065.

Stadler HS, Solursh M (1994) Characterization of the homeobox-containing gene

GH6 identified novel regions of homeobox gene expression in the developing chick embryo. Dev Biol 161:251–262.

Steel KP, Barkway C (1989) Another role for melanocytes: their importance for normal stria vascularis development in the mammalian ear. Development 107:453–463.

Steel KP, Brown SKM (1994) Genes and deafness. Trends Genet 10:428–435.

Steel KP, Harvey G (1992) Development of auditory function in mutant mice. In: Romand R (ed) Development of Auditory and Vestibular systems 2. Amsterdam: Elsevier, pp. 221–241.

Steel KP, Kimberling W (1996) Approaches to understanding the molecular genetics of hearing and deafness. In: Van De Water, TR, Popper AN, Fay RR (eds) Clinical Aspects of Hearing. New York: Springer-Verlag, pp. 10–40.

Steel KP, Smith RJH (1992) Normal hearing in Splotch (Sp/+), the mouse homologue of Waardenburg syndrome type I. Nature Genet 2:75–79.

Stone LM, Finger TE, Tam PPL, Tan S-S (1995) Taste receptor cells arise from local epithelium, not neurogenic ectoderm. Proc Natl Acad Sci USA 92: 1916–1920.

Stoykova A, Gruss P (1994) Role of Pax genes in developing and adult brain as suggested by expression patterns. J Neurosci 14:1395–1412.

Sulik KK, Cotanche DA (1994) Embryology of the ear. In: Toriello H, Choen MM, Gorlin RJ (eds). Hereditary Hearing Loss and Its Syndromes. New York: Oxford University Press, pp. 22–42.

Swanson GJ, Howard M, Lewis J (1990) Epithelial autonomy in the development of the inner ear of a bird embryo. Dev Biol 137:243–257.

Szepsenwol J (1933) Recherches sur les centres organisateurs de vesicules auditives chez des embryons de poulets omphlocephales obtenus experimentalement. Arch Anat Microsc Morphol Exp 29:5–94.

Tabin C (1995) The initiation of the limb bud: growth factors, Hox genes, and retinoids. Cell 80:67–67.

Tachibana M, Wilcox E, Yokotani N, Schneider M, Fex J (1992) Selective amplification and partial sequencing of cDNAs encoding G protein A subunits from cochlear tissues. Hear Res 62:892–898.

Tannahill D, Isaacs, HV, Close MJ, Peters G, Slack JMW (1992) Developmental expression of the Xenopus int-2 (FGF-3) gene: activation by mesodermal and neural induction. Development 115:695–702.

Tassabehji M, Newton VE, Read AP (1994) Waardenburg syndrome type 2 caused by mutations in the human microphthalmia (MITF) gene. Nature Genet 8:251–255.

Tello JF (1931) Le reticule des cellules ciliees du labyrinth chez la souris et son independance des terminaisons nerveuses de la huitieme paire. Trav Lab Rech Biol 27:151–186.

Thaller C, Hofman C, Eichele G (1993) 9-CIS-Retinoic acid, a potent inducer of digit pattern duplications in the chick wing bud. Development 118:957–965.

Tickle C, Alberts BM, Wolpert L, Lee J (1982) Local application of retinoic acid to the limb bud mimics the action of the polarizing region. Nature 296:564–565.

Tilney LG, Cotanche DA, Tilney MS (1992) Actin filaments, stereocilia and hair cells of the bird cochlea. VI. How the number and arrangment of stereocilia are determined. Development 116:213–226.

Torres M, Gomez-Pardo E, Gruss P (1996) Pax2 contributes to inner ear patterning and optic nerve trajectory. Development 122:3381–3391.

Van Bergeijk WA (1967) The evolution of vertebrate hearing. In: Neff WD (ed) Contributions to Sensory Physiology. New York: Academic Press, pp. 1–49.

Van De Water TR (1983) Embryogenesis of the inner ear: "in vitro studies." In: Romand R (ed) Development of Auditory and Vestibular Systems. New York: Academic Press, pp. 337–374.

Van De Water TR, Represa J (1991) Tissue interactions and growth factors that control development of the inner ear. Ann NY Acad Sci 630:116–128.

Van De Water TR, Frenz DA, Giraldez F, Represa J, Lefebvre PP, Rogister B, Moonen G (1992) Growth factors and development of the stato-acoustic system. In: Romand R (ed) Development of Auditory and Vestibular Systems 2. Amsterdam: Elsevier, pp. 1–32.

Von Bartheld CS, Patterson SL, Heuer JG, Wheeler EF, Bothwel M (1991) Expression of nerve growth factor (NGF) receptors in the developing inner ear of chick and rat. Development 113:455–470.

Von Kupffer C (1895) Studien zur vergleichenden Entwicklungsgeschichte des Kopfes der Kranioten. Vol 3. Die Entwicklug der Kopfnerven von *Ammocoetes planeri*. Munich, Lehmann, pp. 80.

Von Kupffer C (1900) Studien zur vergleichenden Entwicklungsgeschichte des Kopfes der Kranioten. Vol 4. Zur Kopfentwicklung von *Bdellostoma*. Munich, Lehmann, pp. 87.

Waddington CH (1937) The determination of the auditory placode in the chick. J Exp Biol 14:232–239.

Warr WB (1992) Organization of olivocochlear efferent systems in mammals. In: Fay RR, Popper AN, Webster DB (eds). The Anatomy of the Mammalian Auditory Pathways. New York: Springer-Verlag, pp. 410–448.

Waterman AJ (1925) The development of the inner ear rudiment of the rabbit embryo in a foreign environment. Am J Anat 63:161–219.

Webb JF, Noden DM (1993) Ectodermal placodes: contributions to the development of the vertebrate head. Am Zool 33:434–447.

Werner G. (1960) Das Labyrinth der Wirbeltiere. Jena: (Germany) Fischer Verlag, pp. 309.

Wever, EG (1974) The evolution of vertebrate hearing. In: Keidel WD, Neff WD (eds) Handbook of Sensory Physiology. Vol. V/1: Auditory System. Berlin: Springer-Verlag, pp. 423–454.

Whitehead MC, Morest DK (1985) The development of innervation patterns in the avian cochlea. Neuroscience 14:255–276.

Whitfield, TT, Granato M, van Eeden FJM, Schach U, Brand M, Furutani-Seiki M, Haffter P, Hammerschmidt M, Heisenberg C-P, Jiang Y-J, Kane DA, Kelsh RN, Mullins MC, Odenthal J, Nüsslein-Volhard C (1996) Mutations affecting development of the zebrafish inner ear and lateral line. Development 123:241–254.

Wilkinson DG, Bhatt S, McMahon AP (1989) Expression pattern of the FGF-related proto-oncogene *int-2* suggests a multiple role in fetal development. Development 105:131–136.

Winklbauer R, Hausen P (1983) Development of the lateral line system in *Xenopus laevis*. I. Normal development and cell movement in the supraorbital system. J Embryol Exp Morphol 76:283–296.

Wu DK, Oh SH (1996) Sensory organ generation in the chick inner ear. J Neurosci 15:6454–6462.

Yamamoto M, Ko LJ, Leonard MW, Beug H, Orkin SH, Enge JD (1990) Activity and tissue-specific expression of the transcription factor NF-E1 multigene family.

Genes Dev 4:1650–1662.

Yntema CL (1939) Self-differentiation of heterotopic ear ectoderm in the embryo of *Amblystoma punctatum*. J Exp Zool 80:1–17.

Yntema CL (1950) An analysis of induction of the ear from foreign ectoderm in the salamander embryo. J Exp Zool 113:211–244.

Yntema CL (1955) Ear and nose. In: Willier BH, Weiss PA, Hamburger V (eds) Analysis of Development. Philadelphia PA: Saunders, pp. 415–428.

Yoo TJ, Cho H, Yamada Y (1991) Hearing impairment in mice with the cmd/cmd (cartilage matrix deficiency) mutant gene. Ann NY Acad Sci 630:265–267.

Zwilling E (1941) The determination of the otic vessicle in *Rana pipiens*. J Exp Zool 86:333–342.

4

Development of Sensory and Neural Structures in the Mammalian Cochlea

R. Pujol, M. Lavigne-Rebillard, and M. Lenoir

1. Introduction

During the last few decades, the structural and functional development of the cochlea has been extensively reviewed (Pujol and Hilding 1973; Rubel 1978; Romand 1983; Eggermont and Bock 1985; Pujol and Uziel 1988; Lim and Rueda 1992). This chapter does not attempt to be a general review but more specifically describes the maturation of sensory hair cells and their neural connections during in vivo development of the mammalian cochlea. An important part of this coverage concerns the ontogeny of the different types of cochlear synapses. Up until now, reviews on cochlear synaptogenesis have been either incomplete (Pujol 1986; Pujol and Sans 1986) or based on in vitro results (Sobkowicz 1992) that do not give a clear understanding of what occurs in vivo. We complement structural findings wherever possible with physiological correlations to present morphology at a functional rather than at a descriptive level.

Before focusing on hair cells and synapses, it is worth summarizing the general trends of cochlear maturation. The coiled mammalian cochlea follows a general gradient of maturation from base to apex: a gradient applicable either to gross morphological development (as reported by Retzius 1884) or to the maturation of more subtle subcellular elements and/or mechanisms (Pujol et al. 1991). The most apical part of the cochlea does not completely finish the maturation process, retaining some immature features into adulthood. This appears to be true from a morphological point of view (we expand on this later in this chapter) and also from a general physiological perspective. At the onset of auditory function, when most of the cochlear structures are still immature, the physiological responses have high thresholds and poor tuning. However, some active properties can be detected quite early at the base of the cochlea, and they mature very quickly (Mills, Norton, and Rubel 1994); conversely, the apical cochlea develops much slower and the most apical part retains primarily passive behavior. Moreover, depending on species, the developmental processes that make the cochlea active could well concern a different extension of the cochlear spiral: more explicitly, adult cats or rats can be regarded as having a "basal" cochlea more extended toward the apex than do adult guinea pigs or humans. At either end of such a scale are the "all-basal" horseshoe bat (*Rhinolophus rouxi*) cochlea (Vater and Lenoir 1992; Vater, Lenoir, and Pujol 1992) and the almost "all-apical" mole rat (*Spalax ehrenbergi*) cochlea (Raphael et al. 1991).

It is possible to give an overall picture of the maturation of sensory and neural structures in the mammalian cochlea and, at the same time, consider data from specific species. Beside obvious differences in gross morphology, such as coiling and total length of the spiral that account for significant differences in the number of sensory cells, the cochleas of all eutherian mammals seem to be remarkably similar at both cellular and subcellular structural levels. Maturation can therefore be compared between species, and, in addition, reliable extrapolations can be made. As an example, the early stages of hair cell differentiation and synaptogenesis are better described in the slow-developing human cochlea than in the fast-developing cochleas of most of common experimental animals. Conversely, because well-preserved tissues are rarely found in late human stages, the last stages of cochlear development are better studied in rodents. To help the reader extract a single rationale from individual species results, Table 4.1 lists and compares the main developmental stages of the most species referred to in this chapter.

2. Development of Cochlear Sensory Hair Cells

2.1 Early Differentiation of Hair Cells

The molecular triggering of hair cell differentiation is still one of the major unsolved problems in cochlear development. Moreover, as mammalian cochlear hair cells are known to lose their proliferative capacities after they enter the differentiation stage, a better knowledge at this stage would be extremely useful to understand normal and abnormal development of the cochlea, as well as the question of regeneration. Before the first signs of hair cell differentiation can be observed (9–10 weeks of gestation in a human fetus), both scanning and transmission electron-microscopic observations indicate the presence of an undifferentiated polystratified sensory epithelium with a field of microvilli and kinocilia at the luminal surface (Fig. 4.1). However, nerve fibers (endings of differentiated spiral ganglion neurons) are already invading this epithelium (Fig. 4.1B) at a very precise location (great epithelial ridge). This could indicate that some postmitotic cells, still not distinguishable from supporting cells, have already started a process of differentiation as sensory cells and begun to interact with nerve fibers (see Section 3.2.1). A few days later, although scanning electron microscopy still does not show the presence of any growing stereocilia among the field of

TABLE 4.1. Timing of the main events (in chronological order) in the development of sensory and neural structures in mammalian cochlea.

Structural events in cochlear development	Timing in different species	Main reference
Terminal mitoses in basal cochlea:		
Spiral ganglion	E12–13, mouse	Ruben 1967
Hair cells	E15–16, mouse	Ruben 1967
Spiral ganglion neuron differentiation	E14, mouse	Ruben 1967
Triggering of hair cell differentiation	???	
Afferent fibers (auditory dendrites) cleary seen within the cochlear epithelium	E15, mouse	Sher 1971
	E16, rat	Gil-Loyzaga and Merchan-Cifuentes 1982
	E32, guinea pig	Pujol et al. 1991
	W9, human	Pujol and Lavigne-Rebillard 1985
Distinguishable IHCs, then OHCs	E17, mouse	Sher 1971
	E17, rat	Gil-Loyzaga and Merchan-Cifuentes 1982
	E34, guinea pig	Pujol, unpublished observations
	W10–11, human	Lavigne-Rebillard and Pujol 1986
First synaptic-like contacts between afferent dendrites and hair cells	E18, rat	Gil-Loyzaga and Merchan-Cifuentes 1982
	E36, guinea pig	Pujol, unpublished observations
	W11–12, human	Lavigne-Rebillard and Pujol 1986
Beginning of stereociliogenesis on IHCs, then on OHCs	E18, rat	Gil-Loyzaga and Merchan-Cifuentes 1982
	E36, guinea pig	Pujol et al. 1991
	W12, human	Lavigne-Rebillard and Pujol 1986
First vesiculated efferent endings in the inner spiral sulcus below IHCs	Birth, rat	Lenoir, Shnerson and Pujol 1980
	Birth, mouse	Shnerson, Devigne, and Pujol 1982
	E40, guinea pig	Pujol, unpublished observations
	W14, human	Pujol and Lavigne-Rebillard 1985
Transient efferent axosomatic synapses with IHCs	P3–10, rat	Lenoir, Shnerson, and Pujol 1980
	P3–10, mouse	Shnerson, Devigne, and Pujol 1982
	P2–8, cat	Pujol, Carlier, and Devigne 1978
	E40–56, guinea pig	Pujol, unpublished observations
Acquisition by the OHCs of the structural characteristics of motile activities	E52–56, guinea pig	Pujol et al. 1991
	P7–16, gerbil	Weaver and Schweitzer 1994
Morphological criteria correlated with the onset of cochlear function (opening of tunnel of Corti, formation of Nuel's spaces)	P3, cat	Pujol and Marty 1970
	P9–10, rat	Lenoir, Shnerson, and Pujol 1980
	P9–10, mouse	Kraus and Aulbach-Kraus 1981
	P9, dog	Pujol and Hilding 1973
	P12, hamster	Pujol and Abonnenc 1977
	P12, gerbil	Weaver, Hoffpauir, and Schweitzer 1994
	E54, guinea pig	Pujol and Hilding 1973
	W18–20, human	Lavigne-Rebillard and Pujol 1990

(*Continued*)

TABLE 4.1. (*Continued*)

Structural events in cochlear development	Timing in different species	Main reference
Formation of efferent axosomatic synapses with OHCs	P6–16, rat	Lenoir, Shnerson, and Pujol 1980
	P7–20, mouse	Shnerson, Devigne, and Pujol 1982
	P6–15, cat	Pujol, Carlier, and Devigne 1978
	E56–P6, guinea pig	Pujol et al. 1991
	W20–30, human	Lavigne-Rebillard and Pujol 1990
Mature looking OHCs and surrounding structures (Deiters cells, medial efferents)	P16–18, rat	Lenoir, Shnerson, and Pujol 1980
	P16–21, mouse	Shnerson, Devigne, and Pujol 1982
	P18, gerbil	Weaver and Schweitzer 1994
	P20, cat	Pujol and Marty 1970
	P6, guinea pig	Pujol et al. 1991
	W30, human	Lavigne-Rebillard and Pujol 1990
End of myelination of spiral ganglion neurons	P120, cat	Romand and Romand 1982

E, embryonic day; P, postnatal day; W, weeks postconception; IHC, inner hair cell; OHC, outer hair cell.

microvilli, transmission electron microscopy enables the early detection, on transmodiolar sections of the same epithelium, of postmitotic cells that can be distinguished from neighboring supporting cells by their slightly different cytoplasmic content and their base surrounded by afferent dendrites (Pujol and Lavigne-Rebillard 1985). These cells obviously correspond to future hair cells.

2.2 Stereociliogenesis

The general pattern of differentiation of stereocilia bundles has been described in the guinea pig cochlea as a four-stage process (Kaltenbach, Falzarano, and Simpson 1994) involving the initial production of stereocilia, differentiation into tall and short populations, formation of distinct ranks, and resorption of supernumerary stereocilia. In the earliest stage, future stereocilia are not very morphologically different from surrounding microvilli but are arranged in round and densely packed clusters (Fig. 4.2). This feature is also observed in the avian basilar papilla (Cotanche and Sulik 1984; Tilney, Cotanche, and Tilney 1992) and the mammalian vestibular end organ (Mbiène, Favre, and Sans 1984).

In the mammalian cochlea, the earliest stages of stereociliary differentiation are similar in both types of hair cells, although the process begins on inner hair cells (IHCs). From the primitive round tuff (Fig. 4.2), a "V-shaped" bundle of stereocilia develops, with the tip of the "V" centered on the kinocilium facing the lateral wall (Fig. 4.3; Sobin and Anniko 1984; Lavigne-Rebillard and Pujol 1986; Lim and Anniko 1985; Kaltenbach, Falzanaro,

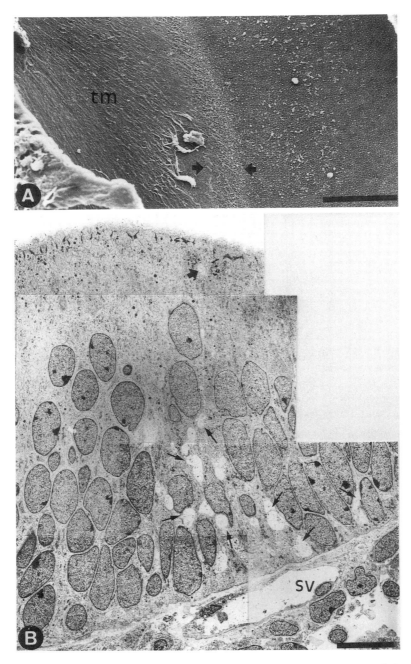

FIGURE 4.1. Scanning (A) and transmission (B) electron micrographs of Corti's epithelium from 10-week-old fetal human cochleae (reprinted from Lavigne-Rebillard and Pujol 1990 with permission). On the surface view (A), the epithelium is undifferentiated. The arrows point to the area of the future organ of Corti. On the modiolar side, the fibrillar material of the nascent tectorial membrane (tm) is seen. On a transverse section (B), nerve fibers (arrows) are seen within the undifferentiated epithelium, above the spiral vessel (SV). One nerve profile (thick arrow) is seen close to the luminal surface. Bars, 50 μm (A); 10 μm (B).

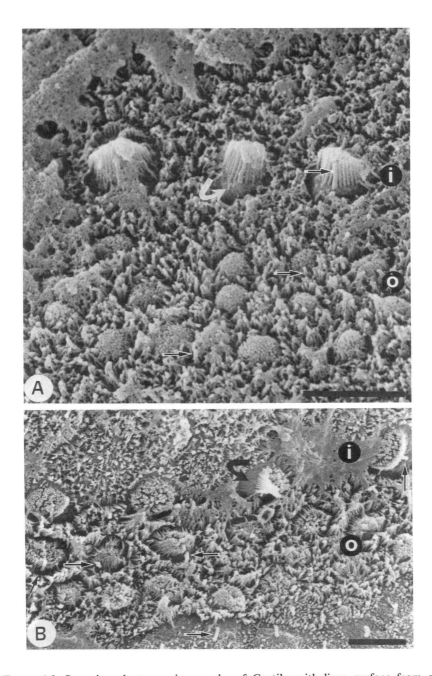

Figure 4.2. Scanning electron micrographs of Corti's epithelium surface from a 12-week-old fetal human cochlea (A) and a rat cochlea at birth (B). Both inner (i) and outer (o) hair cells exhibit bundles of stereocilia associated with a kinocilium (arrows) arising among the numerous microvilli of the surrounding epithelial cells. On the outer hair cells, small stereocilia are densely packed in round bundles. On the inner hair cells, stereocilia are more elongated and their bundle show a V-shape arrangement. Glabrous portions appear on the surface of some inner hair cells (curved arrow). At that stage, the process of ciliogenesis is more advanced at inner hair cells. Bars, 5 μm.

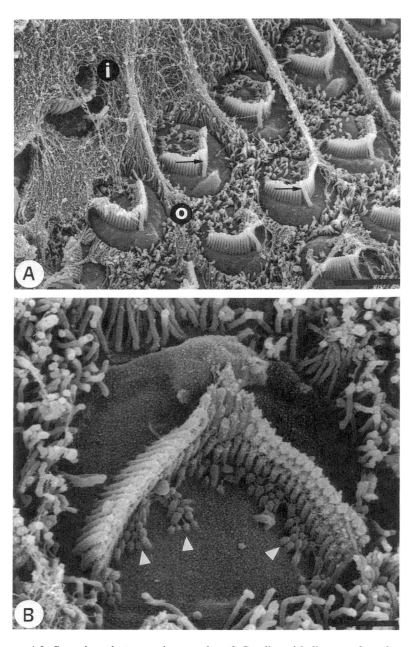

FIGURE 4.3. Scanning electron micrographs of Corti's epithelium surface from a 3-day-old rat cochlea (basal turn). The cuticular plate of both inner (i) and outer (o) hair cells is forming, with a few microvilli remaining on the surface. The kinocilium (arrows) is still present in all hair cells. On the high magnification (B), stereocilia are in typical staircase arrangement and the links between them are visible; the small supernumerary stereocilia (arrowheads) will disappear later on (compare with Fig. 4.4). Bars, 5 μm (A); 1 μm (B).

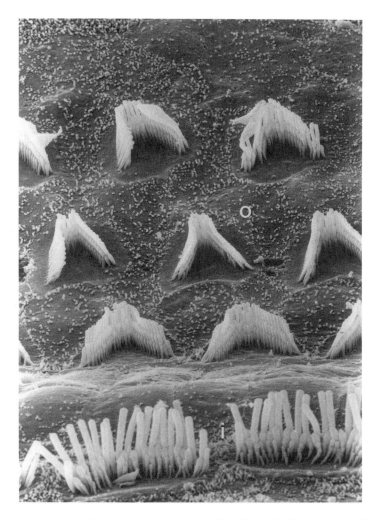

FIGURE 4.4. Scanning electron micrograph of Corti's epithelium surface from a 16-day-old rat cochlea. The stereocilia bundles have an adultlike appearance on both inner (i) and outer (o) hair cells. The surface of the cuticular plate is cleaned from kinocilium, microvilli, and short supernumerary stereocilia. Bar, 10 μm.

and Simpson 1994). On IHCs, the "V-shaped" bundle quickly becomes almost linear. Microvilli and supernumerary stereocilia are eliminated from the developing cuticular plate, and the remaining stereocilia are arranged in three to four parallel rows that are graded in length (Fig. 4.3B; Kaltenbach, Falzanaro, and Simpson 1994), the tallest being located near the kinocilium in the most external row. As soon as the staircase arrangement of stereocilia becomes visible, lateral (cross) and tip links can be observed (Fig. 4.3B;

TABLE 4.2. Timing of the main structural events (in chronological order) in the development of hair cell stereocilia in the mammalian cochlea.

Main structural events in the development of hair cell stereocilia	Species	Main reference
Appearance of stereocilia	Guinea pig IHC and OHC: <E38	Sobin and Anniko 1984
	Human IHC: W11; OHC: W12	Lavigne-Rebillard and Pujol 1986, 1990
	Rat IHC and OHC; <P2	Lenoir, Puel, and Pujol 1987
	Hamster IHC and OHC: <P2	Kaltenbach, Falzarano, and Simpson 1994
	Mouse IHC and OHC: E15	Lim and Anniko 1985
Presense of tip-links	Human IHC and OHC: W20–22	Lavigne-Rebillard and Pujol 1986
	Rat IHC and OHC: P2 (B)	Lenoir, Puel, and Pujol 1987
	Hamster OHC and IHC: P4	Kaltenbach, Falzarano, and Simpson 1994
Resorption of supernumerary stereocilia and microvilli	Mouse OHC: P3 (B)	Lim and Anniko 1985
	Rat OHC: P7 (B); IHC: P13–14 (B)	Lenoir, Shnerson, and Pujol 1987
	Guinea pig OHC: E56 (B)	Pujol et al. 1991
	Hamster OHC: P14 (B-A); IHC P18 (B-A)	Kaltenbach, Falzarano, and Simpson 1994
Resorption of kinocilium	Mouse OHC and IHC: P10 (B), >P14 (A)	Lim and Anniko 1985
Adultlike gross morphology of the bundle	Human OHC and IHC: W20–22 (B-A)	Lavigne-Rebillard and Pujol 1986, 1987
	Rat OHC: P10 (B-A); IHC: P13–14 (B)	Lenoir, Shnerson, and Pujol 1987
	Hamster OHC and IHC: P12–14 (B-A)	Kaltenbach, Falzarano, and Simpson 1994
	Horseshoe bat OHC: before birth (B); IHC; P3 (B)	Vater, Lenoir, and Pujol 1996
End of maturation of actin skeleton	Guinea pig OHC: E52 (B)	Pujol et al. 1991
	Gerbil OHC: P18–19 (B-A)	Weaver, Hoffpauir, and Schweitzer 1994
	Rat OHC: P11 (B-A)	Vago et al. 1996

(Continued)

TABLE 4.2. (*Continued*)

Main structural events in the development of hair cell stereocilia	Species	Main reference
Final length (age and size)	Rat OHC: P12, 2 μm (B), 9 μm (A); IHC: P?, 3 μm (B), 9 μm (A)	Roth and Bruns 1992
	Hamster OHC: P12–14, 1.5 μm (B), 5 μm (A); IHC: P16–18, 4 μm (B), 5 μm (A)	Kaltenbach, Falzarano, and Simpson 1994
	Horseshoe bat OHC: P2, <1 μm (B), 2 μm (A); IHC: P2, 3 μm (B), 4 μm (A)	Vater, Lenoir, and Pujol 1996

B, base of the cochlea; A, apex of the cochlea.

Lavigne-Rebillard and Pujol 1986; Lenoir, Puel, and Pujol 1987; Lim and Rueda 1992). It is interesting to note that tip links, probably functionally related to the mechanical transduction process (Pickles, Comis, and Osborne 1984; Hudspeth 1985) are formed very early, well before the onset of cochlear function. Also before the onset of function, the kinocilium regresses and disappears, first on outer hair cells (OHCs), then on IHCs (Lenoir, Puel, and Pujol 1987). Thus IHCs, although starting to grow their stereocilia before OHCs, get an adultlike cuticular plate surface (Fig. 4.4) slightly later (Kaltenbach, Falzanaro, and Simpson 1994).

The length of stereocilia changes dramatically during maturation (Roth and Bruns 1992; Kaltenbach, Falzanaro, and Simpson 1994), and the final length of the tallest stereocilia appears to be correlated with the final length of the hair cell itself (Strelioff and Flock 1984; Wright 1984). In other words, in a given adult species, IHCs have approximately the same size along the whole cochlear spiral and the length of their stereocilia is also stable. Conversely, as adult OHCs vary (increase) regularly in length from base to apex (Pujol et al. 1992) and in upper coils from the first to the third row, the final length of their stereocilia varies proportionally (Strelioff and Flock 1984; Wright 1984; Vater and Lenoir 1992). The insertion of the tips of the tallest row of OHC stereocilia within the undersurface of the tectorial membrane could well play a role in regulating both the length and the arrangement of the bundles. In fact, this insertion is different between the base and the apex of the cochlea (Lenoir, Puel, and Pujol 1987). At the base, tips are anchored in a smooth layer of thick and dense material covering the undersurface of the tectorial membrane, i.e., Hardesty's membrane (Lim 1972). In more apical regions, Hardesty's membrane is absent, and the stereocilia are in direct contact with large fibers emerging from the main body of the tectorial membrane (Lenoir, Puel, and Pujol 1987; Lavigne-Rebillard and Pujol 1990).

Table 4.2 summarizes and compares in different mammals the timing of the main structural events in the development of cochlear stereocilia.

2.3 Morphology, Arrangement in Rows, and Number of the Hair Cells

In the early stages of development, both types of hair cells are similar in shape: cylindrical but with a slightly swollen base due to the basally located and proportionally large nucleus (one third of the cell volume at this time). Quickly, IHCs increase in size to get their pear-shaped appearance, with the nucleus now located at midheight of the cell. The shape of OHCs remains immature much longer, until the supporting structures (outer pillar and Deiters' cells) elongate and the spaces (tunnel of Corti and Nuel's spaces) are formed, i.e., until the period of the onset of cochlear function (Pujol and Hilding 1973). At this time, OHCs start to show a strict cylindrical shape with a fixed diameter of 6-7 μm while their length evolves depending on basal to apical location as mentioned above. In the adult stage, the length of both the OHC body and OHC stereocilia follows a linear regression highly correlated with frequency tuning: the shorter the lengths, the higher the frequencies (Fig. 4.5; Brundin et al. 1991; Pujol et al. 1992).

When they first appear, hair cells are already arranged in rows (one row of IHCs, three rows of OHCs). The intercellular spacing of hair cells within a single row is wide at the beginning, then reduces due to two probable factors: continuing hair cell production and an increase in hair cell diameter (Kaltenbach and Falzarano 1994). This also partly accounts for the increase in length and width of the cochlear spiral during the same developmental period (Kaltenbach and Falzarano 1994). Hair cells in the very apical part of the cochlea are arranged differently at all stages of development (Fig. 4.6; Lavigne-Rebillard and Pujol 1986; Lenoir, Puel, and Pujol 1987; Kaltenbach and Falzarano, 1994; Zhou and Pickles 1994). Even in adult cochleae (Fig. 4.6B), although the row of IHCs can be recognized almost to the end of the spiral, the OHCs appear to be more disorganized.

The presence at some point of in vivo development of supernumerary hair cells (both IHCs and OHCs) has been mainly described in human cochleae (Bredberg 1968; Kawabata and Nomura 1978; Tanaka, Sakai, and Terayama 1979; Igarashi 1980; Lavigne-Rebillard and Pujol 1986, 1990). The question has to be asked as to why this phenomenon, so obvious in humans where two rows of IHCs and four to six rows of OHCs are seen (Fig 4.7; Pujol and Lavigne-Rebillard 1995), has not been reported for experimental mammalian species. One explanation could be the difference in species; another that transient supernumerary hair cells are frequently seen in humans because the rate of development is much slower than that of common laboratory animals. Supporting this idea are the recent findings

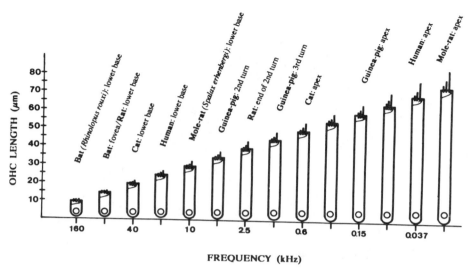

FIGURE 4.5. Correlation between outer hair cell (OHC) length across adult mammalian cochleae and estimated frequency (reprinted from Pujol et al. 1992 with kind permission from Elsevier Science Ltd, The Boulevard, Langford Lane, Kidlington OX5 1B, UK.). Note the remarkable stability in outer hair cell cylindrical shape and diameter; only the length of the cell body and stereocilia varies. It seems that, whatever the species, a given frequency corresponds to a fixed length.

(Abdouh, Després, and Romand 1994) of hair cell overproduction in the developing mouse cochlea in vitro, a condition known to slow down development. Thus, an overproduction of hair cells followed by a downregulation could be a "normal" developmental phenomenon. Understanding the molecular mechanisms underlying the overproduction of hair cells both in vivo and in vitro, where this overproduction seems to be enhanced by retinoic acid (Kelley et al. 1993), would appear to be of vital importance in understanding the challenging question of regeneration.

2.4 Late OHC Differentiation

Late development of OHCs is well known and has been tentatively correlated with coincidental striking changes in the functional properties of the cochlea, such as increasing sensitivity gain and frequency selectivity (Pujol, Carlier, and Lenoir 1980). Apart from their previously discussed elongation (see Section 2.3), OHCs undergo other drastic changes at a subcellular level. At the onset of functioning, an adultlike distribution of microfilaments and intermediate filaments is acquired, with an accumulation of actin in the cuticular plate accompanied by a disappearance of cytokeratin (Raphael et al. 1987). Other changes involve mitochondrial and endoplasmic reticulum distribution and the related formation of laminated

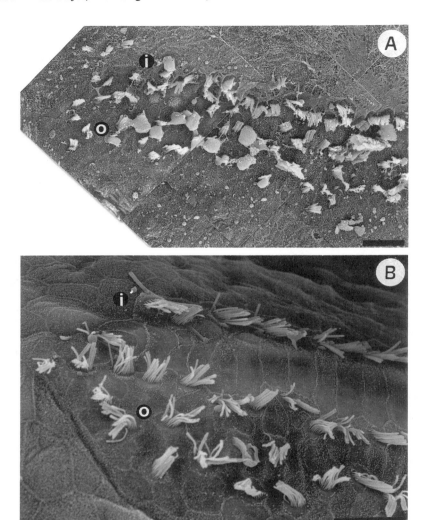

FIGURE 4.6. Scanning electron micrographs of Corti's epithelium surface at the extreme apex of the cochlea from a 20-week-old human fetus (A) and an adult rat (B: reprinted from Lavigne-Rebillard, Cousillas, and Pujol 1985 with permission of Wiley-Liss, Inc., a subsidiary of John Wiley and Sons, Inc.). In the fetal cochlea (A), both inner (i) and outer (o) hair cells are disorganized. In the adult cochlea (B), this disarrangement persists at outer hair cells, whereas the row of inner hair cells is well organized. Bars, 20 μm.

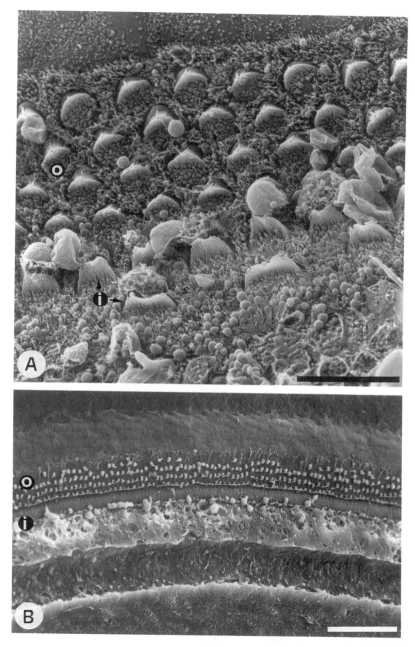

FIGURE 4.7. Scanning electron micrographs of Corti's epithelium surface in the basal cochlea from a 15-week-old (A; reprinted from Lavigne-Rebillard and Pujol 1986 with permission) and a 22-week-old (B) human fetus. Four rows of outer hair cells (o) can be observed at both stages; note the presence of a second row of extra inner hair cells (i) at the earliest stage (A). Bars, 10 μm (A); 50 μm (B).

cisternae and sub-plasma membrane complex (Fig. 8; Pujol et al. 1991; Lim and Rueda 1992; Weaver and Schweitzer 1994). Temporal correlations have been established between these changes in the subcellular organization of OHCs and either the development of in vitro motile properties (Pujol et al. 1991; He, Evans, and Dallos 1994) or the shift of the place code (Weaver and Schweitzer 1994). It is noteworthy that, in the functionnally hyperspecialized cochlea of the horseshoe bat, an hyperspecialized sub-plasma membrane complex (Fig. 4.8F) is seen at the final stage of development (Vater, Lenoir, and Pujol 1992, 1997). Altogether, the OHCs seem to follow a two-stage developmental process, as if the cell was differentiating twice: first as a normal sensory cell (IHC-like differentiation) and second as a totally new type of cell with both sensory and motile properties. These two steps may depend on two sets of genetic and epigenetic factors. In several mouse mutants with strong developmental defects such as deafness (*dn/dn;* Pujol et al. 1983) or shaker 1 (*sh1/sh1;* Shnerson et al. 1983), the OHCs degenerate during the first step of differentiation. Similarly, in rat pups deprived of thyroid hormone (Uziel et al. 1981, 1983), OHCs never seem to acquire the second stage of differentiation.

2.5 Markers of Developing Hair Cells

Hair cell maturation has been recently investigated with immunocytochemical localization of a variety of cytoplasmic and cell-surface markers that prove to be specifically expressed at certain stages of development. Neuron-specific enolase has been shown to label both the spiral ganglion neurons

---

$\longrightarrow$

FIGURE 4.8. Outer hair cell cytological maturation. A, B, and C: 37-day-old guinea pig fetal cochlea. At this stage, stereociliogenesis is just starting: the immature cuticular plate is covered by microvilli, short stereocilia (curved arrows), and a kinocilium (B, open arrow). Above the outer hair cell (o) nucleus, mitochondria and smooth endoplasmic reticulum are abundant (A and B). Neither the sub-plasma membrane nor the laminated cistern complexes are visible. At nuclear level and below, two adjacent outer hair cells (o1 and o2) are in direct contact (C). D: 47-day-old guinea pig fetal cochlea. The laminated cistern (arrowheads) is forming, although the lateral wall of the outer hair cell is still in contact with a Deiters cell (d). E and F: final stage in the maturation of outer hair cell lateral wall in a guinea pig cochlea at birth (E) and in a young adult bat (F). In E, the laminated cistern and sub-plasma membrane complexes are typical of an adult mammalian outer hair cell, with classical "pillars" (thin arrows) linking the plasma membrane and the cistern of smooth endoplasmic reticulum (arrowheads). In F, the hyperspecialized complex of the adult bat outer hair cell (see also Vater, Lenoir, and Pujol 1992, 1997) is formed by "pillars" (thin arrows) regularly connected to cylindrical and circumferential structures (thick arrows). n, Nuel's space. Bars, 2 μm (A); 1 μm (B and C); 0.5 μm (D, E and F).

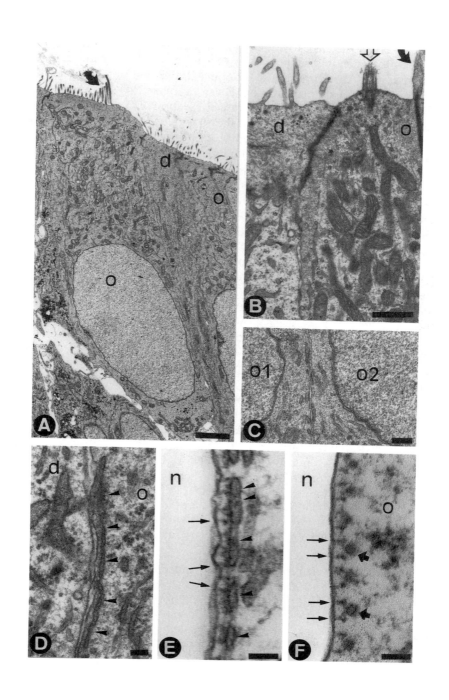

and the IHCs in the adult guinea pig cochlea (Altschuler et al. 1985). The same experiment repeated in developing rat (Dechesne and Pujol 1986) and mouse (Whitlon and Sobkowicz 1988) cochleae showed that both IHCs and OHCs were labeled as early as embryonic day 19. However, the labeling on OHCs subsequently disappeared by the end of the first postnatal week. It would appear, therefore, that developing OHCs, but not IHCs, lost some primitive "neuronal" properties; this can be related to the second step of differentiation of OHCs just discussed (Section 2.4). The same type of developmental patterning has been shown with calretinin immunoreactivity (Dechesne, Rabejac, and Desmadryl 1994), whereas calbindin, another calcium-binding protein, labels both types of cochlear hair cells at all stages of development (Dechesne and Thomasset 1988). Other markers have been shown to be transiently expressed in both types of hair cells during development. Indeed NFM, a neurofilament protein, is expressed in mouse cochlea IHCs and OHCs from embryonic day 19 to postnatal day 20 (Hasko et al. 1990). Antibodies against blood-group (HBO) cell-surface glycoconjugates also label rat cochlear hair cells as soon as they differentiate, while immunostaining disappears during the second postnatal week (Gil-Loyzaga et al. 1989); H antigens are transiently expressed during stereociliogenesis, and B antigens specifically label both types of hair cells up to the onset of functioning (day 10 in the rat). It is worthy of note that these transient expressions appear to be modulated by thyroid hormone (Gil-Loyzaga et al. 1994) because the immunoreactivity lasts much longer in hypothyroid rat cochleae that have a delayed and abnormal development (Uziel et al. 1981, 1983). Microfilaments and intermediate filaments of the hair cell cytoskeleton can also be used as developmental markers. Cytokeratin immunoreactivity is only expressed in hair cells at very early stage and is then restricted to supporting cells (Raphael et al. 1987). Actin and fodrin immunoreactivities are good markers of the cuticular plate and stereocilia formation (Raphael et al. 1987; Pirvola, Lehtonen, and Ylikoski 1991); a significant increase in actin labeling of the OHC lateral wall has been correlated to the onset of cochlear function (Weaver, Hoffpauir, and Schweitzer 1994).

3. Development of Hair Cell Innervation

Adult mammalian cochlear hair cells are innervated by four types of nerve fibers (see Spoendlin 1969, 1979; Warr and Guinan 1979; Kiang et al. 1982; Pujol and Lenoir, 1986). IHCs send their auditory message to the cochlear nuclei via radial afferents: unbranched peripheral dendrites of type I ganglion neurons (95% of the spiral ganglion population). In turn, the lateral efferent axons from the ipsilateral superior olive bring a postsynaptic regulation to the radial afferent dendrites below IHCs. On the other side of the cochlea, OHCs are connected to the auditory brain stem via the spiral afferents: convergent dendrites of the small unmyelinated type II ganglion neurons (5% of the spiral ganglion). The medial efferent axons from contralateral (75%) and ipsilateral (25%) medial olivary nuclei synapse directly with OHCs. The morphology and distribution of synapses (see Pujol and Lenoir 1986; Liberman, Dodds, and Pierce 1990) as well as the localization of most neurotransmitters (see Eybalin 1993) are now well established for the adult mammalian cochlea. During development, the pattern of hair cell innervation undergoes some drastic changes that deserve special mention.

3.1 Development of the Spiral Ganglion

The onset of spiral ganglion neuron differentiation is very early and before the differentiation of hair cells. The placode origin of the statoacoustic ganglion is now generally accepted (see Rubel 1978), although for many years it was contended that the neural crest also contributed to its formation (see Deol 1967). A recent study of the mouse embryo (Wikström and Anniko 1987) still supports the notion that cells migrating from the neural crest could play a role, albeit minor, in the formation of the spiral ganglion. Are these neural crest cells the origin of a small and specific population of neurons? Due to the strong characteristic differences between mature type I and type II neurons, it is tempting to suggest a different origin of these cells and hence to hypothesize the neural crest origin of type II cells. However, the difficulty of morphologically recognizing the types of ganglion cells at early stages of development (Nakai and Hilding 1968; Anniko 1983; Wikström and Anniko 1987; Simmons et al. 1991) prevents this line of argument from being further explored until such time as this question can be resolved by using other characteristics. In fact, type I and type II neurons can only be distinguished late in development when the course of their peripheral dendrites can be tracked to the different types of hair cells (Perkins and Morest 1975; Simmons et al. 1991) or when the first signs of myelination can be seen on type I cells (Pujol and Hilding 1973; Romand et al. 1980; Romand and Romand 1982; Schwartz, Parakkal, and Gulley 1983). Even immunocytochemistry, which in adulthood helps to differentiate the two types of neurons, is not particularly useful at the early stages. Calretinin immunoreactivity specific to type I neurons appears at about the time of birth in the mouse cochlea (Dechesne, Rabejac, and Desmadryl 1994), far too late to pick up the first stage of neuronal differentiation. Other markers, such as α-enolase (Dechesne and Keller 1996), neurofilament protein (NF 160; Hafidi and Romand 1989), and peripherin (Hafidi, Després, and Romand 1993), appear to be type II specific in adults, whereas immunoreactivity is seen on both types of ganglion neurons earlier on.

Myelination of type I neurons is a good index of late spiral ganglion development; in kittens, this process has been shown to extend up to 3–6 months (Romand and Romand 1982). This contrasts with all other structural and/or physiological criteria of development of the kitten cochlea that appears to be fully mature at postnatal day 20 (Pujol and Marty 1970; Pujol, Carlier, and Devigne 1978; Pujol, Carlier, and Lenoir 1980). Another criterium of late development in the spiral ganglion has been reported in the cochleae of human neonates where the proportion of type II vs. type I cells is significantly greater than in adults (Chiong, Burgess, and Nadol 1993). Although not reported in animal studies, this finding is perhaps related to the transient extranumerary OHCs well evidenced in the developing human cochlea (see Section 2.4).

3.2 Development of the Afferent Innervation in the Organ of Corti

3.2.1 Early Stages of Radial Afferent Innervation

Radial nerve fibers, i.e., peripheral dendrites of type I spiral ganglion neurons, have been reported to invade the cochlear epithelium very early, even before morphological differentiation of hair cells can be observed. At this stage, some nerve endings can be seen close to the luminal surface of the epithelium (Fig. 4.1B; Pujol and Lavigne-Rebillard 1985). It has also been suggested that the invasion of an undifferentiated cochlear epithelium by afferent fibers is very early (Tello 1931), as evidenced during in vivo development in mouse (Sher 1971), rat (Gil-Loyzaga and Merchan-Cifuentes 1982), and human fetus (Pujol and Lavigne-Rebillard 1985) cochleae. Although similar observations have been made in the neighboring vestibular epithelium (Sans and Dechesne 1985), the role of nerve fibers in triggering inner ear hair cell differentiation is questionable (Pujol and Sans 1986) and inconsistent with in vitro results (Van De Water 1976). However, it is probable that afferent nerves have an early influence on the cytodifferentiation and development of hair cells. This is indicated by the result that hair cells in vitro deprived of their ganglion do not express B and H antigens (Gil-Loyzaga, Remezal, and Oriol 1992), cell-surface glycoconjugates that have proved to be specific markers of cytodifferentiations of these cells (see Section 1.5). Nevertheless, as soon as hair cells differentiate, the nerve endings stop spreading within the epithelium to concentrate on the base of the newly differentiated sensory cells (Gil-Loyzaga and Merchan-Cifuentes 1982; Pujol and Lavigne-Rebillard 1985).

A characteristic of radial afferents early in development is the dramatic sprouting of dendritic endings below the IHCs, a process that does not occur in the adult configuration. This sprouting has been seen with light microscopy as well as Golgi and silver staining techniques (Lorente de No 1937; Perkins and Morest 1975; Gil-Loyzaga and Merchan-Cifuentes 1982; Ginzberg and Morest 1984). The sprouting can be better seen in cochleae where there is dendritic swelling resulting from exposure to excitotoxicity, such as after a slight ischemia (Fig. 4.9; Lavigne-Rebillard and Pujol 1990) or after a kainic acid injection (Fig. 4.10A; Gil-Loyzaga and Pujol 1990). During normal development, this primitive branched stage, also evidenced by transmission electron microscopy (Fig. 4.11A), is followed by a pruning mechanism that results in the adult unbranched configuration, where the only branching to be seen is at the most apical part of the cochlea (Liberman, Dodds, and Pierce 1990). This is another sign of "immaturity" of the extreme apex. Strikingly, sprouting has been reported to reoccur in adult pathological cochleae after mechanical damage to the eighth nerve (Spoendlin 1985) or to the organ of Corti (Sobkowicz and Slapnick 1992). In a recent study reporting radial dendrite regeneration in adult guinea pig cochleae after excitotoxic damage (Puel et al. 1995), we observed branching of regrowing radial dendrites (Fig. 4.11B) on their way to reconnect the IHCs.

The question arising from these findings concerning the branching of early radial afferent innervation of the organ of Corti is, do these radial afferents (or some of their branches) also reach the OHCs at this stage of development? Early light-microscopic studies of hair cell innervation development (Retzius 1884; Lorente de No 1937) reported "some fibers innervating both types of cells." However, these reports were unable to clearly differentiate between the type of ganglion cell sending branches to both IHCs and OHCs. Later on, the same type of branching was reported in a Golgi study (Perkins and Morest 1975) where the authors favored a type II (spiral afferent) branching. Our results, obtained with kainic acid in rat cochleae (Pujol et al. 1985; Gil-Loyzaga and Pujol 1990), clearly show that the OHCs are innervated early on by the same type of kainate-sensitive dendrites as IHCs (Fig. 4.10), an observation consistent with transient radial branching (Pujol 1986; Gil-Loyzaga and Pujol 1990). Such a transient radial innervation of OHCs has been recently confirmed by horseradish peroxidase (HRP) studies in gerbil (Etcheler 1992) and hamster (Simmons 1994). The temporary radial afferent branching on immature OHCs disappears just before the onset of function. This roughly coincides in time with the formation of medial efferent synapses (Lenoir, Shnerson, and Pujol 1980; Pujol 1986) and also with the acquisition by OHCs of their motile properties (Pujol et al. 1991). It appears that the regression of radial afferents from the OHC base is not fully dependent on either the arrival of efferents or the onset of sound-induced activity (Etcheler 1992). Regression is more likely to be linked to the intrinsic late developmental changes in the OHCs discussed in Sections 2.4 and 2.5 as being "a second step in OHC differentiation," when the IHC-like immature OHC cell switches to being a sensory-motile type of cell.

3.2.2 Radial Afferent Synaptogenesis

Radial afferent synaptogenesis begins very early in the mammalian cochlea. As soon as the sensory cells start to differentiate, they are connected with

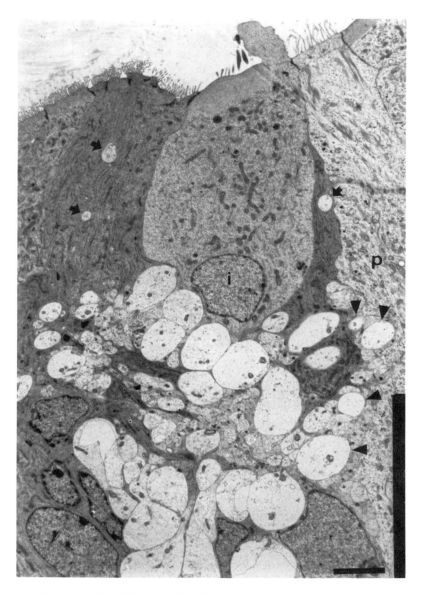

FIGURE 4.9. Inner hair cell (i) in a 3-day-old rat cochlea after 20 minutes of ischemia. Due to the ischemia-related excitotoxicity, all type I radial afferent dendrites are swollen (clear profiles with sparse cytoplasmic content). A cytoplasmic extrusion (due to ischemia) is seen behind the cuticular plate of the inner hair cell. Note the exuberant number of afferent endings surrounding the inner hair cell base; characteristic of this immature stage, some swollen afferents are seen in ectopic positions (arrows) or invading (arrowheads) the pillar cell (p), as if they were crossing to the outer hair cell side. Bar, 2 μm.

nerve endings of type I spiral ganglion neurons. Membrane thickening is immediately observed, followed by the formation of presynaptic specializations (Pujol and Hilding 1973; Pujol and Lavigne-Rebillard 1985). Thus, at least the ultrastructural features for synapses appear in the cochlea long before the onset of function. The slow-developing human fetus enables the very first stages of cochlear synaptogenesis to be precisely described. Classic afferent synapses, with synaptic densities and presynaptic bodies surrounded by vesicles, can be seen in a 12-week-old fetal cochlea, first at IHC and then a few days later at OHC bases (Pujol and Lavigne-Rebillard 1985; Lavigne-Rebillard and Pujol 1990). At this stage, stereocilia are just beginning to grow, and the onset of cochlear function is still 6–8 weeks ahead! For a short period of time, these afferent synapses follow the same pattern of development in both types of cells, although the number of endings connecting sensory cells is always much higher at IHCs than at OHCs. Certain transient characteristics of immature afferent synapses have been reported (Pujol, Carlier, and Devigne 1979); examples include filopodia-like endings making long and/or multiple contacts with the hair cell, postsynaptic endings containing different sizes of developmental vesicles, multiple (Figs. 4.10B, 4.12A) and/or ectopic presynaptic bodies (see also Sobkowicz et al. 1982; Sobkowicz 1992). Some of these criteria reappear at IHCs in adult cochleae during posttraumatic (excitotoxic; see Section 3.2.3) neosynaptogenesis (Fig. 4.12C, D; Puel et al. 1995).

Concerning IHCs, the late stage of radial afferent synaptogenesis involves the disappearance of the transient features discussed immediately above. A few days after the onset of function, the IHC/radial afferent synapse reaches its classic adult configuration (Liberman, Dodds, and Pierce 1990). Namely, unbranched dendrites terminate on IHCs with a single bouton now devoid of vesicles and, on the opposite side within the IHC, a single presynaptic body can be seen (Fig. 4.12B).

3.2.3 Spiral Afferent Synaptogenesis

There is little to be said about spiral afferent synaptogenesis. Very early in development (12 weeks in human fetal cochlea), fibers with a denser cytoplasmic content than radial afferents have been seen to spiral between Deiters' cells and to contact newly differentiated OHCs (Pujol and Lavigne-Rebillard 1985; Lavigne-Rebillard and Pujol 1990). Although it is difficult to clearly recognize the type of endings at this time (see Section 3.1; type I vs. type II ganglion neurons) when the OHC is almost exclusively surrounded by afferent endings (Fig. 4.13), there is indirect evidence provided by excitotoxic preparations. In the adult cochlea, an excess of glutamate (due to either direct application of a glutamate agonist, local ischemia, or noise trauma) causes acute and specific damage (called excitotoxicity) to the type I auditory dendrites, the type II afferents never being affected (Pujol et al. 1985; Pujol, Puel, and Eybalin 1992). If this is also true during

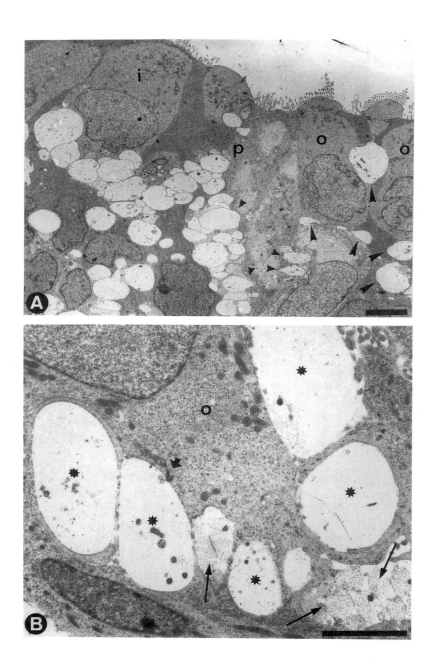

development, the presence of both swollen and unswollen afferents below OHCs, either after ischemia (Pujol and Lavigne-Rebillard 1985) or after kainate exposure (Fig. 4.10B; Pujol et al. 1985; Gil-Loyzaga and Pujol 1990), suggests that the unswollen endings found at this time belong to spiral fibers. It is conceivable (Simmons et al. 1991) that spiral afferents from type II neurons are innervating the cochlea as early as radial ones and that they acquire typical properties such as a specific microfilament content during later development (Hafidi and Romand 1989; Hafidi, Després, and Romand 1993). Moreover, early in development, the transient branching of type II endings to IHCs has been recently hypothesized (Hafidi, Després, and Romand 1993). Thus a common feature of the development for all afferents seems to be an exuberant and transient branching to both types of hair cells preceding the final (adult) stage of specific innervation patterning: type I/IHCs and type II/OHCs.

3.3 Development of Efferent Innervation

3.3.1 Early Arrival of Efferents in the Cochlea

The early arrival of efferents within the developing otocyst has been suggested by pioneer works (Tello 1931; Lorente de No 1937). Recently, methods using a fluorescent carbocyanide dye (DiI) have confirmed their very early arrival. By embryonic day 12 in the mouse, some axons from neurons of the superior olivary complex have been traced to the vestibular and cochlear components of the otocyst (Fritzsch and Nichols 1993). This result means that efferents and afferents (see Tello 1931; Sher 1971) can be seen at about the same time in the primitive cochlear epithelium, well before the first structurally recognizable sign of hair cell differentiation. The use of the same DiI technique in the postnatal rat has provided complementary

←
_____

FIGURE 4.10. Probable radial afferent innervation of immature outer hair cells. During the immature stage, both inner (i) and outer (o) hair cells are innervated by the same type of fibers sensitive to excitotoxicity: these fibers probably belong to the radial system that shows exhuberant branching at that time (see also Figs. 4.9 and 4.11). A: 1-day-old rat cochlea after kainic acid treatment (see also Gil-Loyzaga and Pujol 1990). The same type of swollen profiles are seen below the inner hair cell, crossing (small arrowheads) the pillar cells (p) and reaching (large arrowheads) the outer hair cells. B: 5-day-old rat cochlea (reprinted from Pujol et al. 1985 with kind permission from Elsevier Science–NL, Sara Burgerhartstraat 25, 1055 KV Amsterdam, The Netherlands). The base of an outer hair cell is only surrounded by afferents and most of these profiles are swollen (asterisks) due to kainic acid exposure. Note a double synaptic body (large arrow) opposite one of these swollen afferents. Some afferents are not swollen and may belong to the spiral system (arrows). Bars: 50 μm (A); 5 μm (B).

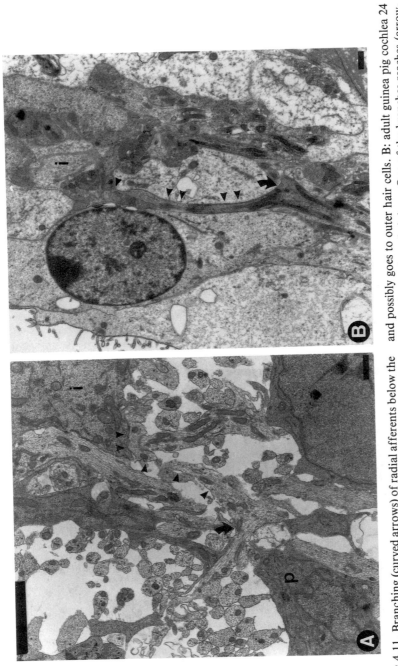

FIGURE 4.11. Branching (curved arrows) of radial afferents below the inner hair cell (i). A: 37-day-old guinea pig fetal cochlea. While one branch goes (arrowheads) to the inner hair cell base, where it can be followed in serial sections, the other branch crosses the pillar cell (p) and possibly goes to outer hair cells. B: adult guinea pig cochlea 24 hours after excitotoxic injury. One of the branches reaches (arrowheads) the inner hair cell. Bars, 1 μm.

precisions about the maturation of the two efferent systems and their sequential projection to both the IHC and OHC areas. At birth, the number of efferent neurons and their projection to the cochlea are not significantly different from adult organization (Robertson, Harvey, and Cole 1989). However, although efferent fibers enter the organ of Corti at birth (Robertson, Harvey, and Cole 1989), they first stay at the IHC level and they only invade the OHC region progressively from postnatal day 3 to 11 (Cole and Robertson 1992). This sequential pattern of development is confirmed by HRP studies in postnatal gerbils (Simmons et al. 1990), and by all data on efferent synaptogenesis.

3.3.2 Development of Efferent Synapses at IHCs

Below adult cochlear IHCs, the efferents correspond to highly branched terminals of thin unmyelinated axons from lateral superior olivary neurons. Terminals, and/or en passant vesiculated endings from this so-called lateral efferent system (see Warr 1992) synapse mainly with type I radial auditory dendrites (Fig. 4.12B). Although each radial auditory dendrite seems to be contacted, the number of synapses is greater with dendrites of modiolar high-threshold fibers (Liberman 1980). Such a morphological synaptic arrangement can be interpreted as bringing some postsynaptic modulation (or protection; see Pujol 1994) to IHC-auditory nerve synapses.

During development, the first vesiculated endings have been recognized quite early in the inner spiral sulcus of all studied species (see Pujol 1986). Synaptic contacts between these vesiculated efferents and radial afferent fibers have been reported in 14-week-old human fetuses (Lavigne-Rebillard and Pujol 1988) and in 1-day-old rats (Lenoir, Shnerson, and Pujol 1980) and mice (Shnerson, Devigne, and Pujol 1982) well before the onset of cochlear function. A possible role of this early lateral efferent innervation, as strongly suggested in adult cochleae (see Pujol 1994), could be the early protection of radial auditory dendrites against excitotoxicity, which indeed affects these dendrites very early in development (Figs. 4.9, 4.10A; Gil-Loyzaga and Pujol 1990).

A characteristic of the immature IHC efferent innervation is the surprisingly high number of direct contacts between vesiculated endings and the IHC itself, most of them showing typical synaptic differentiations such as a postsynaptic cistern in the IHC (Fig. 4.12A, 4.14A). In adult cochleae, apart from the most apical portion, these direct contacts between lateral vesiculated endings and IHC basal membrane are sparse. In addition, these contacts can very rarely be described as "synapses" in serial sections (Liberman 1980). An interesting explanation has been proposed from an HRP reconstruction of efferent fibers in the postnatal hamster cochlea (Simmons et al. 1990): medial efferents transiently innervate IHCs before crossing the tunnel and forming synapses with OHCs. Whatever the nature (lateral or medial) of these efferent axosomatic synapses with immature IHCs, they should have a functional significance because (1) mRNAs for

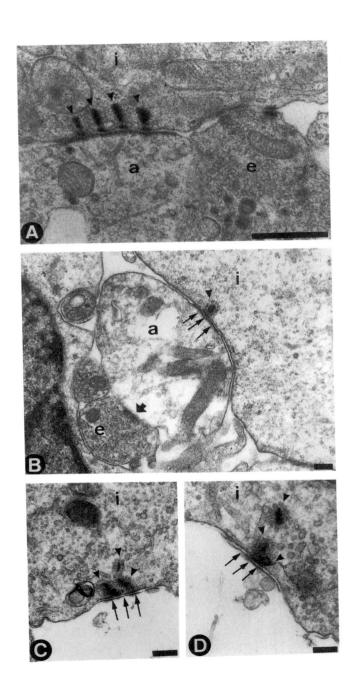

some cholinergic receptors (α9, Elgoyen et al. 1994; m3, Safieddine et al. 1996) have been found in IHCs and (2) axosomatic synapses reappear transiently in adulthood after the IHC has been disconnected from auditory dendrites by excitotoxic injury (see Sections 3.2.2 and 3.2.3 and Fig. 4.14B; Puel et al. 1995). These findings suggest that IHCs may receive certain information from cholinergic efferents that could be useful during development in the setting of their final pattern of innervation and, in adulthood, in the processes of recovery of this pattern after damage.

3.3.3 Development of Efferent Synapses at OHCs

Below adult cochlear OHCs, the efferents endings correspond to branched terminals of the myelinated axons from medial superior olivary neurons (see Warr 1992). These medial efferents form mainly axosomatic synapses with the OHCs, the size and number of synapses decreasing at the apex of the cochlea.

Compared with IHCs (see Section 3.3.2), efferent synapses mature much later at OHCs (see Pujol 1986). The direct axosomatic contacts between medial efferents and the basal pole of OHCs do not develop before the time of the onset of cochlear function, although efferents can be seen in the area of OHCs a few days earlier (Fig. 4.13). Typical efferent synapses with a well-developed postsynaptic cistern are not found until the second postnatal week in the mouse (Kikuchi and Hilding 1965; Shnerson, Devigne, and Pujol 1982) and the rat (Lenoir, Shnerson, and Pujol 1980), until embryonic day 55 in the guinea pig (Pujol et al. 1991), or until weeks 20–22 in human fetus (Lavigne-Rebillard and Pujol 1988). In all studied species, the formation of medial efferent synapses with OHCs immediately precedes or directly coincides with the onset of cochlear function even though there is no causal relationship between development of efferents and the onset of function. The adult appearance of these efferent synapses has been described as one of the last events in the maturation of cochlear sensory and neural structures (Pujol, Carlier, and Lenoir 1980).

$\leftarrow$

FIGURE 4.12. Synaptic pole of the inner hair cell (i) early in development (A: 37-day-old guinea pig fetal cochlea), at the end of development (B: young adult guinea pig cochlea), and 3 hours after excitotoxic injury (C and D: adult guinea pig cochleas). A: the immature stage is characterized by multiple presynaptic bodies (arrowheads) facing an afferent profile (a) and by a direct efferent (e) axosomatic synapse. B: adult pattern of a synaptic complex below an inner hair cell. A single presynaptic body (arrowhead) is facing postsynaptic membrane density (small arrows) within the radial afferent bouton (a); one efferent (e) makes a typical axodendritic synapse (thick arrow) with the afferent ending. C and D: in adulthood, 3 hours after an excitotoxic injury, the postsynaptic afferent profile has blown out. Multiple and/or ectopic presynaptic bodies (arrowheads) are seen facing remnants of the postsynaptic membrane (small arrows). Bars, 0.5 μm.

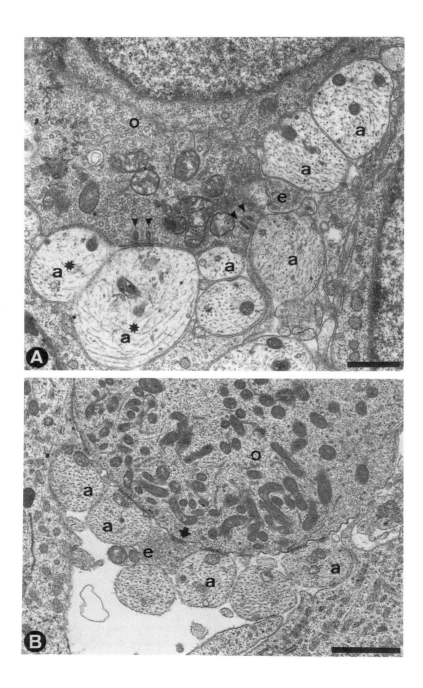

Returning to the idea that the completion of certain maturation processes differs from base to apex (see Sections 1. and 3.2), the immature aspect of some of the few medial efferent synapses with apical OHCs is worth noting. At the extreme apex of the cochlea, such synapses, if any (Fig. 4.15A), are very rare. In the apical turn (Fig. 4.15B), presynaptic varicosities are smaller than at the base of the cochlea (Fig. 4.15C) and the postsynaptic cistern is incompletely developed. It is also in apical coils that axodendritic en passant synapses between efferent varicosities and spiral afferents are frequently encountered (Pujol and Lenoir 1986). Such synapses are very rare in basal portions of the cochlea but are very common during development at a stage that precedes the establishment of direct axosomatic efferent synapses with the OHCs (Lenoir, Shnerson, and Pujol 1980; Pujol 1986).

3.4 Development of Cochlear Neurotransmitters and/or Neuromodulators

3.4.1 Afferent Neurotransmitters

Very little can be said about the development of glutamate, clearly the only reliable putative neurotransmitter of cochlear hair cells (see Eybalin 1993). Although immunocytochemistry has not been used for glutamate or its receptors in the immature cochlea, there are, however, two findings that indirectly support the assumption that the main afferent neurotransmitter is probably present very early. First, all morphological characteristics of afferent synapses appear very early in development, as soon as contacts between afferent dendrites and the differentiating hair cells have been made (see Section 3.2.2). Clear regularly sized microvesicles surrounding a synaptic body have been observed on the presynaptic side of the hair cell as early as week 12 in human fetuses (Pujol and Lavigne-Rebillard 1985) or at birth in different altricial mammals (Pujol, Carlier, and Devigne 1978; Lenoir, Shnerson, and Pujol 1980; Shnerson, Devigne, and Pujol 1982). Do these synaptic vesicles already contain some neurotransmitter (glutamate) at this very early stage? An indirect positive answer is given by the swollen

←——————————————————————————————

FIGURE 4.13. Early stages of synaptogenesis at the base outer hair cells (o). A: 5-day-old rat cochlea (a few days before the onset of cochlear function). The outer hair cell, still surrounded by numerous afferents (a), is contacted by a small efferent (e). Multiple synaptic bodies (arrowheads) are seen facing two afferent profiles. Some of these afferent profiles are swollen (a*), with a clear cytoplasmic content; they could belong to the radial system (see also Fig. 4.10A). B: 52-day-old guinea pig fetal cochlea (at the time of the onset of cochlear function). One efferent (e) begins to form an axosomatic synapse with an uncompleted postsynaptic cistern (thick arrow) in the outer hair cell. Bars: 0.5 μm.

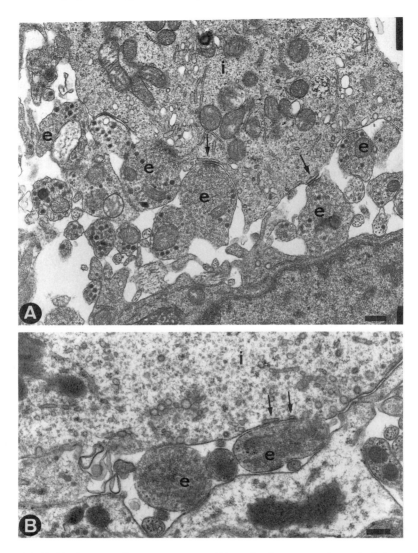

FIGURE 4.14. Both at an immature stage (A) and after an excitotoxic injury in adulthood (B) efferents (e) are in direct contact with the inner hair cell (i). A: in a 47-day-old guinea pig fetal cochlea, several efferents contact the inner hair cell, some of them form typical axosomatic synapses (arrows). B (Reprinted from Puel et al. 1995 with permission): In an adult guinea pig cochlea, one day after an excitotoxic injury that has blown out all afferent dendrites, two efferents are seen in direct contact with the inner hair cell; arrows point to postsynaptic cisterns. Bars: 0.5 μm.

appearance of endings (see Figs. 4.9 and 4.10) that indicate a sensitivity to ischemia-induced excitotoxicity linked to glutamate (Pujol et al. 1992). Also, direct glutamate-agonist excitotoxicity has been demonstrated early in rat cochleae (Pujol et al. 1985; Gil-Loyzaga and Pujol 1990). The weight of evidence thus indicates early glutamate release together with an early development of glutamate receptors on postsynaptic fibers. Another question arises, if glutamate and glutamate receptors are present in the developing cochlea well before the onset of auditory potentials, what is their role? A trophic role may be envisaged in setting the final configuration of the IHC-auditory nerve synapses. For instance, in the central nervous system during developmental or neosynaptogenesis, N-methyl-D-aspartate (NMDA) receptors have been shown to play a major role in synapse elimination and stabilization (Rabacchi et al. 1992) and in the pruning of exuberant dendritic branches (Kozlowski, Jones, and Schallert 1994). This idea is supported by the recent finding that NMDA-receptor mRNA is transiently overexpressed during the auditory nerve regeneration process after an excitotoxic injury (Puel et al. 1995).

3.4.2 Efferent Neurotransmitters

Not all neurotransmitters and/or neuromodulators recognized as being expressed in olivocochlear efferent neurons and localized in vesiculated endings of the organ of Corti (see Eybalin 1993) have been investigated during the developmental period. Available data on the development of cochlear neurotransmitters and receptors are summarized on Table 4.3. To date, most of these data come from light-microscopic immunocytochemistry and concern acetylcholine (ACh), γ-aminobutyric acid (GABA), calcitonin gene-related peptide (CGRP), and enkephalins.

ACh is considered as being the main neurotransmitter for both lateral and medial efferents (Eybalin 1993). The best indication for its presence in efferent cochlear synapses, either in the adult or during development, results from immunocytochemistry with its synthesizing enzyme choline acetyltransferase (ChAT). Cholinesterase data cannot easily be taken into account due to the poor specificity of the reaction, especially during the developmental stages. In the postnatal rat cochlea (Merchan-Peŕez et al. 1994), ChAT immunoreactivity is already present within the inner spiral bundle (below IHC) at the earliest stage that has been examined, i.e., day 1. A few days later (from day 3 on), reactivity also begins to develop below OHCs where it reaches its adult pattern of distribution by day 15 (the time of morphologically mature medial efferent-OHC synapses; Lenoir, Shnerson, and Pujol 1980). The early presence of ChAT in the cochlea, well before the onset of cochlear function, could support a regulative function of ACh in cochlear neuritogenesis, as previously reported for other parts of the nervous system (Mattson 1988).

GABA immunoreactivity has been identified in adult efferent presy-

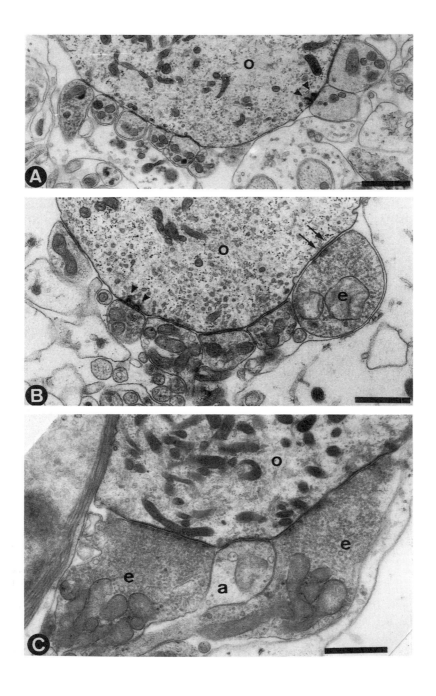

napses of the lateral, as well as of the medial, systems (although limited to the apex of the cochlea in that later case; see Eybalin 1993). A study in the postnatal rat (Merchan-Pérez et al. 1990b, 1993) shows GABA-like immunoreactivity appearing first (day 3) within the inner spiral bundle, then (day 9) below OHCs. This timing is comparable with that reported for the mouse cochlea (Whitlon and Sobkowicz 1989). Again, the early presence of GABA, at least within the inner spiral bundle, could indicate a role for this substance in cochlear synaptogenesis, as demonstrated in cervical ganglion (Wolff, Joo, and Dames 1978).

CGRP is known to be colocalized in efferent cholinergic neurons and synapses (see Eybalin 1993). Immunocytochemical detection of CGRP always appears in postnatal rat cochlea (Merchan-Pérez, Gil-Loyzaga, and Eybalin 1990a) at day 4 in the inner spiral bundle and at days 6–7 below OHCs. This appearance coincides with the development of mature-looking efferent synapses at both rat cochlea sites (Lenoir, Shnerson, and Pujol 1980). This timing is consistent with a possible role for CGRP in the formation and maintenance of ACh synapses, as has been shown at the neuromuscular junction (Sala et al. 1995).

[Met]enkephalin and related peptides are recognized as being neurotransmitters and/or neuromodulators for lateral efferents (see Eybalin 1993). During development, a guinea pig study (Gil-Loyzaga, Cupo, and Eybalin 1988) has shown the early (embryonic day 50) appearance of enkephalin-like immunoreactivity in the inner spiral bundle, at about the time of appearance of the onset of cochlear function in this animal (see Pujol and Hilding 1973).

Altogether, these data support the duality of efferents (Warr 1992) and sequential maturation shown by morphological studies: lateral efferent synapses below IHCs appearing before medial efferent synapses with OHCs (see Sections: 3.3.2 and 3.3.3). In addition, a developmental study of synaptophysin, a membrane protein specific to synaptic vesicles that can be tentatively correlated with the development of efferent synapses, also agrees with this timing. In the postnatal rat cochlea (Gil-Loyzaga and Pujol 1988), synaptophysin reactivity is first observed (day 3) in the inner spiral plexus, then (day 10) below OHCs.

←—————————————————————————————————————

FIGURE 4.15. Synaptic pole of adult outer hair cells (o) from different levels of guinea pig cochlea: extreme apex (A), 4th turn (B), and 2nd turn (C). A: only afferent endings are in contact with the outer hair cell; a double synaptic body (arrowheads) is seen facing one of these afferents. B: slightly more basally, an efferent ending (e) is seen facing an incomplete postsynaptic cistern (small arrows) in the outer hair cell; all other endings are afferents and a double presynaptic body (arrowheads) is seen opposite one of them. C: two huge presynaptic efferent endings (e) form typical axosomatic synapses with the outer hair cell; a small afferent profile (a) is in between. Bars: 1 μm.

TABLE 4.3. Development of cochlear neurotransmitters.

NT	Reference	Methods	Results (species, timing, localization)
Glutamate	Pujol et al. 1985	KA excitotoxicity (TEM)	Rat P0-6: IHC and OHC levels; adult: IHC level
	Gil-Loyzaga and Pujol 1990	KA excitotoxicity (TEM)	Rat E19–P3: IHC and OHC levels
	Lefebvre et al. 1991	KA excitotoxicity (in vitro)	Rat P5: ganglion neurons
	Janssen, Schweitzer, and Jensen 1991	Glu excitotoxicity	Rat P22–adult: ganglion neurons
	Luo, Brumm, and Regan 1995	In situ GluR mRNAs	Rat E16–17: + +GluR2,3
ACh	Emmerling and Sobkowicz 1988	BcAs: AChE	Mouse birth-P20: ↑
	Sobkowicz and Emmerling 1989	Im, ImEM AChE (in vitro)	Mouse birth: IHC level; P4-12: OHC level
	Bartolami et al. 1990	Bc2m (muscarinic receptor)	Rat P1–25: IP^3 ↑, peaks at P12, then ↓
	Roth, Dannhof, and Bruns 1991*	ImEM ChAT	Rat P20: OHCs
	Bartolami et al. 1993	Bdg MR^3	Rat P4–adult: affinity ↑ and number ↓
	Merchan-Pérez et al. 1994*	Im ChAT	Rat P1: IHC level; P3–15: OHC level
	Glowatzki et al. 1995	RTPCR $\alpha^9 R$	Rat P4: IHCs and OHCs
GABA	Wittlon and Sobkowicz 1989	Im GABA (in vitro)	Mouse P2: IHC level; P6–10: OHC level
	Merchan-Pérez, Gil-Loyzaga, and Eybalin 1990a	Im GAD	Rat birth: IHC level; P15: OHC level
	Merchan-Pérez et al. 1993	Im GABA	Rat P3: IHC level; P9: OHC level
Dopamine	Gil-Loyzaga and Parés Herbute 1986	HPLC DA	Rat birth-P30: DA ↑
Enkephalins	Gil-Loyzaga, Cupo, and Eybalin 1988	Im MET, MET8	Guinea pig E50: IHC level
CGRP	Tohyama et al. 1989	Im CGRP (surface preparation)	Rat P8: IHC level; P19: OHC level
	Merchan-Pérez, Gil-Loyzaga, and Eybalin 1990a	Im CGRP (sections)	Rat P4: IHC level; P6: OHC level

The dates for immunocytochemistry (light microscopy [Im]; light and electron microscopies [ImEMI]) or in situ hybridization indicate the early appearance of the reactivity. For other applied methods (excitotoxicity; biochemical assays [BcAs]; biochemistry of second messenger [Bc2m]; receptor binding [Bdg]; reverse-transcription polymerase chain reaction [RTPCR]; high-performance liquid chromatography [HPLC]; dates directly refer to the reported results. NT, neurotransmitter; TEM, transmission electron microscopy; Glu, glutamate; GluR, glutamate receptor (R2,3, ionotropic non-N-methyl-D-aspartate types 2 and 3); Ach, acetylcholine; ChAT, choline acetyltransferase; AChE, acetylcholine esterase; MR^3, muscarinic receptor type 3; IP^3, inositol triphosphate; $\alpha^9 R$, cholinergic α^9-receptor; GABA, γ-aminobutyric acid; GAD, glutamic acid decarboxylase; DA, dopamine; MET, Met-enkephalin; MET8, Met-enkephalin-Arg[6]-Gly[7]-Leu[8]; CGRP, calcitonin gene-related peptide; ↑, increase; ↓, decrease. Level means that reactivity is seen not in the cell itself but in the neighboring neurites. *Discrepant results due to different technical approaches and/or antibody specificities.

4. Conclusion and Perspective

Remarkable progress has been made in the last two decades in understanding the structure and function of the mammalian cochlea. In the 1970s, the development of sensory and neural structures in the cochlea was already well described (at least at cellular and subcellular levels), whereas the active mechanisms and role of OHCs had yet to be specified. During this period, questions were frequently raised concerning the adult cochlear organization and function that are now regarded as premonitory. For instance, the sequential development of efferents at IHC and then at OHC levels suggested a dual organization, later described by Warr and colleagues as lateral and medial systems (Warr, 1992; Warr and Guinan, 1979). Furthermore, the late maturation of OHCs and the changes occurring in their innervation pattern (with a loss of most of their connections to the brain) suggested that these cells evolve toward a very atypical type of sensory receptor.

The 1980s (decade of the active mechanisms) witnessed a period almost entirely devoted to the adult cochlea. Major physiological advances were made: recordings from single hair cells, discovery of otoacoustic emissions, of OHC motilities, etc. Another big step was made in neurochemistry, with the characterization and localization of most of the neuroactive substances used at afferent and efferent cochlear synapses. In the developmental area, apart from the brand new "place code principle" (see Chapter 5), progress was made to confirm and complement these advances: development of otoacoustic emissions (see Chapter 5) and development of neurotransmitters (see Section 3.4 and Table 4.3).

Only recently, the discovery of hair cell regeneration in the avian model reawakened interest in developmental studies. It quickly became clear that a new dimension of developmental studies (at cellular and molecular biology levels) was essential for the better understanding of hair cell differentiation processes, cell-to-cell interactions, formation of synapses, role of growth factors, etc. Similarly, the general boom in genetics has affected the cochlea, and there is an increasing demand for molecular biology of the development. Schematically, one could summarize the areas where we are still most naive and where major contributions should come in the future.

(1) Determination of the fate of cells within the growing otocyst. Are they precursors or progenitors of hair cells, of supporting cells? Why and when do these cells become quiescent and start differentiating? Interestingly, one point may prove to be crucial in this area, what makes the late differentiation of OHCs so different from that of IHCs? Are these two types of sensory cells coming from two types of progenitors? Which genetic and/or epigenetic factors drive the very specific late step of differentiation of the OHCs toward a motile type?

(2) Reciprocal interactions between hair cells and neurons. The real

influence of hair cells on synaptogenesis and, conversely, of synapses on hair cell maturation has still to be specified. Furthermore, there are certain other questions that will probably be solved in the same area, such as the early roles of some neurotransmitters and receptors and the role of nerve growth factors. It is well known in other areas of the nervous system that some neurotransmitters and some of their receptors play a specific developmental role, such as in the pruning or retraction of dendrites and in the stabilization or elimination of synapses. This is also probably true in the cochlea where the adult pattern of IHC and OHC innervation is regulated by different neuroactive substances. Two recent findings suggest a promising outlook for this area of research: the role of certain nerve growth factors (BNDF, NT3) and their receptors (trkB, trkC) in the patterning of cochlear innervation (see Chapter 3) and the influence of NMDA-glutamate receptors in the posttraumatic neosynaptogenesis below IHC (see Section 3.4.1).

5. Summary

There are two general gradients in the differentiation and maturation of cochlear hair cells and their neural connections. One is the classic base-to-apex gradient that applies to most of the criteria and means that at each maturation stage the midbasal turn is more advanced than the second turn, and the second turn is more advanced than upper turns. The second gradient is from IHCs to OHCs, meaning that IHCs differentiate and develop first. This does not necessarily imply that IHCs are the first to achieve all adult characteristics. For example, the completion of the ciliogenesis process is first achieved at OHCs.

The development of neural connections in the mammalian cochlea exhibit some general features that are classically found in the nervous system during the synaptogenesis process: transient stages of exuberant branching, pruning, and remodeling of synaptic connections. The general and characteristic features of cochlear synaptogenesis can be summarized as follows (see also Figs. 4.16 and 4.17):

(1) Afferents (peripheral endings of the spiral ganglion neurons), and possibly efferents (axonal endings of olivocochlear neurons), invade the developing otocyst very early, before morphologically detectable hair cell differentiation.

(2) The formation of synapses between hair cells and afferents begins well before hair cells are completely differentiated and well before they start functioning. These first synapses are characterized by multiple contacts from the same nerve endings and multiple presynaptic bodies in the hair cell (Figs. 4.16, Ia and 4.17, Oa).

(3) The pattern of IHC innervation changes during development; a transient stage (Fig. 4.16, Ia) is characterized by an exuberant branching of

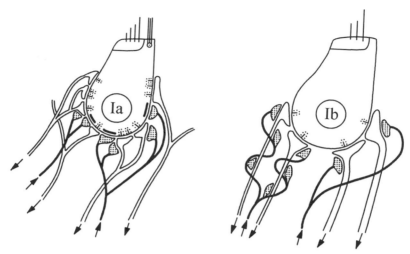

FIGURE 4.16. Schematic representation of the synaptogenesis at an inner hair cell level in the mammalian cochlea. Afferents are represented by open fibers with an outward arrow. Efferents are represented by filled fibers with vesiculated profiles and inward arrow. The inner hair cell immature stage (Ia) is characterized by a branching of radial afferents that make multiple or elongated contacts with cell. Multiple synaptic bodies are often seen. The efferents make both axodendritic and axosomatic contacts (with a postsynaptic cistern in the inner hair cell). In adulthood (Ib), the typical modiolar (thin) vs. pillar (thick) radial fibers are represented with their respective axodendritic efferent synapses.

afferent dendrites and by the presence of numerous axosomatic efferent synapses. In the final stage (Fig. 4.16, Ib), afferents are unbranched and almost all lateral efferent synapses are axodendritic.

(4) Some characteristics of developmental IHC synaptogenesis are once again found in adulthood during a posttraumatic (excitotoxic) degeneration-regeneration process of afferent neurites.

(5) The developmental changes in the pattern of OHC innervation are even more dramatic. The OHCs are first exclusively surrounded by radial and spiral afferent fibers (Fig. 4.17, Oa). Subsequently, they lose their connections with radial afferents and get connected by medial efferent endings (Fig. 4.17, Ob, Oc).

Overall, it looks as if OHCs undergo two distinct stages of differentiation or were differentiated twice: first as a "normal" sensory cell, paralleling IHC differentiation and then as a totally new type of cell having sensory and motile properties. The late and profound changes in both OHC morphology and innervation pattern reflect this double differentiation.

In most common mammalian species (i.e., not species specialized toward very high or very low frequency hearing), the most apical portion of the cochlea still has certain features in adulthood, suggesting that the matura-

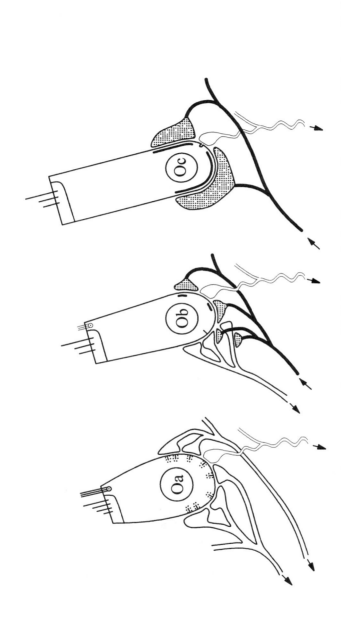

FIGURE 4.17. Schematic representation of the synaptogenesis at an outer hair cell level in the mammalian cochlea. Afferents are represented by open fibers (straight: radial; waived: spiral) with an outward arrow. Efferents are represented by filled fibers with vesiculated profiles and inward arrow. The immature outer hair cell (Oa) is only innervated by afferents, probably of two types: radial (branches from the inner hair cell radial afferents?) or spiral. Multiple synaptic bodies are often seen at this stage. At the intermediate stage (Ob), efferents (medial) arrive, synapse with afferent dendrites, and begin to form typical axosomatic synapses. At the mature (Oc) stage (apart from the apical cochlea that is not schematized here), only spiral afferent and medial efferents forming huge axosomatic synapses are seen.

tion process has not been completely achieved. These features include afferent branching, presence of multiple synaptic bodies, direct efferent contacts with the IHC, and few, if any, axosomatic medial efferent synapses with the OHC. These immature-like morphological characteristics of "apical" cochlear innervation could account for its physiological properties. For instance, apical OHCs might have kept some "IHC-like sensory properties"? Depending on species, the "apical" cochlea may be restricted to the very apex (as in the rat cochlea) or be more extended (as in the human cochlea).

Acknowledgements. A large amount of the work reported in this review has been done in our laboratory either in Marseille or in Montpellier. It is a pleasure to thank the different past or present co-workers, technicians, or students whose names appear on the respective references. In the preparation of manuscript, authors are particularly indebted to Pierre Sibleyras and Paul Paulet for photographic work, to Régine Leduc for drawings, and to Ghyslaine Humbert for skillful art preparation. Thanks also to George Tate and Michèle Paolucci for editing the work.

References

Abdouh A, Despres G, Romand R (1994) Histochemical and scanning electron microscopic studies of supernumerary hair cells in embryonic rat cochlea in vitro. Brain Res 660:181–191.

Altschuler RA, Reeks KA, Marangos PJ, Fex J (1985) Neuron-specific enolase-like immunoreactivity in inner hair cells but not outer hair cells in the guinea pig organ of Corti. Brain Res 327:379–384.

Anniko M (1983) Early development and maturation of the spiral ganglion. Acta Otolaryngol 95:263–276.

Bartolami S, Guiramand J, Lenoir M, Pujol R, Récasens M (1990) Carbachol-induced inositol phosphate formation during rat cochlea development. Hear Res 47:229–234.

Bartolami S, Planche M, Pujol R (1993) Characterization of muscarinic binding sites in the adult and developing rat cochlea. Neurochem Int 23:419–425.

Bredberg G (1968) Cellular pattern and nerve supply of the human organ of Corti. Acta Otolaryngol Suppl 236:1–135.

Brundin L, Flock A, Khanna SM, Ulfendahl M (1991) Frequency-specific position shift in the guinea pig organ of Corti. Neurosci Lett 128:77–80.

Cole KS, Robertson D (1992) Early efferent innervation of the developing rat cochlea studied with a carbocyanine dye. Brain Res 575:223–230.

Cotanche DA, Sulik KK (1984) The development of stereociliary bundles in the cochlear duct of chick embryos. Dev Brain Res 16:181–193.

Chiong CM, Burgess BJ, Nadol JB (1993) Postnatal maturation of human spiral ganglion cells: light and electron microscopic observations. Hear Res 67:211–219.

Dechesne CJ, Keller A (1996) Differential α enolase immunoreactivity in the two neuron types of the rat spiral ganglion during postnatal development. Comparison with neurofilament protein immunoreactivity. Aud Neurosci 2:33–46.

Dechesne CJ, Pujol R (1986) Neuron-specific enolase immunoreactivity in the developing mouse cochlea. Hear Res 21:87–90.

Dechesne CJ, Thomasset M (1988) Calbindin (CaBP 28 kDa) appearance and distribution during development of the mouse inner ear. Dev Brain Res 40:233–242.

Dechesne CJ, Rabejac D, Desmadryl G (1994) Development of calretinin immunoreactivity in the mouse inner ear. J Comp Neurol 346:517–529.

Deol MS (1967) The neural crest and the acoustic ganglion. J Embryol Exp Morphol 17:533–541.

Eggermont JJ, Bock GR (1985) Normal and abnormal development of hearing and its clinical implications. Acta Otolaryngol Suppl 421, pp. 1–128.

Elgoyhen AB, Johnson DS, Boulter J, Vetter DE, Heinemann S (1994) α9: An acetylcholine receptor with novel pharmacological properties expressed in rat cochlear hair cells. Cell 79:705–715.

Emmerling MR, Sobkowicz HM (1988) Differentiation and distribution of acetylcholinesterase molecular forms in the mouse cochlea. Hear Res 32:137–146.

Etcheler SM (1992) Developmental segregation in the afferent projections to mammalian auditory hair cells. Proc Natl Acad Sci USA 89:6324–6327.

Eybalin M (1993) Neurotransmitters and neuromodulators of the mammalian cochlea. Physiol Rev 73:309–373.

Fritzsch B, Nichols DH (1993) DiI reveals a prenatal arrival of efferents at the differentiating otocyst of mice. Hear Res 65:51–60.

Gil-Loyzaga P, Merchan-Cifuentes JA (1982) Histogenesis y desarrollo del receptor auditivo. In: Merchan-Cifuentes M (ed) El Oido Interno. Salamanca, Spain: Univ Salamanca Press, pp. 85–133.

Gil-Loyzaga P, Parés-Herbute N (1989) HPLC detection of dopamine and noradrenaline in the cochlea of adult and developing rats. Dev Brain Res 48:157–160.

Gil-Loyzaga P, Pujol R (1988) Synaptophysin in the developing cochlea. Int J Dev Neurosci 6:155–160.

Gil-Loyzaga P, Pujol R (1990) Neurotoxicity of kainic acid in the rat cochlea during early developmental stages. Eur Arch Otorhinolaryngol 248:40–48.

Gil-Loyzaga P, Cupo A, Eybalin M (1988) Met-enkephalin and Met-enkephalin-Arg[6]-Gly[7]-Leu[8] immunofluorescence in the developing guinea-pig organ of Corti. Dev Brain Res 42:142–145.

Gil-Loyzaga P, Pujol R, Mollicone R, Dalix AM, Oriol R (1989) Appearance of B and H blood-group antigens in developing cochlear hair cells. Cell Tissue Res 257:17–21.

Gil-Loyzaga P, Remezal M, Oriol R (1992) Neuronal influence on B and H human blood-group antigen expression in rat cochlear cultures. Cell Tissue Res 269:13–20.

Gil-Loyzaga P, Remezal M, Mollicone R, Ibanez A, Oriol R (1994) H and B human blood-group antigen expression in cochlear hair cells is modulated by thyroxine. Cell Tissue Res 276:239–243.

Ginzberg RD, Morest DK (1984) Fine structure of cochlear innervation in the cat. Hear Res 14:109–127.

Glowatzki E, Wild K, Brändle U, Fakler G, Fakler B, Zenner HP, Ruppersberg JR (1995) Cell-specific expression of the α9 n-Ach receptor subunit in auditory hair cells revealed by single-cell RT-PCR. Proc R Soc Lond Biol Sci 262:141–147.

Hafidi A, Romand R (1989) First appearance of type II neurons during ontogenesis in the spiral ganglion of the rat. An immunocytochemical study. Dev Brain Res 48:143–149.

Hafidi A, Després, G, Romand R (1993) Ontogenesis of type II spiral ganglion neurons during development: peripherin immunohistochemistry. Int J Dev Neurosci 11:507–512.

Hasko JA, Richardson GP, Russell IJ, Shaw G (1990) Transient expression of neurofilament protein during hair cell development in the mouse cochlea. Hear Res 45:63–74.

He DZZ, Evans BN, Dallos P (1994) First appearance and development of electromotility in neonatal gerbil outer hair cells. Hear Res 78:77–90.

Hudspeth AJ (1985) The cellular basis of hearing: the biophysics of hair cells. Science 230:745–752.

Igarashi Y (1980) Cochlea of the human fetus: a scanning electron microscope study. Arch Histol Jpn 43:195–209.

Janssen R, Schweitzer L, Jensen KF (1991) Glutamate neurotoxicity in the developing rat cochlea: physiological and morphological approaches. Brain Res 552:255–264.

Kaltenbach JA, Falzarano PR (1994) Postnatal development of the hamster cochlea. I. Growth of hair cells and the organ of Corti. J Comp Neurol 340:87–97.

Kaltenbach JA, Falzarano PR, Simpson TH (1994) Postnatal development of the hamster cochlea. II. Growth and differentiation of stereocilia bundles. J Comp Neurol 350:187–198.

Kawabata I, Nomura Y (1978) Extra-internal hair cells. A scanning electron microscopic study. Acta Otolaryngol 85:342–348.

Kelley MW, Xu XM, Wagner MA, Warchol ME, Corwin JT (1993) The developing organ of Corti contains retinoic acid and forms supernumerary hair cells in response to exogenous retinoic acid in culture. Development 119:1041–1053.

Kiang NYS, Rho JM, Northrop CC, Liberman MC, Ryugo DK (1982) Hair-cell innervation by spiral ganglion cells in adult cats. Science 217:175–177.

Kikuchi K, Hilding D (1965) The development of the organ of Corti in the mouse. Acta Otolaryngol 60:207–222.

Kozlowski DA, Jones TA, Schallert T (1994) Pruning of dendrites and restoration of function after brain damage: role of the NMDA receptor. Restorative Neurol Neurosci 7:119–126.

Kraus HJ, Aulbach-Kraus K (1981) Morphological changes in the cochlea of the mouse after the onset of hearing. Hear Res 4:89–102.

Lavigne-Rebillard M, Pujol R (1986) Development of the auditory hair cell surface in human fetuses. A scanning electron microscopy study. Anat Embryol 174:369–377.

Lavigne-Rebillard M, Pujol R (1988) Hair cell innervation in the fetal human cochlea. Acta Otolaryngol 105:398–402.

Lavigne-Rebillard M, Pujol R (1990) Auditory hair cells in human fetuses: synaptogenesis and ciliogenesis. J Electron Microsc Techn 15:115–122.

Lavigne-Rebillard M, Cousillas H, Pujol R (1985) The very distal part of the basilar papilla in the chicken: a morphological approach. J Comp Neurol 238:340–347.

Lefebvre PP, Weber T, Leprince P, Rigo JM, Delrée P, Rogister B, Moonen G (1991) Kainate and NMDA toxicity for cultured developing and adult rat spiral ganglion neurons: further evidence for a glutamatergic excitatory neurotransmission at the inner hair cell synapse. Brain Res 555:75–83.

Lenoir M, Shnerson A, Pujol R (1980) Cochlear receptor development in the rat with emphasis on synaptogenesis. Anat Embryol 160:253–262.

Lenoir M, Puel JL, Pujol R (1987) Stereocilia and tectorial membrane development in the rat cochlea. A SEM study. Anat Embryol 175:477–487.

Liberman MC (1980) Efferent synapses in the inner hair cell area of the cat cochlea: an electron microscopic study of serial sections. Hear Res 3:189–204.

Liberman MC, Dodds LW, Pierce S (1990) Afferent and efferent innervation of the cat cochlea: quantitative analysis with light and electron microscopy. J Comp Neurol 301:443–460.

Lim DJ (1972) Fine morphology of the tectorial membrane. Its relationship to the organ of Corti. Archiv Otolaryngol 96:199–215.

Lim DJ, Anniko M (1985) Developmental morphology of the mouse inner ear. A scanning electron microscopic observation. Acta Otolaryngol Suppl 422, pp. 1–69.

Lim DJ, Rueda J (1992) Structural development of the cochlea. In: Romand (ed) Development of Auditory and Vestibular Systems 2. Amsterdam: Elsevier, pp. 33–58.

Lorente de No R (1937) Sensory endings in the cochlea. Laryngoscope 47:373–377.

Luo L, Brumm D, Ryan AF (1995) Distribution of non-NMDA glutamate receptor mRNAs in the developing rat cochlea. J Comp Neurol 361:372–382.

Mattson MP (1988) Neurotransmitters in the regulation of neuronal cytoarchitecture. Brain Res Rev 13:179–212.

Mbiène JP, Favre D, Sans A (1984) The pattern of ciliary development of the fetal mouse vestibular receptors: a qualitative and quantitative SEM study. Anat Embryol 170:229–238.

Merchan-Pérez A, Gil-Loyzaga P, Eybalin M (1990a) Immunocytochemical detection of calcitonin gene-related peptide in the postnatal developing rat cochlea. Int J Dev Neurosci 8:603–612.

Merchan-Pérez A, Gil-Loyzaga P, Eybalin M (1990b) Immunocytochemical detection of glutamate decarboxylase in the postnatal developing rat organ of Corti. Int J Dev Neurosci 8:613–620.

Merchan-Pérez A, Gil-Loyzaga P, Lopez-Sanchez J, Eybalin M, Valderrama FJ (1993) Ontogeny of γ-aminobutyric acid in efferent fibers to the rat cochlea. Dev Brain Res 76:33–41.

Merchan-Pérez A, Gil-Loyzaga P, Eybalin M, Fernandez-Mateos P, Bartolomé MV (1994) Choline-acetyltransferase-like immunoreactivity in the organ of Corti of the rat during postnatal development. Dev Brain Res 82:29–34.

Mills DM, Norton SJ, Rubel EW (1994) Development of active and passive mechanics in the mammalian cochlea. Aud Neurosci 1:77–99.

Nakai Y, Hilding D (1968) Cochlear development. Some electron microscopic observations of maturation of hair cells, spiral ganglion and Reissner's membrane. Acta Otolaryngol 66:369–385.

Perkins RE, Morest DK (1975) A study of cochlear innervation patterns in cats and rats with the Golgi method and Nomarski optics. J Comp Neurol 163:129–158.

Pickles JO, Comis SD, Osborne MP (1984) Morphology and cross-linkage of stereocilia in the guinea pig labyrinth examined without the use of osmium as a fixative. Cell Tissue Res 237:43–48.

Pirvola U, Lehtonen E, Ylikoski J (1991) Spatiotemporal development of cochlear innervation and hair cell differentiation in the rat. Hear Res 52:345–355.

Puel JL, Safieddine S, Gervais d'Aldin C, Eybalin M, Pujol R (1995) Synaptic regeneration and functional recovery after excitotoxic injury in the guinea pig cochlea. C R Acad Sci Ser III Sci Vie 318:67–75.

Pujol R (1986) Synaptic plasticity in the developing cochlea. In: Ruben RW, Van De Water TR, Rubel EW (eds) The Biology of Change in Otolaryngology. New York: Elsevier, pp. 47–54.

Pujol R (1994) Lateral and medial efferents: a double neurochemical mechanism to protect and regulate inner and outer hair cell function in the cochlea. Br J Audiol 28:185–191.

Pujol R, Abonnenc M (1977) Receptor maturation and synaptogenesis in the golden hamster cochlea. Arch Otorhinolaryngol 217:1–12.

Pujol R, Hilding D (1973) Anatomy and physiology of the onset of auditory function. Acta Otolaryngol 76:1–11.

Pujol R, Lavigne-Rebillard M (1985) Early stages of innervation and sensory cell differentiation in the human fetal organ of Corti. Acta Otolaryngol Suppl 423:43–50.

Pujol R, Lavigne-Rebillard M (1995) Sensory and neural structure in the developing human cochlea. Int J Pediatr Otorhinolaryngol 32:S177–S182.

Pujol R, Lenoir M (1986) The four types of synapses in the organ of Corti. In: Altschuler R, Bobbin R, Hoffman D (eds) Neurobiology of Hearing: The Cochlea. New York: Raven Press, pp. 161–172.

Pujol R, Marty R (1970) Postnatal maturation in the cochlea of the cat. J Comp Neurol 139:115–126.

Pujol R, Sans A (1986) Synaptogenesis in the mammalian inner ear. In: Aslin R (ed) Advances in Neural and Behavioral Development. Norwood, NJ: Ablex Press, pp. 1–18.

Pujol R, Uziel A (1988) Auditory development: peripheral aspects. In: Meisami E, Timiras PS (eds). Handbook of Human Growth and Developmental Biology. Vol. I, part B. Boca Raton, FL: CRC Press, pp. 109–130.

Pujol R, Carlier E, Devigne C (1978) Different patterns of cochlear innervation during the development in the kitten. J Comp Neurol 117:529–536.

Pujol R, Carlier E, Devigne C (1979) Significance of presynaptic formations in the very early stages of cochlear synaptogenesis. Neurosci Lett 15:97–102.

Pujol R, Carlier E, Lenoir M (1980) Ontogenetic approach to inner and outer hair cells functions. Hear Res 2:423–430.

Pujol R, Shnerson A, Lenoir M, Deol MS (1983) Early degeneration of sensory and ganglion cells in the inner ear of mice with uncomplicated genetic deafness (dn): preliminary observations. Hear Res 12:57–63.

Pujol R, Lenoir M, Robertson D, Eybalin M, Johnstone BM (1985) Kainic acid selectively alters auditory dendrites connected with cochlear inner hair cells. Hear Res 18:145–151.

Pujol R, Zajic G, Dulon D, Raphael Y, Altschuler RA, Schacht J (1991) First appearance and development of motile properties in outer hair cells isolated from guinea-pig cochlea. Hear Res 57:129–141.

Pujol R, Lenoir M, Ladrech S, Tribillac F, Rebillard G (1992) Correlation between the length of outer hair cells and the frequency coding of the cochlea. In: Cazals Y, Demany L, Horner KC (eds) Auditory Physiology and Perception. Oxford, UK: Pergamon Press, pp. 45–52.

Pujol R, Puel JL, Eybalin M (1992) Implication of non-NMDA and NMDA receptors in cochlear ischemia. NeuroReport 3:299–302.

Rabacchi S, Bailly Y, Delhaye-Bouchaud N, Mariani J (1992) Involvement of the N-methyl-D-aspartate (NMDA) receptor in synapse elimination during cerebellar development. Science 256:1823–1825.

Raphael Y, Marshak G, Barash A, Geiger B (1987) Modulation of intermediate filament expression in the developing cochlear epithelium. Differentiation 35:151–162.

Raphael Y, Lenoir M, Wroblewski R, Pujol R (1991) The sensory epithelium and its innervation in the mole rat cochlea. J Comp Neurol 314:367–382.

Retzius G (1884) Gehörorgan des Wirbeltiere. II Das Gehörorgan der Reptilien, der Vögel, und der Säugetiere. Stockholm: Samson and Wallin.

Robertson D, Harvey AR, Cole KS (1989) Postnatal development of the efferent innervation of the rat cochlea. Dev Brain Res 47:197–207.

Romand R (1983) Development of the cochlea. In: Romand R (ed) Development of Auditory and Vestibular Systems. New York: Academic Press, pp. 47–88.

Romand R, Romand MR (1982) Myelination kinetics of spiral ganglion cells in kitten. J Comp Neurol 204:1–5.

Romand R, Romand MR, Mulle C, Marty R (1980) Early stages of myelination in the spiral ganglion cells of the kitten during development. Acta Otolaryngol 90:391–397.

Roth B, Bruns V (1992) Postnatal development of the rat organ of Corti. Anat Embryol 185:571–581.

Roth B, Dannhof B, Bruns V (1991) ChAT-like immunoreactivity of olivocochlear fibers on rat outer hair cells during the postnatal development. Anat Embryol 183:483–489.

Rubel EW (1978) Ontogeny of structure and function in vertebrate auditory system. In: Jacobson M (ed) Handbook of Sensory Physiology. Vol. IX. Development of Sensory Systems. Berlin: Springer-Verlag, pp. 135–237.

Ruben RJ (1967) Development of the inner ear of the mouse: a radioautographic study of terminal mitoses. Acta Otolaryngol Suppl 220, pp. 1–44.

Safieddine S, Bartolami S, Wenthold RJ, Eybalin M (1996) Pre- and postsynaptic M3 muscarinic receptor mRNAs in the rodent peripheral auditory system. Mol Brain Res 40:127–135.

Sala C, Andreose JS, Fumagalli G, Løhmo T (1995) Calcitonin gene-related peptide: possible role in formation and maintenance of neuromuscular junctions. J Neurosci 15:520–528.

Sans A, Dechesne C (1985) Early development of vestibular receptors in human embryos: An electron microscopic study. Acta Otolaryngol Suppl 423:51–58.

Schwartz AM, Parakkal M, Gulley RL (1983) Postnatal development of spiral ganglion cells in the rat. Am J Anat 167:33–41.

Sher AE (1971) The embryonic and postnatal development of the inner ear of the mouse. Acta Otolaryngol Suppl 285, pp. 1–77.

Shnerson A, Devigne C, Pujol R (1982) Age-related changes in the C57BL/6J mouse cochlea. II. Ultrastructural findings. Dev Brain Res 2:77–78.

Shnerson A, Lenoir M, Van De Water TR, Pujol R (1983) The pattern of sensori-neural degeneration in the cochlea of the deaf Shaker-1 mouse: ultrastructural observations. Dev Brain Res 9:305–315.

Simmons DD (1994) A transient afferent innervation of outer hair cells in the postnatal cochlea. NeuroReport 5:1309–1312.

Simmons DD, Manson-Gieseke L, Hendrix TW, McCarter S (1990) Reconstructions of efferent fibers in the postnatal hamster cochlea. Hear Res 49:127–140.

Simmons DD, Manson-Gieseke L, Hendrix TW, Morris K, Williams SJ (1991) Postnatal maturation of spiral ganglion neurons: a horseradish peroxidase study.

Hear Res 55:81–91.

Sobin A, Anniko M (1984) Early development of cochlear hair cell stereociliary surface morphology. Arch Otorhinolaryngol 241:55–64.

Sobkowicz HM (1992) The development of innervation in the organ of Corti In: Romand R (ed) Development of Auditory and Vestibular Systems 2. Amsterdam: Elsevier, pp. 59–100.

Sobkowicz HM, Emmerling MR (1989) Development of acetylcholinesterase-positive neuronal pathways in the cochlea of the mouse. J Neurocytol 18:209–224.

Sobkowicz HM, Slapnick SM (1992) Neuronal sprouting and synapse formation in response to injury in the mouse organ of Corti in culture. Int J Dev Neurosci 10:545–566.

Sobkowicz HM, Rose JE, Scott GE, Slapnick SM (1982) Ribbon synapses in the developing intact and cultured organ of Corti in the mouse. J Neurosci 2:942–957.

Spoendlin H (1969) Innervation patterns of the organ of Corti of the cat. Acta Otolaryngol 67:239–254.

Spoendlin H (1979) Neural connections of the outer hair cell system. Acta Otolaryngol 87:381–387.

Spoendlin H (1985) Nerve proliferation in the cochlea. In: Nomura Y (ed) Hearing Loss and Dizziness. Tokyo: Igaku-Shoin, pp. 68–82.

Strelioff D, Flock A (1984) Stiffness of sensory-cell hair bundles in the isolated guinea pig cochlea. Hear Res 15:19–28.

Tanaka K, Sakai N, Terayama Y (1979) Organ of Corti in the human fetus. Scanning and transmission electron microscope studies. Ann Otol Rhinol Laryngol 88:749–758.

Tello JF (1931) Le réticule des cellules ciliées du labyrinthe chez la souris et son indépendance des terminaisons nerveuses de la VIIIe paire. Trav Lab Rech Biol Univ Madrid 27:151–186.

Tilney LG, Cotanche DA, Tilney MS (1992) Actin filaments, stereocilia and hair cells of the bird cochlea. IV. How the number and arrangement of stereocilia are determined. Development 116:213–226.

Tohyama Y, Kiyama H, Kitajiri M, Yamashita T, Kumazawa T, Tohyama M (1989) Ontogeny of calcitonin gene-related peptide in the organ of Corti of the rat. Dev Brain Res 45:309–312.

Uziel A, Gabrion J, Ohresser M, Legrand C (1981) Effects of hypothyroidism on the structural development of the organ of Corti in the rat. Acta Otolaryngol 92:469–480.

Uziel A, Pujol R, Legrand C, Legrand J (1983) Cochlear synaptogenesis in the hypothyroid rat. Dev Brain Res 7:295–301.

Vago R, Ripoll C, Tournebize R, Lenoir M (1996) Distribution of actin and tubulin in outer hair cells isolated from developing rat cochlea: a quantitative study. Eur J Cell Biol 69:308–315.

Van De Water TR (1976) Effects of removal of the statoacoustic ganglion complex upon the growing otocyst. Ann Otol Rhinol Laryngol 85:Suppl 33:1–32.

Vater M, Lenoir M (1992) Ultrastructure of the horseshoe bat's organ of Corti. I. Scanning electron microscopy. J Comp Neurol 318:367–379.

Vater M, Lenoir M, Pujol R (1992) Ultrastructure of the horseshoe bat organ of Corti. II. Transmission electron microscopy. J Comp Neurol 318:380–391.

Vater M, Lenoir M, Pujol R (1997) Development of the organ of Corti in horseshoe bats: scanning and transmission electron microscopy. J Comp Neurol 377:520–534.

Warr WB (1992) Organization of olivocochlear efferent systems in mammals. In: Webster DB, Popper AN, Fay RR (eds) Mammalian Auditory Pathway: Neuroanatomy. New York: Springer-Verlag, pp. 410–448.

Warr WB, Guinan JJ (1979) Efferent innervation of the organ of Corti: two separate systems. Brain Res 173:152–155.

Weaver SP, Schweitzer L (1994) Development of gerbil outer hair cells after the onset of cochlear function: an ultrastructural study. Hear Res 72:44–52.

Weaver SP, Hoffpauir J, Schweitzer L (1994) Distribution of actin in developing outer hair cells in the gerbil. Hear Res 72:181–188.

Whitlon DS, Sobkowicz HM (1988) Neuron-specific enolase during the development of the organ of Corti. Int J Dev Neurosci 6:77–87.

Whitlon DS, Sobkowicz HM (1989) GABA-like immunoreactivity in the cochlea of the developing mouse. J Neurocytol 18:505–518.

Wikström SO, Anniko M (1987) Early development of the stato-acoustic and facial ganglia. Acta Otolaryngol 104:166–174.

Wolff JR, Joo F, Dames W (1978) Plasticity in dendrites shown by continuous GABA administration in superior cervical ganglion of adult rat. Nature 274:72–74.

Wright A (1984) Dimensions of the cochlear stereocilia in man and the guinea pig. Hear Res 13:89–98.

Zhou SL, Pickles JO (1994) Early hair-cell degeneration in the extreme apex of the guinea pig cochlea. Hear Res 79:147–160.

5

The Development of Cochlear Function

Rudolf Rübsamen and William R. Lippe

1. Introduction

The cochlea is the window through which the central auditory system views its acoustic environment. The transduction of air borne sound by the hair cells and the neural encoding at the periphery place constraints on the acoustic features that are available for further processing by auditory neurons in the brain. At birth, the cochlea in most altricial mammals is still very immature. It is effectively unresponsive to sound and generates little sustained (spontaneous) activity. During the first month after the onset of hearing, significant changes occur in cochlear functioning, changes that are reflected in both the overall level and spatiotemporal pattern of nerve impulses that are transmitted centrally over the auditory nerve. The task of determining the extent to which maturational changes in auditory perception, spontaneous activity, and central responses to sound originate within the cochlea and to what degree these changes reflect the development of central synaptic processes remains a formidable challenge.

Probably the most significant advance in our understanding of cochlear functioning during the past 20 years has been the discovery of active cochlear mechanisms (Kemp 1978; Dallos 1992). It is now generally accepted among cochlear physiologists that two closely related characteristics of audition in adult mammals, namely, high sensitivity and high frequency resolution, are due to the operation of a "cochlear amplifier" (Davis 1983; Dallos 1992). This theory postulates that acoustic stimulation of the cochlea produces a cascade of events, jointly termed mechanoelectrical signal transduction, that alters the membrane potential of both the inner (IHCs) and outer hair cells (OHCs). The change in membrane potential in the OHCs causes these cells to undergo a motile response that injects mechanical energy back into the basilar membrane via a reverse process of electromechanical transduction, thereby amplifying basilar membrane motion. Support for this theory comes from a variety of experiments. Direct measurements of basilar membrane motion yield gain functions of basilar membrane displacement that are dependent on the physiological status of the OHCs (Johnstone, Pattuzi, and Yates 1986; Ruggero and Rich 1991). More convincing evidence for the amplification of basilar membrane vibration comes from indirect measurements of inner ear activity, namely, from the occurrence of spontaneous (Wilson 1980) and acoustically evoked otoacoustic emissions (Kemp 1978; Whitehead et al. 1996) and from the fact that acoustic emissions can also be elicited by direct electrical alternating-current stimulation of the cochlea (Hubbard and Mountain 1983).

In light of these new findings showing the importance of OHC-mediated active processes during cochlear stimulus transduction in adult animals, it would seem promising to reevaluate what we know about the maturation of auditory responsiveness in young postnatal mammals and to ask what implications the active processes have for our understanding of the development of cochlear function. For example, it has repeatedly been shown in various mammalian species that the sensitivity and frequency selectivity of auditory nerve fibers increase concomitantly during early postnatal life (Rubel et al. 1985; Walsh and McGee 1986; Rübsamen 1992; Walsh and Romand 1992). The observation that cochlear damage in adults, particularly to OHCs, causes a reduction in sensitivity and frequency selectivity to a level comparable to that seen in early development suggests that OHC-mediated processes underlie the acquisition of adultlike tuning and sensitivity (Walsh and McGee 1990). However, to date, there have been no direct recordings in vivo from hair cells in maturing mammals. What we know about the maturation of stimulus transduction and the contribution of active processes comes mainly from a variety of other, more accessible measures of cochlear physiology. These measures include, for example, the endocochlear potential (ECP), the cochlear microphonic (CM), the summating potential (SP), and distortion-product otoacoustic emissions (DPOAEs). Four types of neuronal events might also be used as indicators of cochlear functioning. These include the compound action potential (CAP) and the auditory nerve response, both of which reflect the discharge of first-order cochlear afferents, as wells as the auditory evoked potential recorded from within the brain stem and the scalp-recorded auditory brain stem response, both of which predominantly reflect the activity of central auditory neurons. However, these measures must be treated cautiously when used to interpret cochlear development. Neuronal responses are unquestionably affected by developmental processes within the cochlea and thus will reflect cochlear development. However, developmental changes in neuronal responses might also reflect the development of afferent synaptic transmission or the maturation of central synaptic processes.

A number of recent reviews have focused on the structural development of the cochlea (e.g., Pujol, Lavigne-Rebillard, and Uziel 1991; Pujol and Lavigne-Rebillard 1992) or have presented a more generalized description of the maturation of the cochlea (Walsh and Romand 1992; Henley and Rybak 1995). The present chapter will focus on selected topics concerning the development of cochlear function. The first part of the chapter concentrates mainly on four different direct measurements of cochlear physiological processes: the ECP, the CM, electromotility in OHCs, and DPOAEs. These measurements provide information about the development of signal transduction and the contribution of various components of the inner ear to active cochlear amplification. Next, we examine the characteristics of spontaneous activity in the developing auditory pathway both before and after the onset of hearing and consider the idea that sustained neural firing could play a trophic or instructive role during development. Finally, we review the experimental studies on the development of cochlear tonotopy and examine the unresolved issues that still remain almost 20 years after the hypothesis of a shifting place code was initially proposed. The chapter will concentrate on precocial mammals. Findings from altricial mammals and birds are only considered when relevant.

The text provides an overview of the material. The readers are referred to the detailed tables accompanying each section that describe the results of individual studies.

2. Endocochlear Potential (Table 5.1)

The stria vascularis gives rise to an electrochemical gradient within the cochlea and an associated "silent" (or standing) current that flows out of the scala media and across the organ of Corti (Brownell 1990; Zidanic and Brownell 1990). The potential difference between the endolymph and the perilymph, the ECP, typically measures 75–100 mV in mature mammals. Because most of the hair cell is located within the perilymphatic space, with only the stereocilia and apical surface of the hair cell extending into the endolymph, the total electrical gradient across the stereociliary membrane of the hair cells amounts to 120–160 mV. Acoustic stimulation results in a displacement of the cochlear partition that causes a tilting of the hair cell stereocilia. The deflection of the stereocilia, in turn, causes a change in the hair cell's membrane potential (receptor potential) that is mediated by mechanically gated ion channels believed to be located at or near the tips of the stereocilia (Hudspeth and Corey 1977; Shotwell, Jakobs, and Hudspeth 1981; Jaramillo and Hudspeth 1991). The receptor potential is thought to not only trigger the release of neurotransmitter from the hair cell but also to drive the OHC's force-generating electromotile response (Brownell et al. 1985; Brownell 1990). The motile response of the OHC is part of a positive mechanical feedback process that increases the movement of the basilar membrane near threshold (Geisler 1986; Dallos 1992).

In adult mammals, the feedback amplification of basilar membrane motion evoked by low-level signals critically depends on the ECP. At moderate stimulus levels, drug-induced lowering of the ECP reduces the amplitude of basilar membrane displacement, particularly at the characteristic frequency (Ruggero and Rich 1991). Lowering the ECP also causes a

TABLE 5.1. Development of the endocochlear potential.

Species (Reference)	Location	Age	Results/Comments
Rabbit (1)	Cochlea, scala media	0–37 DAB	Electrophysiology. ECP appears at 5 DAB and increases rapidly in subsequent days. At 15 DAB, the adult value of ~ 80 mV is reached. Development of the ECP closely parallels the development of the CM.
Mouse (2, 3)	Cochlea, scala media	0–14 DAB	At birth, the ionic composition is similar in endo- and perilymphatic spaces. Adultlike composition of endolymph is reached at 6–8 DAB. The maturation of endolympathic composition correlates well with the morphological maturation of the stria vascularis.
Mouse (4)	Cochlea, scala media	E8 to 10 DAB	X-ray microanalysis. Endolymphatic potassium concentration increases between 4 and 8 DAB. The ionic composition of the endolymph becomes adultlike before the ECP matures to its adult level.
Mouse (5, 6)	Cochlea	0–21 DAB	Histochemistry, X-ray microanalysis. High levels of Na^+-K^+-ATPase and adenylate cyclase occur in the stria at the contraluminal side of marginal cell membranes. Increase in the adenylate cyclase content occurs ~ 1 day before and in parallel with the rise of the potassium content in endolymph. Maturation of endolymph parallels the morphological maturation of the fine structure of the stria vascularis and the increase in strial adenylate cyclase content.
Mouse (14)	Cochlea, scala media		Electrophysiology. ECP reaches its adultlike level at 16 DAB.
Mouse (17)	Cochlea	E1 to 120 DAB	The development of enzymatic activity of adenylate cyclase up to P8 supports the assumption that adenylate cyclase contributes to the regulation of endolymphatic fluid.

Species	Structure	Age	Description
Rat (11–13)	Cochlea, scala media	8–60 DAB	Electrophysiology. ECP in animals 8–28 DAB is more susceptible to furosemide ototoxicity than that in animals 30 DAB and older. Rapid increase in the level of the ECP occurs at 11–13 DAB; adultlike values are reached at 17 DAB. Development of the ECP and CAP are inversely related; the ECP level increases as the CAP threshold declines.
Rat (7)	Cochlea, scala media	8 DAB to adult	Electrophysiology. At 8 DAB, the level of the ECP is low; the maximum rate of increase occurs between 11 and 16 DAB; at 13–14 DAB, the rate of increase is ~1mV/h. Rapid potential increase occurs simultaneously in all three cochlear turns. Endolymphatic ion concentration remains unchanged during the phase of rapid increase in the ECP.
Rat (16)	Cochlea	1–30 DAB	In situ hybridization. Expression of Na^+-K^+-ATPase α 1-subunit mRNA in the developing rat cochlea. The adult expression level is attained between P11 and P14.
Guinea pig (9)	Cochlea, scala media	0 DAB to adult	Electrophysiology. Simultaneous measurements of the ECP and K^+, Na^+, and Cl^- ions in the endolymph.
Gerbil (10)	Cochlea, scala media	0–20 DAB	X-ray microanalysis. Adultlike ionic composition in the endolymph is established before the appearance of a measurable ECP.
Gerbil (15, 18)	Cochlea, scala media	0–60 DAB	Electrophysiology. A positive ECP (2–3 mV) is first observed at 10 DAB; thereafter, the ECP increases and reaches +75 to +85 mV at 20 DAB.
Cat (8)	Cochlea	0 DAB to adult	Electrophysiology. At P1, the ECP is ~10–15 mV. Subsequently, the rate of increase is ~1.5 mV/day during the first 1.5 weeks and ~3.5 mV/day during the next 2 weeks; adultlike values are reached by the end of the first month.

DAB, days after birth; ECP, endocochlear potential; ATPase, adenosinetriphosphatase; E, embryonic day; P, postnatal days (mammal); days after hatching (birds). 1, Ånggard 1965; 2, Anniko, Wroblewski, and Wersäll 1979; 3, Anniko and Wroblewski 1986; 4, Anniko and Nordemar 1980; 5, Anniko 1985; 6, Anniko and Wroblewski 1986; 7, Bosher and Warren 1971; 8, Fernández and Hinojosa 1974; 9, Komune et al. 1993; 10, Ryan and Woolf 1983; 11, Rybak et al. 1991; 12, Rybak, Weberg, and Whitworth 1991; 13, Rybak, Whitworth, and Scott 1992; 14, Steel and Barkway 1989; 15, Woolf, Ryan, and Harris 1986; 16, Yao et al. 1994; 17, Zajic, Anniko, and Schacht 1983; 18, McGuirt, Schmiedt, and Schulte 1995.

reduction in the amplitude of evoked otoacoustic emissions (Mills, Norton, and Rubel 1993, 1994) and increases the threshold of afferent neuronal responsiveness (Evans and Klinke 1982; Sewell 1984a,b; Rübsamen, Mills, and Rubel 1995). In effect, lowering the ECP reduces the "gain" of the cochlear amplifier. Because of the ECP's role in controlling the gain of the cochlear amplifier, an analysis of the postnatal maturation of the ECP is critical for understanding the development of auditory sensitivity.

The ECP itself is generated primarily by energy-dependent ion pumps in the stria vascularis, although other processes may contribute as well (Dallos 1981; Komune et al. 1993). The pumps also contribute to maintaining the characteristic ionic composition of the cochlear endolymph. In mature animals, the endolymph of the scala media has a high concentration of K^+ and a low concentration of Na^+ (guinea pig: 156.9 ± 4.6 mM K^+; 1.5 ± 0.3 mM Na^+). In contrast, the ionic content of perilymph in the scala vestibuli and the scala tympani has a low level of K^+ and a high level of Na^+ (4.4 ± 0.4 mM K^+; 153.3 ± 4.9 mM Na^+) (Komune et al. 1993).

The development of the ionic composition of the endolymph has been studied in several different species of rodents (mouse, rat, gerbil) and in kittens. In rodents, it is generally reported that, at birth, the ionic composition of the endolymph and perilymph is similar. During subsequent maturation, the K^+ concentration in the endolymph increases to reach adult values by the end of the first postnatal week (Fig. 5.1) (mouse pups: Anniko, Wroblewski, and Wersäll 1979; Anniko and Nordemar 1980; Anniko and Wroblewski 1986; rat: Bosher and Warren 1971; gerbils: Ryan and Woolf 1983). The increase in K^+ concentration parallels the morphological maturation of the fine structure of the stria vascularis (Anniko, Wroblewski, and Wersäll 1979) and also coincides with the rise of adenylate cyclase in the marginal cells of the stria vascularis, an enzyme that is thought to be one of the major constituents of the regulatory system controlling inner ear electrolyte balance (Anniko 1985; Anniko and Wroblewski 1986; Zajic, Anniko, and Schacht 1983).

However, by postnatal days 8–10 (P8–10), when the ionic composition of the endolymph is the same as in adults, the ECP is still very immature. It is between P12 and P16–17 that the ECP increases to the adult value of 75–100 mV (Fig. 1) (rat: Bosher and Warren 1971; Rybak, Whitworth, and Scott 1992; mouse pups: Steel and Barkway 1989; gerbil: Woolf, Ryan, and Harris 1986; McGuirt, Schmiedt, and Schulte 1995). The later development of the ECP is paralleled by an increase in Na^+-K^+-adenosine triphosphatase activity in the stria vascularis (Kuijpers 1974), and it is tempting to speculate that the increase in the ECP might be directly related to the increased activity of this enzyme.

Bosher and Warren (1971) speculated that a low resistance of tissue in the scala media might be the reason why the ECP in mouse pups is still at an immature level at a time when the ionic composition of the endolymph is adultlike. If this is true, the delayed development of the ECP relative to the

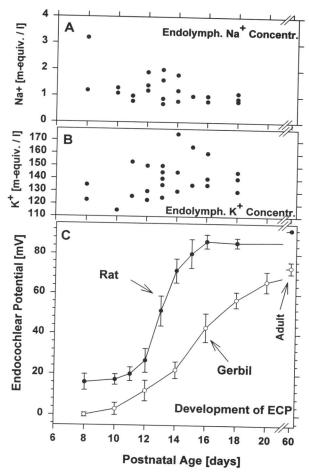

FIGURE 5.1. Developmental changes in the ionic concentration (concentr.) of sodium (Na$^+$) (A) and potassium (K$^+$) (B) in the endolymph of rat pups between postnatal days 8–20. (Modified from Bosher and Warren 1971, The Physiological Society.) C: development of the endocochlear potential (ECP) in young rats and gerbils between the 8th postnatal day (4 days before the onset of acoustically evoked cochlear responses) and postnatal days 18–20, when threshold levels become adultlike. Adult ECP values are shown at postnatal day 60. Note that the ECP continues to increase even after the concentrations of Na$^+$ and K$^+$ are adultlike. Values are means ± SD. (Data from Bosher and Warren 1971 [rat] and Woolf, Ryan, and Harris 1986 [gerbil]).

endolymph would suggest that an adultlike composition of the endolymph is necessary, but not sufficient, for the establishment of the adult ECP. Indeed, experiments in the adult chinchilla and guinea pig have shown that the ECP can be substantially reduced without significantly altering the ionic concentration of the endolymph. This is observed after systemic injection of the ototoxic diuretic furosemide, which has little effect on the concentration of Na^+ and K^+ but which transiently shifts the ECP to negative values (Brusilow 1976; Rybak and Morizono 1982).

Generally, the maturation of the ECP in the kitten is quite similar. However, in this species, the overall time course of development is prolonged. The ECP starts to increase during the first days after birth, but adult values are only reached after 1 month (Fernández and Hinojosa 1974).

3. Outer Hair Cell Electromotility

A variety of stimuli can induce either fast or slow force-generating length changes in OHCs, even when the hair cells are isolated from the organ of Corti and studied in vitro. The slow length changes of OHCs that accompany the tilting of the cuticular plate at the base of the stereocilia are mediated by an influx of Ca^{2+} through voltage-operated Ca^{2+} channels (Lewis and Hudspeth 1983; Zenner 1986a,b; Hudspeth 1989). The slow motility seems to be based on an actin-myosin interaction and is dependent on calmodulin, inositol 1,4,5-trisphosphate, and adenosine triphosphate (Schacht and Zenner 1987; Slepecky, Ulfendahl, and Flock 1988; Slepecky and Ulfendahl 1992). The possibility that slow length changes might play a role in modifying the overall sensitivity of the cochlea has been discussed previously (Zenner 1986a,b; Zenner, Zimmermann, and Gitter 1988). Thus, it is possible that the development of slow motility might contribute to the postnatal increase in auditory sensitivity. However, to date, there are no experimental physiological data that directly address this question. Circumstantial evidence for this idea comes from investigations of the morphological development of OHCs. These studies revealed a significant increase in actin labeling in the region of the hair cell cortical cytoskeleton at the onset of acoustically evoked cochlear responses (Weaver, Hoffpauir, and Schweitzer 1994).

Fast length changes of OHCs can be elicited by direct electrical stimulation (Brownell et al. 1985; Ashmore 1987; Evans and Dallos 1993) and are thought to be due to voltage-sensitive force generators associated with the hair cell's plasma membrane (Holley 1991; Holley, Kalinec, and Kachar 1992). In vivo fast length changes are believed to be driven by the cell's receptor potential (Brownell et al. 1985; Dallos 1992). The receptor potential itself results from the modulation of the standing ionic current (mainly K^+), which runs from the endolymph through mechanically gated channels at the tips of the hair cell stereocilia into the hair cell cytosol through voltage-gated channels at the hair cell base and into the perilymph (Dallos 1992). Recordings from auditory nerve fibers in vivo have shown that cochlear feedforward amplification, as indicated by low thresholds and sharp tuning, critically depends on the integrity of OHCs (Dallos and Harris 1978; Liberman and Dodds 1984b).

The best way to assess the contribution of OHCs to the maturation of cochlear functioning would be to record intracellularly from hair cells at young ages and to directly measure basilar membrane vibration developmentally. To date, such measurements have not been made. What is available are data obtained from in vitro studies of isolated OHCs in guinea pig embryos (Pujol et al. 1991) and young gerbils (He, Evans, and Dallos 1994). In guinea pigs, electrically evoked (fast) and Ca^{2+}-induced (slow) motility in isolated OHCs are first observed prenatally. Because of its long gestational period and mature state at the time of birth, the guinea pig is not often used in investigations of auditory system development. Therefore, only a limited amount of data is available in this species for relating the observations on the development of hair cell motility to the overall maturation of the cochlea. These data show that OHC motility in guinea pigs first appears between gestational days 50 and 60, the same period when CMs and CAPs can first be recorded in vivo (Romand 1971). The observations of Pujol and co-workers (1991) suggest that fast electromotility is first found in hair cells in the basal cochlear turn (high frequency) and only a few days later in more apical regions. Because of methodological reasons discussed by the investigators, the earlier onset of electromotility in basal vs. apical cells in the guinea pig must be interpreted cautiously, although similar findings have been reported in the gerbil (He, Evans, and Dallos 1994). If confirmed, this finding would be consistent with the basal-to-apical gradient of cochlear differentiation that is generally observed in altricial mammals (see Section 8).

An animal model for which we have more developmental data is the Mongolian gerbil. In this species, electromotile responses can first be measured from a small fraction of isolated OHCs on P7. The proportion of responsive cells increases steadily until P12, when neural activity can first be evoked by acoustic stimulation (He, Evans, and Dallos 1994) (Fig. 5.2). Although all OHCs from the apical and basal cochlear turns show motile responses by P12, the maximal absolute response amplitude continues to increase up to approximately P16–17, when mature values are approached (Fig. 5.3). The increase in response amplitude results only, in part, from the motile apparatus associated with the hair cell membrane. The continued growth (elongation) of the OHCs also contributes. This is shown by examining the relationship between the maximum current-induced length and the hair cell's resting (nominal) length. This analysis suggests that the "effectiveness" of the motile apparatus in basal and apical turn OHCs continues to increase only to P13 and P14, respectively. This is also the time when the threshold of the electrically evoked motile response becomes adultlike.

These observations have implications for the development of active cochlear processes and their contribution to cochlear functioning. If the

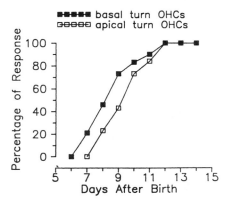

FIGURE 5.2. Development of electromotility in gerbil outer hair cells. The percentage of outer hair cells (OHCs) that exhibit motilie responses to direct transcellular electrical stimulation in vitro is plotted as a function of age. Note that, at 12 days after birth (the onset of acoustically evoked responses), all OHCs show motile responses. (From He, Evans, and Dallos 1994, Elsevier Science Publishers.)

absolute magnitude of the electromotile responses is a valid indicator of the OHC's contribution to active cochlear processes, then the active mechanism in the gerbil would not be mature until at least P17. Through as a yet unknown mechanism, OHC motility is linked to a layered system of subsurface cisternae located next the OHC's lateral plasma membrane (Dieler, Shehata-Dieler, and Brownell 1991; Shehata, Brownell, and Dieler 1991). In gerbils, a single layer of cisternae was observed at P10, the youngest age investigated. Subsequently, up to as many as 5 more layers are added. Because this development is not complete until 5 days after the OHCs show their maximum motile response (P22) (Weaver and Schweitzer 1994), maturational changes in the dynamics of OHC motility cannot result from the increased number of cisternal layers.

A better correlation is found between the development of OHC motility and maturation of the ECP, both of which reach maturity by P16–17 (see above). In adults, cochlear sensitivity has been shown to be closely related to both OHC motility, as inferred from measurements of tone-evoked basilar membrane vibration, and the level of the ECP (Ruggero and Rich 1991). Thus it seems reasonable to assume that both factors contribute significantly to the improvement of auditory thresholds during development. A critical test of this assumption will require direct measurement of cochlear feedforward amplification in immature animals. To date, only measurements that serve as indirect indicators of developmental changes in active processes during mechanoelectrical stimulus transduction are available. Two of these "indirect" physiological measures are the CM and evoked otoacoustic emissions.

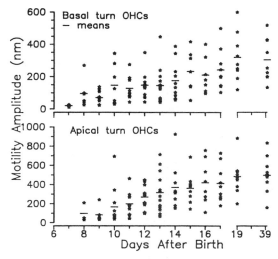

FIGURE 5.3. The maximal electromotile response amplitude of individual gerbil OHCs is plotted as a function of age. The maximal response amplitude was defined as the response observed at the highest stimulus level (280 mV) to a series of voltage pulses of progressively increasing amplitude. The horizontal bars indicate mean values at the respective ages. Although all OHCs show motile responses by 12 days after birth (see Fig. 5.2), response amplitude continues to increase until 16–17 days after birth when mature levels are approached. (From He, Evans, and Dallos 1994, Elsevier Science Publishers.)

4. Cochlear Microphonic (Table 5.2)

One type of cochlear potential that is linked to OHC activity, and which thus might serve as an indicator of active processes during stimulus transduction, is the CM (Adrian 1931). Evidence that the CM is generated by OHCs comes from the finding that destruction of OHC stereocilia by the ototoxic aminoglycoside kanamycin causes an almost complete loss of the CM response (Dallos 1973). Furthermore, in deaf mutant mice (deafness mutation *dn/dn*) in which hair cells are distorted and degenerated, the ECP exhibits a normal course of postnatal development, but a measurable CM is absent (Bock and Steel 1983). CMs are a favorite of auditory physiologists because these signals make it possible to measure OHC activity from extracellular or even extracochlear recordings (primarily at the round window). However, despite the frequent use of CM measurements, a generally accepted model for the generation of these signals is still lacking, and there is some uncertainty of how to interpret changes in CM amplitude across a range of stimulus frequencies (Pickles 1988).

It is generally believed that the CM reflects relative shifts in voltage between the endolymph and the perilymph due to ion flux through the tips

TABLE 5.2. Development of the cochlear microphonic.

Species (Reference)	Location	Age	Results/Comments
Rat (3)	Scalp-recorded CMs	7-70 DAB	CM amplitude reaches adult levels by 14 DAB, whereas the ECP continues to increase until 20 DAB. CM latency matures by 17 DAB.
Rat (9)	CM, round window recording	8-30 DAB	CMs first recorded at 8 DAB, 2-3 days before the CAP. CM amplitude and CM sensitivity are adultlike by P15.
Gerbil (2, 4, 5)	Intracochlear CM recording	8-90 DAB	Sound-evoked CMs can be recorded as early as 12-14 DAB, at which time evoked activity is restricted to frequencies <5 kHz.
Gerbil (11, 12)	CM, round window recording	10 DAB to adulthood	CMs are first recorded at 12 DAB. CM threshold and maximum CM response develop in parallel across different frequencies and achieve adultlike levels by 18 DAB (threshold) and by 30 DAB (maximum response).
Gerbil (10)	CM, round window recording	8 DAB to adulthood	With direct mechanical stimulation of the stapes, CM can first be elicited at 10 DAB, 2 days earlier than with acoustic stimulation. Bypassing the middle ear by direct stimulation of the stapes provides evidence that the CM is adultlike by P16.
Cat (7)	Scalp-recorded CM	1-54 DAB	CMs can be recorded as early as P1, 1-2 days before auditory brain stem responses.
Cat (8)	CM, round window recording	1-30 DAB	CM amplitude reaches maturity by the fourth postnatal week.

CM, cochlear microphonic. 1, Bock and Steel 1983; 2, Arjmand, Harris, and Dallos 1988; 3, Church, Williams, and Holloway 1984; 4, Finck, Schneck, and Hartman 1972; 6, Harris and Dallos 1984; 7, Mair, Elverland, and Laukli 1978; 8, Moore 1981; 9, Uziel, Romand, and Marot 1981; 10, Woolf and Ryan 1988; 11, Woolf and Ryan 1984; 12, McGuirt, Schmiedt, and Schulte 1995.

of the OHCs. If this is true, then the CM should serve as a good indicator of OHC activity and could provide information about feedforward amplification during cochlear stimulus transduction. From measurements of CMs during development, it should be possible to determine both the time of onset of acoustically generated OHC transduction currents and the subsequent increase in absolute level of these currents. The latter measurement relates directly to the gain of the cochlear amplifier.

In this respect, studies reporting sound-evoked CMs in mouse and rat pups as early as P8 should be regarded with caution (Crowley and Hepp-Reymond 1966; Uziel, Romand, and Marot 1981). At this age, only a very low level positive ECP is measurable in either of these species (Fig. 5.1C). More recent investigations in gerbils report the first CMs at P10-12 (Woolf and Ryan 1984, 1988) and at P13 (McGuirt, Schmiedt, and Schulte 1995). At this time, the ECP measures 0-20 mV. Thereafter, the "threshold" of the CM for frequencies between 0.1 and 10 kHz decreases rapidly by 70-80 dB and reaches adult levels at P18 (Fig. 5.4). The developmental improvement in threshold parallels almost exactly the development of the ECP, which also reaches adult values at P18-20 (Woolf, Ryan, and Harris 1986; McGuirt, Schmiedt, and Schulte 1995). If the CM amplitude depends on the level of the ECP, then the developmental increase in the ECP should cause a parallel shift in the entire CM input-output function, in addition to affecting the CM threshold. That is, the CM amplitude at respective stimulus frequency-intensity combinations should increase during the same period, as should the maximum amplitude of the CM. This has indeed been shown (Woolf and Ryan 1984, 1988; McGuirt, Schmiedt, and Schulte 1995) (Fig. 5.5). Surprisingly, the CM amplitude continues to increase after P20 and at least up to the end of the first postnatal month, although the ECP remains stable. In rat pups and kittens, which differ from each other in the dynamics of morphological and physiological cochlear maturation, a similar correlation between the development of ECP amplitude and CM thresholds is found, and in both species the maximum CM amplitude continues to increase over a prolonged period (rat ECP: Bosher and Warren 1971; rat CM: Uziel, Romand, and Marot 1981; cat ECP: Fernández and Hinojosa 1974; cat CM: Mair, Elverland, and Laukli 1978; Moore 1981). The reason for this prolonged development of CM amplitude is unknown.

As mentioned above, one advantage of CM recordings is the possibility of investigating OHC activity from extracochlear locations (mostly the round window) (Finck, Schneck, and Hartman 1972; Moore 1981; Uziel, Romand, and Marot 1981; Woolf and Ryan 1984, 1988) and even from scalp recordings (Mair, Elverland, and Laukli 1978; Laukli and Mair 1981; Church, Williams, and Holloway 1984). The disadvantage of this type of farfield measurement is the poor spatial resolution of these recordings, i.e., the CM signals recorded from outside the cochlea reflect activity integrated over a large area of the basilar membrane. Round window recordings of the CM will, irrespective of the stimulus frequency, mainly reflect the activity

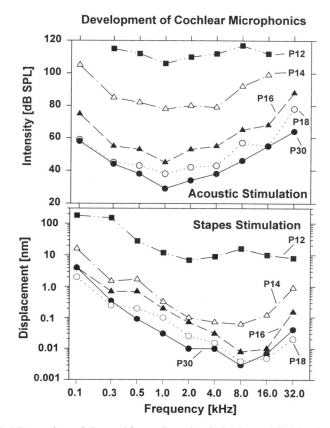

FIGURE 5.4. Maturation of the cochlear microphonic in the gerbil. The development of cochlear microphonic threshold to acoustic stimulation and direct mechanical stimulation of the stapes is plotted as a function of stimulus frequency and age. The ordinates indicate the sound intensity and stapes displacement required to produce a criterion level cochlear microphonic of 1.5 μV rms. Chronological age is indicated in postnatal days (P). SPL, sound pressure level. (Data from Table 1 in Woolf and Ryan 1984 [top] and Table 1 in Woolf and Ryan 1988 [bottom].)

of hair cells located near the basal end of the cochlea (Tasaki, Davis, and Legoiux 1952; Pickles 1988). This is why, in studies of the development of the CM, input-output functions are never saturating at intensities above 50 to 60 dB sound pressure level (SPL), as would be expected if the signals reflect the focal activity at the representation site of a specific characteristic frequency (CF). Instead, the signals show an almost linear increase in amplitude up to stimulus intensities of 100 dB SPL or higher (Moore 1981; Woolf and Ryan 1984, 1988; McGuirt, Schmiedt, and Schulte 1995). Comparable linear, nonsaturating off-CF input-output functions have also been seen in direct measurements of basilar membrane motion (Johnstone,

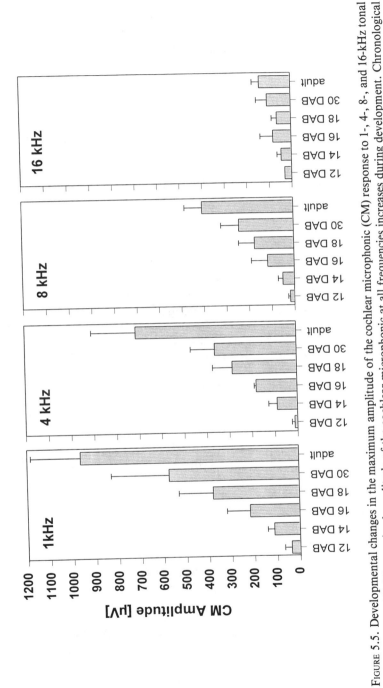

FIGURE 5.5. Developmental changes in the maximum amplitude of the cochlear microphonic (CM) response to 1-, 4-, 8-, and 16-kHz tonal stimuli in the gerbil. The maximal amplitude of the cochlear microphonic at all frequencies increases during development. Chronological age is given in days after birth (DAB). Values are means ± SD. (Data from Table 2 in Woolf and Ryan 1984.)

Patuzzi, and Yates 1986; Ruggero and Rich 1991) and in OHC receptor potentials (Cody and Russell 1987). More detailed information about the development of the OHC system, especially about differential development along the basal-to-apical extent of the basilar membrane, could be obtained from intracochlear recordings of CMs. However, studies using this approach have focused on the question of developmental changes in the cochlear frequency map (Harris and Dallos 1984; Arjmand, Harris, and Dallos 1988). This topic is discussed in Section 8.

5. Distortion Product Otoacoustic Emissions (Table 5.3)

The DPOAE is another type of cochlear-generated signal that is useful for evaluating the contribution of OHC force-generating electromotility to stimulus transduction. These acoustic signals, which can be recorded in the external ear canal, are generated within the cochlea in response to simultaneous stimulation of the same ear by two different frequency tones (Probst, Lonsbury-Martin, and Martin 1991). DPOAEs consist of two components: a vulnerable DPOAE, which is typically evoked by stimulus intensities up to 55 to 65 dB SPL, and a nonvulnerable component, which is generated at higher stimulus levels. The vulnerable DPOAE depends on an intact OHC system (Long and Tubis 1988; Brown, McDowell, and Forge 1989) and on a positive ECP (Brownell 1990; Mills, Norton, and Rubel 1994), whereas the nonvulnerable DPOAE does not. The vulnerable component of the DPOAE is a primary manifestation of the active amplification of the traveling wave by the OHCs. Thus the vulnerable DPOAE (or "active" emission) provides, in a highly frequency-specific fashion, information about active cochlear mechanisms. An advantage of DPOAE measurements is that they provide a noninvasive probe for evaluating cochlear functioning. Therefore, in the near future, DPOAE measurements will become a standard tool in clinical audiology, and particularly in pediatric audiology, because their acquisition does not depend on the patient's cooperation (Probst, Lonsbury-Martin, and Martin 1991; Norton 1992; Probst, Harris, and Hauser 1993). There have been an increasing number of reports on otoacoustic emissions in infants, including preterm babies (Bargones and Burns 1988; Bonfils et al. 1990; Norton and Widen 1990; Burns et al. 1992; Chuang, Gerber, and Thornton 1993; Morlet et al. 1993). However, because these studies focus primarily on clinical issues and, in most cases, the infants were examined as late as 1–6 months after hearing onset (seventh postconceptual month), these studies will not be considered in the present review.

Evoked otoacoustic emissions have been used to examine the development of stimulus transduction along the basilar membrane in the rat and gerbil (Lenoir and Puel 1987; Henley et al. 1989; Norton, Bargones, and Rubel 1991; Mills, Norton, and Rubel 1994; Mills and Rubel 1996). The fact that emissions provide highly selective frequency information makes them especially well suited for such investigations. In both rodent species,

TABLE 5.3. Development of evoked otoacoustic emissions.

Species (Reference)	Location	Age	Results/Comments
Rat (1)	Outer ear canal	11–40 DAB	Measurement of DPOAEs to $2f_1 - f_2 = 3$, 5, and 7 kHz. First responses at 12 DAB to 5 and 7 kHz and at 14 DAB to 3 kHz. Adultlike responses by 18 DAB for 3 kHz and by 28 DAB for 7 kHz. Threshold values ranged from 90 dB SPL at 16 DAB to 65–85 db SPL in adults. Likely that responses are passive rather than active emissions.
Rat (2)	Outer ear canal	12–51 DAB	Measurement of DPOAEs to $2f_1 - f_2 = 2.8$–8.0 kHz. Responses first detected at 14 DAB to 4.0–8.0 kHz and 18 DAB to 2.8 kHz. input-output functions for higher frequencies are adultlike by third to fourth postnatal week.
Gerbil (3)	Outer ear canal	13–102 DAB	Measurement of DPOAEs to $f_2 = 1.3$–13.0 kHz and $f_2/f_1 = 1.3$. Orderly progression in the maturation of emissions: responses to high frequencies (13 kHz) appear first at 13–14 DAB; responses to middle and high frequencies mature earlier than responses to low frequencies (1.3 kHz). Low-frequency responses do not appear until 18–19 DAB and are not mature until 30 DAB.
Gerbil (4, 5)	Outer ear canal	14–60 DAB	Measurement of DPOAEs to $f_2 = 0.5$–48 kHz and $f_2/f_1 = 1.28$. Measurements were made before and after systemic injection of the ototoxic drug furosemide; this allowed the gain of the cochlear amplifier to be estimated and active vs. passive emissions to be distinguished. At 14 DAB, active emissions are evoked by stimulus frequencies from 2.0–20 kHz but not by lower (0.5 to 1.0 kHz) and higher frequencies (24 to 48 kHz). At 14 DAB, the gain of the cochlear amplifier is at adult levels for midfrequencies. Cochlear amplifier gain at lower and higher frequencies reaches mature levels by ~25 DAB. After 22 DAB, a reduction of cochlear amplification occurs in the mid- and high-frequency range, which results in flattening of the audiogram during development. Passive emission thresholds improve during development, particularly at higher frequencies.

DPOAE, distortion-product otoacoustic emission; f, frequency; SPL, sound pressure level. 1, Lenoir and Puel 1987; 2, Henley et al. 1989; 3, Norton, Bargones, and Rubel 1991; 4, Mills, Norton, and Rubel 1994; 5, Mills and Rubel 1996.

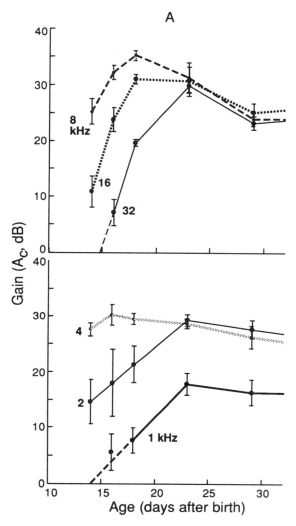

FIGURE 5.6. Development of active and passive cochlear mechanics in the neonatal gerbil. The cochlear amplifier gain and the threshold of the passive mechanical response were estimated by measuring the amplitude of the cubic distortion tone (CDT) emission before and after systemic injection of furosemide. See the inset in "B" for the definition of terms. A: cochlear amplifier gain (A_C) is plotted as a function of age for different stimulus frequencies. Note that the gain at midfrequencies (4–8 kHz) is already at or near mature levels by 14 days after birth. The dashed lines at 1 and 32 kHz indicate the estimated gain because emissions at these frequences were not measurable above the noise level at 14 days after birth. Values are means ± SE. B: development of the passive "threshold" (L_p) at different

(Continued)

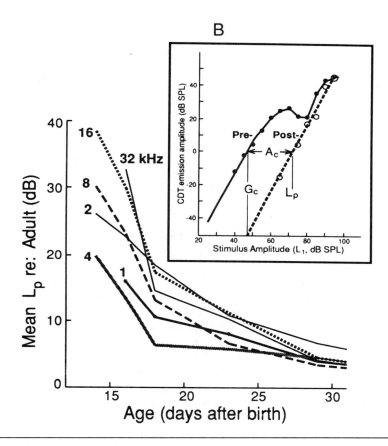

stimulus frequencies. The values shown at each age are the improvements in threshold that occur between that age and adulthood (42–46 days after birth). Note that passive thresholds improve most markedly at the higher stimulus frequencies. The inset shows typical CDT emission input-output functions before (Pre-) and after (Post-) injection of furosemide. Furosemide injection causes a transient inactivation of the active process that, in turn, results in the elimination of the active component of the CDT emission and the linearization of the input-output function. L_1 is the amplitude of the lower of the 2 primary stimulus frequency tones. A_c is defined as the horizontal difference between the parallel portions of the input-output functions. This measure of cochlear amplifier gain represents the improvement in threshold with the cochlear amplifier functional compared to nonfunctional. G_c, an alternative estimate of cochlear amplifier gain, is defined as the vertical distance between the parallel portions of the input-output functions at low stimulus levels. L_p is the stimulus level required to produce a CDT emission amplitude equal to 0 dB SPL after furosemide injection. (Modified from Mills and Rubel 1996, American Institute of Physics.)

DPOAEs were first detected at P12–13. This coincides with the time when positive ECPs and CMs can be measured in these animals. Thus all three types of measurements give an indication of the onset of OHC-mediated cochlear amplification.

All reports agree that emissions develop, in a frequency-specific fashion, between P18 and P30. However, conclusions regarding the spatiotemporal gradient of DPOAE development differ among studies. Although Lenoir and Puel (1987) suggest that DPOAEs develop along an apical-to-basal gradient, Henley et al. (1989) and Norton, Bargones, and Rubel (1991) report the opposite. There are several methodological factors that might account for the discrepancy in results. For example, the conclusions in the 2 latter studies are based on lower intensities of acoustic stimulation and a wider range of stimulus frequencies than were used by Lenoir and Puel (1987). Furthermore, not all DPOAEs can serve as an indicator of active cochlear processes. A reliable estimate of the development of cochlear amplification is only possible after differentiating between the *vulnerable* (*active*) and *nonvulnerable* (*passive*) components of the DPOAE. Norton, Bargones, and Rubel (1991) attempted to do this by comparing pre- and postmortem DPOAEs in young gerbils. The vulnerable DPOAE was observed as early as P15. Subsequently, in the period up to P30, the level difference between the vulnerable and nonvulnerable DPOAE components measured at low- to midstimulus intensities (50 to 60 dB SPL) steadily increased up to 40–45 dB. The progressive increase in the magnitude of this difference was interpreted to reflect the maturation of the gain of the cochlear amplifier. In a recent series of investigations, Mills, Norton, and Rubel (1994) and Mills and Rubel (1996) produced a transient reduction in the gain of the cochlear amplifier in young gerbils by lowering the ECP through systemic injection of the ototoxic diuretic furosemide (Fig. 5.6). These studies confirmed that the cochlear amplifier is effective from as early as P14–15 and onward. In mid-to-basal positions along the basilar membrane, where 2–16 kHz is represented, the cochlear amplifier gain is between 10 and 25 dB. By P22–23, the gain increases up to 25–35 dB. At lower stimulus frequencies, cochlear amplification is first observed at P17 (for 1 kHz) and P19 (for 0.5 kHz), and maximum gain values are reached at P19 and P22, respectively. Interestingly, after P22, a noticeable reduction in the gain of the cochlear amplifier occurs in the frequency range of 2–32 kHz.

These experiments suggest that the postnatal increase of cochlear amplification and the positional shift of frequency representation along the cochlea are not the consequence of the same developmental modifications in inner ear structure and function, as hypothesized by Romand (1987). The increase in gain, which is mirrored by an increase in the sensitivity of cochlear afferents (cat: Walsh and McGee 1987; gerbil: Müller 1991, 1996) and central auditory neurons (gerbil: Woolf and Ryan 1985; bat: Rübsamen, Neuweiler, and Marimuthu 1989; Rübsamen and Schäfer 1990a), may be due to the development of the OHCs and, in part, to the increase in the ECP (Mills, Norton, and Rubel 1994), whereas the shift in cochlear high-frequency representation may result from developmental changes in the passive mechanical properties of the basilar membrane (also see Rübsamen 1992). This is discussed further in Section 8.

6. Inner Hair Cells (Tables 5.4 and 5.5)

Cochlear output is transmitted centrally through the activity of type I auditory nerve fibers. In adults, these fibers constitute 95–98% of the cochlear afferents and form synapses exclusively with IHCs. Direct recordings from individual IHCs in developing animals would reveal how closely the development of IHC physiology correlates with the maturation of other aspects of cochlear function discussed in the sections 2 through 5 above. Additionally, such investigations might reveal the extent to which the functional development of IHCs is a limiting factor in the maturation of cochlear functioning as a whole. But, as is the case with OHCs, the development of IHC physiology has only been measured indirectly. However, these measurements are even "more indirect" than measures such as the CM and DPOAE which have been used to assess OHC functioning. All measurements that have been used as indicators of the development of IHC functioning reflect the activity of the cochlear afferents rather than the activity of the receptor cells themselves.

The specific measurements that might provide insight into the functional development of IHCs include the (1) CAP, (2) spontaneous and driven activity of auditory nerve fibers, and (3) the activity of neurons in the central auditory nuclei. This rank order also represents the increasing distance from the IHCs, on which our interest is focused.

Evidence for a direct link between the CAP and hair cell functioning comes from studies in a mice strain in which the hair cells do not develop normally (deafness mutation *dn/dn*) (Bock and Steel 1983). From as early as P6, all hair cells exhibit signs of degeneration, and by adulthood, few hair cells remain intact. Other inner ear structures, including Reissner's membrane, the stria vascularis, and the spiral ganglion neurons, show no obvious signs of pathology. These mice show a normal maturation of the ECP but no discernable CAPs, even after the first postnatal month.

In normal mouse, rat, and gerbil pups, CAPs can first be recorded at P12–13 (Shnerson and Pujol 1981; Uziel, Romand, and Marot 1981; Rybak, Whitworth, and Scott 1992; McGuirt, Schmiedt, and Schulte 1995). This is the same time when reliable measurements of CMs are first obtained (see Section 4), which is consistent with the idea that the onset of function occurs simultaneously in the IHCs and OHCs. Subsequent development is characterized by a steady increase in the amplitude and a decrease in the threshold and response latency of the CAP (Uziel, Romand, and Marot 1981; Rybank, Whitworth, and Scott 1992). It has been suggested that the increase in CAP amplitude reflects an increase in the synchronization of afferent neuronal discharges, as indicated by the improvement in the phase-locked responses of afferent nerve fibers that occurs over the same period of development (Moore 1981; Kettner, Feng, and Brugge 1985). This emphasizes the fact that the conclusive interpretation of CAP data requires that these data be related to unit recordings from single auditory nerve

TABLE 5.4. Development of inner and outer hair cells.

Species (Reference)	Location	Age	Results/Comments
Rat (1)	Cochlea	E1–P5	Immunocytochemistry. Ingrowing nerve fibers may influence the differentiation of the apical cytoskeleton in OHCs. Alternatively, fiber ingrowth and cytoskeletal differentiation may be controlled by common external signals.
Hamster (2, 3)	OHC, IHC	0–20 DAB	Transmission electron microscopy. OHC stereocilia attain adultlike sizes by 14 DAB. Tip links are found as early as P4. IHCs and OHCs reach adultlike sizes at 16–18 DAB (onset of acoustically evoked neural activity). The period of hair cell growth extends beyond the period of cochlear growth.
Gerbil (4)	OHC	0–20 DAB	Immunocytochemistry. OHCs first express the facilitated glucose transporter (GLUT-5) at 10 DAB. Adultlike pattern of immunolabeling is reached at 16 DAB. GLUT-5 is thought to support OHC function by facilitating uptake of glucose, which enables hair cells to sustain a high level of metabolic activity in a relatively anaerobic environment.
Gerbil (5)	OHC	0–30 DAB	Transmission electron microscopy. Ultrastructural characteristics of OHCs change after the onset of hearing (12 DAB) and up to 21 DAB: the distribution of organelles changes, and more subsurface cisternae are added.
Gerbil (6)	OHC	0–30 DAB	Immunocyochemistry. Actin labeling in the OHC cytoskeleton increases at the onset of cochlear function (12 DAB). Subsequently, during the period when auditory thresholds decrease, actin labeling is maintained at the same level found at 12 DAB. This contrasts with the gradual development of subsurface cisternae that occurs between 12 and 21 DAB.

OHC, outer hair cell; IHC, inner hair cell. 1, Pirvola, Lehtonen, and Ylikoski 1991; 2, Kaltenbach, Falzarano, and Simpson 1994; 3, Kaltenbach and Falzarano 1994; 4, Nakazawa, Spicer, and Schulte 1995; 5, Weaver and Schweitzer 1994; 6, Weaver, Hoffpauir, and Schweitzer 1994.

TABLE 5.5. Development of compound action potential.

Species (Reference)	Location	Age	Results/Comments
Mouse (2)	VIII nerve CAP	12–50 DAB	CAP tuning (2.0–35 kHz). Sensitivity to tones and threshold tuning curve sharpness (Q10) increase from 12 to 20 DAB. Response latencies attain minimum values by 20 DAB. The intensity functions are essentially identical in 12- and 16-day-old mice.
Mouse (3)	Round window CAP	6–40 DAB	No discernible CAP can be recorded in deafness mutant mice that show normal development of the ECP but that have distorted and degenerated hair cells.
Mouse (5)	Round window CAP	8–30 DAB	CAP first detected at 11 DAB to 6-kHz stimuli.
Rat (1)	Round window CAP	8 DAB to adult	Click stimulation. CAP first appears at 11–12 DAB. Threshold of the CAP then decreases and reaches adult values in the fourth week; the amplitude and threshold sensitivity of both the CM and CAP increase progressively with age as N1 latency decreases. All the parameters are within the adult range by the 15th day for the CM and by the 4th week for the CAP.
Rat (4)	Round window CAP	10–30 DAB	CAP first recorded at 13 DAB. Development of the CAP and ECP are inversely correlated; the CAP threshold declines as the level of the ECP increases.

CAP, compound action potential. 1, Uziel, Romand, and Marot 1981; 2, Shnerson and Pujol 1981; 3, Bock and Steel 1983; 4, Rybak, Whitworth, and Scott 1992; 5, Harvey and Steel 1992.

fibers. Recordings from individual afferent fibers are especially promising for understanding the development of IHC function because a single auditory nerve fiber is postsynaptic to only a single IHC, and deafferented auditory nerve fibers do not show any spontaneous activity. Thus it can be generally stated that the activity of mature auditory nerve fibers reliably reflects the activity of a specific set of IHCs. In relating developmental changes in auditory nerve fiber activity to the maturation of IHC function, one must, of course, consider other possible causes for changes in afferent activity. These might include, for example, the maturation of the mechanically gated ion channels at the tips of the OHC stereocilia, changes in the effectiveness of transmitter release by IHCs, or maturational changes in the spiral ganglion cells themselves. Despite these qualifications, to date recordings from single auditory nerve fibers provide the most reliable source of information about the functional development of IHCs.

7. Spontaneous Activity (Table 5.6)

The majority of auditory nerve fibers in adult mammals discharge in the absence of any intentional sound stimulation. This spontaneous activity is believed to be triggered by postsynaptic potentials resulting from a random, quantal release of neurotransmitter from the base of the hair cell. Several lines of evidence indicate that spontaneous activity is not a response to uncontrolled, low-level acoustic stimulation (Kiang et al. 1965; Manley and Robertson 1976; Furukawa and Matsuura 1978; Robertson and Johnstone 1979a; Crawford and Fettiplace 1980; Sewell 1984a). It is likely that the maintained discharge of mammalian ventral cochlear nucleus neurons and second-order magnocellularis neurons in birds also originates peripherally because this activity is eliminated after cochlear removal or injection of tetrodotoxin into the cochlear perilymph (Koerber, Pfeiffer, and Kiang 1966; Born and Rubel 1988; Born, Durham, and Rubel 1991; Lippe 1994). In addition, spontaneous activity in the ventral cochlear nucleus, as indicated by 2-deoxyglucose utilization, is not observed in mutant mice lacking hair cells (Durham, Rubel, and Steel 1989). In contrast, neurons in the dorsal cochlear nucleus exhibit an intrinsic spontaneous firing that arises independently of synaptic input (Koerber, Pfeiffer, and Kiang 1966; Hirsch and Oertel 1988; Charpak et al. 1989; Waller and Godfrey 1994).

It is uniformly reported that the level of spontaneous activity in the auditory nerve as well as in the central auditory nuclei is much lower in developing animals than in the adult (Figs. 5.7 and 5.8). In mammals, the development of spontaneous activity in the auditory nerve has been studied quantitatively primarily in the cat (Romand 1984; Walsh and McGee 1987; Walsh and Romand 1992) and to a lesser extent in the gerbil (Müller 1996). Spontaneous discharge rates of adult feline auditory nerve fibers have a bimodal distribution such that ~50% of the fibers discharge at rates <10 spikes/s, and the remainder have a broad temporal distribution with maximum rates of 100–120 spikes/s (Kiang et al. 1965; Liberman 1978). The range of spontaneous firing rates in the auditory nerve of other adult mammalian species is similar (Manley and Robertson 1976; Schmiedt 1989). At the onset of hearing on P2 in the cat, few spontaneously active auditory nerve fibers are encountered, and these generally discharge at <1 spike/s. During development, the number of fibers lacking spontaneous activity decreases and the frequency distribution of spontaneous discharge rate shifts to include higher values. According to Romand (1984) on P6 approximately one-third of those fibers that can be activated by sound fire spontaneously at rates of <1/s, and rates >20/s are not encountered. By day 20, fibers with rates <1/s constitute only slightly more than 15% of the population and rates as high as 60/s are observed. The distribution of spontaneous rates continues to shift but doesn't become adultlike until after P40. Comparable developmental increases in the percentage of spontaneously discharging neurons and in the rate of discharge have been described for the feline cochlear nucleus and auditory cortex, the auditory nerve and cochlear nucleus in the gerbil, and the inferior colliculus in the mouse (Brugge, Javel, and Kitzes 1978; Shnerson and Willott 1979; Brugge and O'Connor 1984; Woolf and Ryan 1985; Eggermont 1991; Müller 1996).

The reason for the low level of ongoing activity in immature mammals is not known. The spontaneous discharge rate of auditory nerve fibers in adult mammals is correlated with the level of the ECP (Sewell 1984a). Hypoxia and systemic injection of the ototoxic diuretic furosemide cause the ECP and the spontaneous discharge rate of auditory nerve fibers to decrease simultaneously (Manley and Robertson 1976; Sewell 1984a). This finding is consistent with the hypothesis that the random release of neurotransmitter underlying the generation of spontaneous spike activity depends on the transmembrane potential of the hair cell and thus, indirectly, on the resting current that normally flows through the hair cells. The magnitude of the resting current, in turn, is partially determined by the level of the ECP. The ECP is low at the onset of hearing and increases during development (see Section 2). Thus it is probable that the low level of the ECP in developing animals accounts, in part, for the low level of spontaneous activity.

Additional factors may contribute as well. Consistent with this is the observation that the overall mean spontaneous discharge rate in the auditory nerve of the cat, and possibly in the ventral cochlear nucleus of the gerbil as well, continues to change even after the ECP has reached an adult level (cat: cf. Romand 1984 with Fernández and Hinojosa 1974; gerbil: cf. Woolf, Ryan, and Harris 1986 with Woolf and Ryan 1985). Differences in the spontaneous discharge rate among single auditory nerve fibers in adult animals are correlated with several aspects of neuronal morphology, including fiber diameter, synaptic morphology, synaptic location around the circumference of the IHC-, mitochondrial content, and the integrity of IHC stereocilia. Smaller diameter fibers typically exhibit lower spontaneous

TABLE 5.6. Development of spontaneous activity.

Species (Reference)	Location	Age	Results/Comments
Cat (1)	Auditory nerve	2–200 DAB	Anesthetic: ketamine + sodium pentobarbital. Discharge rate increases during first postnatal month, reaching adult level sometime after P40. Before 14 DAB, mean discharge rate averages ~10 spikes/s, and few neurons discharge faster than 20 spikes/s. This contrasts with adults in which rates up to 100 spikes/s occur and where rate shows a bimodal distribution with low and high spontaneous rates found. During the first 12 DAB, auditory nerve fibers and some cochlear nucleus neurons discharge rhythmically to long-duration stimuli; in some neurons, rhythmic bursting at a lower bursting rate continues briefly after the termination of stimulation.
Cat (2, 3)	Auditory nerve	2–40 DAB	Anesthetic: sodium pentobarbital. Mean spontaneous discharge rate of acoustically responsive fibers increases rapidly from P2 (<5 spikes/s) to P20 (20 spikes/s) and subsequently more slowly to P40 (28 spikes/s) and maturity (45 spikes/s). Percentage of fibers with spike rates < 1.0/s decreases from ~38% on P2 to 15% on P40; over the same period, the percentage of fibers with spike rates > 20/s increases from 0 to 40%. Number of fibers showing no spontaneous activity decreases with age. Interspike interval histogram with single peak and exponential decay appears about P14.
Cat (4)	AVCN	4–45 DAB, adult	Anesthetic: ketamine + sodium pentobarbital. Spontaneous activity is rare during first 2 weeks, and rates are usually less than a few spikes per second; highest discharge rate observed was 25–30 spikes/s at 10 DAB.

Cat (5)	PVCN, DCN	5–52 DAB	Anesthetic: ketamine + sodium pentobarbital. During first week, 25% of neurons exhibit spontaneous activity; maximum rate is 9 spikes/s. Maximum discharge rate is < 34 spikes/s in cells recorded during first month. Some classes of immature neurons show rhythmic bursting to long-duration stimulation; mode of interburst interval is 100–140 ms; interburst interval decreases with increasing intensity of stimulation; bursting is rarely seen after 10–12 DAB. Spontaneous discharges don't show rhythmic pattern.
Cat (6)	Primary auditory cortex	9–53 DAB	Anesthetic: ketamine + sodium pentobarbital. Mean spontaneous discharge rate increases from 0.2 to 0.75 spikes/s between 15 and 53 DAB. Highest rate of ~3 spikes/s is observed in animals older than 27 DAB. Low discharge rates occur at all ages. Upper range of discharge rate increases during development.
Gerbil (7)	Auditory neve	12–14 DAB, young adult (2–4 weeks)	Anesthetic: sodium pentobarbital. In younger animals, frequency distribution of spontaneous rate is unimodal, with a peak at <10 spikes/s; maximum spontaneous rate was 50 spikes/s; 77% of fibers had spontaneous rates below 10 spikes/s. Distribution of spontaneous rate in older animals is bimodal, with maximums at <10 and 50–80 spikes/s; spontaneous rates range from 0 to 163 spikes/s.
Gerbil (8)	Cochlear nucleus	10–90 DAB	Anesthetic: ketamine + sodium pentobarbital. Spontaneous activity observed on P10, 2 days before the onset of sound-evoked activity. At 12 DAB, many units don't respond to sound but are spontaneously active. All quantitative data from units that respond to sound. Mean discharge rate increases monotonically. Percentage of units with spontaneous rates > 5 spikes/s increases from 12% on 12 DAB to 73% in adults; during the same period, maximum discharge rate increases from ~33 to 108 spikes/s. Percentage of units lacking spontaneous discharge decreases from 30% at 12 DAB to 9% at 18–30 DAB. Between 12 and 18 DAB, mean discharge rate and range increase.

(Continued)

TABLE 5.6. (*Continued*)

Species (Reference)	Location	Age	Results/Comments
Gerbil (9)	Inferior colliculus	9–13 DAB	Anesthetic: ketamine + chloral hydrate. Spontaneous unit discharges at all ages. Discharge pattern is irregular; alterations between period of activity and quiescence are common; mean discharge rate is 0.4 spikes/s; mean maximum discharge rate is 36.7 spikes/s.
Tammar wallaby (10)	Auditory nerve, cochlear nucleus	PD 94–122	Anesthetic: ketamine + xylazine. Spontaneous firing occurs as early as PD94 (46–56 days before the first measurable tone-evoked CAP). Eighty-four percent of units discharge in rhythmic bursts. Modal interburst interval is 0.6–0.8 s.
Chicken (11, 12)	Nucleus magnocellularis and nucleus laminaris. Single- and multiunit activity	E14 to P21	Anesthetic: ketamine. Spontaneous multiunit activity occurs in rhythmic bursts during embryonic development. Overall bursting rate increases progressively from 0.20 Hz on E14 to 0.48 Hz on E18; bursting gives way to an adultlike, steady level of firing on E19. Bursting rate is progressively faster at higher frequency locations along the tonotopic axis. Bursting most probably originates within the basilar papilla.
Chicken (13)	Cochlear ganglion	P2, P21	Anesthetic: sodium pentobarbital + chloral hydrate. Spontaneous activity is similar at both ages. Mean discharge rate is 20.5 spikes/s at P2 and 22.9 spikes/s at P21. Spontaneous rate has a unimodal distribution. About 15% of fibers, and only those with CFs < 1 kHz, show preferred intervals in their interspike interval histograms, which is indicative of electrical tuning in the hair cells.

Bat (14)	Inferior colliculus. Single- and multiunit activity	2–35 DAB	Anesthetic: ketamine + xylazine. Spontaneous discharges are observed 4–5 days before the appearance of tone-evoked unit discharges. Early spontaneous firing in the foveal region, where high frequencies will later be represented, frequently shows a rhythmic bursting pattern. Rhythmic bursting continues after hearing onset and is replaced by continuous background activity at about the time that tonotopy first appears in this region. Data are not presented.
Mouse (15)	Inferior colliculus. Single- and multiunit activity	9–20 DAB	Anesthetic: sodium pentobarbital. Spontaneous discharges are observed on 9 DAB, 1 day before the onset of tone-evoked unit responses. Activity occurs in the ventral (high-frequency) region. Data are not presented.
Mouse (16)	Inferior colliculus	12–17 DAB	Anesthetic: chlorprothixene + sodium pentobarbital. Percentage of spontaneously discharging units, mean firing rate, and maximum firing rate in the ventrolateral region of the central nucleus increase from 3%, 0.2 spikes/s, and 0.2 spikes/s, respectively, at 13–14 DAB to 48%, 18 spikes/s, and 70 spikes/s at 15–17 DAB. Spontaneous activity is not at adult levels by 15–17 DAB. Maturation of spontaneous activity differs between the ventrolateral and other peripheral regions of the inferior colliculus.

AVCN, anteroventral cochlear nucleus; PVCN, posteroventral cochlear nucleus; DCN, dorsal cochlear nucleus; CF, characteristic frequency. 1, Walsh and McGee 1978; 2, Romand 1984; 3, Romand and Dauzat 1981; 4, Brugge, Javel, and Kitzes 1978; 5, Brugge and O'Connor 1984; 6, Eggermont 1991; 7, Müller 1996; 8, Woolf and Ryan 1985; 9, Kotak and Sanes 1995; 10, Gummer and Mark 1994; 11, Lippe 1994; 12, Lippe 1995; 13, Manley et al. 1991; 14, Rübsamen and Schäfer 1990a; 15, Romand and Ehret 1990; 16, Shnerson and Willott 1979.

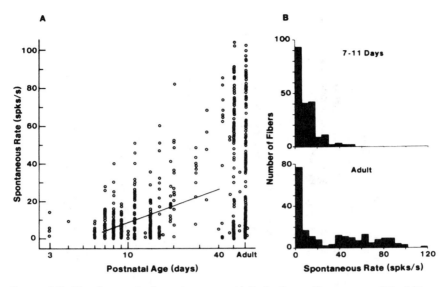

FIGURE 5.7. Development of spontaneous activity in the auditory nerve of the kitten. A: spontaneous discharge rates of individual auditory nerve fibers are shown from approximately the onset of hearing on P3 through adulthood. A least squares linear regression line is fitted to all values from animals younger than 40 postnatal days. B: frequency distributions of the spontaneous discharge rates of auditory nerve fibers between 7–11 postnatal days and adulthood are compared. spks, spikes. (From Walsh and McGee 1987, Elsevier Science Publishers.)

rates (Liberman 1980a,b, 1982; Liberman and Dodds 1984a; Liberman and Oliver 1984). All these morphological properties undergo considerable change during development. The potential contribution of these and other factors, including immaturities in neurotransmitter systems, the pattern of hair cell innervation, the relationship of hair cell stereocilia to the tectorial membrane, and immaturities in spiral ganglion neurons and their processes, have been discussed previously (Brugge 1986; Walsh and McGee 1987). Two methodological considerations may also be relevant. First, almost all observations of spontaneous activity in immature mammals have been made using barbiturate anesthesia, which can greatly suppress the overall level of spontaneous activity and affect the temporal pattern of discharge (Bock and Webster 1974; Webster and Aitkin 1975; Young and Brownell 1976; Brownell, Manis, and Ritz 1979; Anastasio, Correia, and Perachio 1985). It is possible that immature animals are more susceptible to barbiturate anesthesia, and to the effects of anesthesia in general, than are adults. Second, extracellular recording of unit discharges does not detect subthreshold synaptic transmission, which is likely more frequent in immature neural networks.

Relatively little is known about the prevalence or characteristics of spontaneous activity in the auditory pathway before hearing onset. Elec-

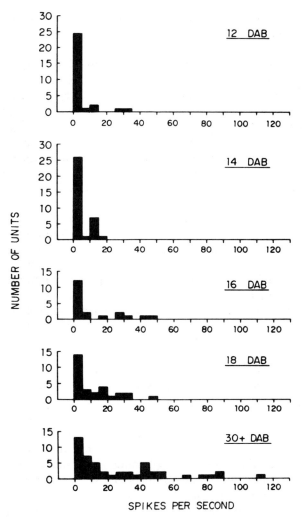

FIGURE 5.8. Development of spontaneous activity in the cochlear nucleus of the neonatal gerbil. Changes in the frequency distribution of spontaneous discharge rate of ventral cochlear nucleus neurons from the onset of hearing at 12 DAB through maturity (30+ DAB) are shown. (From Woolf and Ryan 1985, Elsevier Science Publishers.)

trical stimulation studies in vivo and in isolated brainstem slice preparations have shown that synaptic transmission between the auditory nerve and the cochlear nucleus (cat, rat, and mouse) and between the lateral superior olive and its afferents (rat and gerbil) can be evoked as early as 2–14 days before the time that air borne sound is first capable of influencing postsynaptic function (Marty and Thomas 1963; Pujol and Marty 1968; Tokimoto,

Osako, and Matsuura 1977; Mysliveček 1983; Romand, Despers, and Giry 1987; Wu and Oertel 1987; Sanes 1993; Kandler and Friauf 1995). Kandler and Friauf (1995) suggested that the auditory pathway in the rat is capable of transmitting neuronal activity from the cochlear nucleus to the inferior colliculus more than 2 weeks before the onset of hearing.

Evidence for the occurrence of early spontaneous activity comes from the relatively few studies that have reported low levels of ongoing activity in the auditory pathway before hearing onset. Spontaneous discharges have been recorded in the inferior colliculus of the mouse (Romand and Ehret 1990) and the rufous horseshoe bat (*Rhinolophus rouxi*) (Rübsamen and Schäfer 1990a) 1 and 5 days, respectively, before the onset of sound-evoked activity and in the cochlear nucleus of the gerbil on P10 (Woolf and Ryan 1985) and as early as P7 (Rübsamen and Lippe, unpublished observations). Spontaneously active neurons are also infrequently encountered in electrode penetrations through the inferior colliculus of the gerbil at P9–11 (Kotak and Sanes 1995). The discharge pattern tends to be irregular, with brief periods of activity alternating with periods of silence. The average discharge rate of most neurons is <0.5 spikes/s, although maximum discharge rates are significantly higher (mean = 37 spikes/s), reflecting the high discharge rate that occurs during intermittent bursts of activity. Presently, it is not known if the early spontaneous activity in the central auditory nuclei is generated intrinsically or originates in the cochlea. However, prepotentials are frequently evident in extracellular recordings of the spontaneous discharge of ventral cochlear nucleus neurons in the gerbil well before the onset of hearing (Rübsamen and Lippe, unpublished observations) (Fig. 5.9). This suggests that spontaneous activity of cochlear origin is most probably being generated and transmitted centrally at these early ages.

Direct evidence for early cochlear generated spontaneous firing comes from electrophysiological recordings in the tammar wallaby (*Macropus eugenii*), an Australian marsupial (Gummer and Mark 1994). Cochlear nucleus neurons and auditory nerve fibers discharge spontaneously as early as pouch days 94–122 (PD94–122), which is before the opening of the external auditory meatus (PD125–130) and well before tone-evoked CAPs can first be detected on PD130–140. Correlational analysis of the temporal discharge pattern revealed that ~85% of the units fired in a rhythmic pattern such that bursts of action potentials were followed by brief periods (0.6–0.8 s) of little or no activity.

Spontaneous rhythmic discharges have also been observed in the auditory midbrain of the horseshoe bat (*Rhinolophus rouxi*) before the onset of hearing (Rübsamen and Schäfer 1990a) and stimulated, but not spontaneous, rhythms occur in the kitten auditory system during the first 2 postnatal weeks (see Sanes and Walsh, Chapter 6). A very robust rhythmic pattern of spontaneous firing also occurs in the auditory pathway of the prehatching chick (Lippe 1994, 1995) (Fig. 5.10). From as early as embryonic day 14 (E14), shortly after the onset of synaptic transmission, and through E18, 3

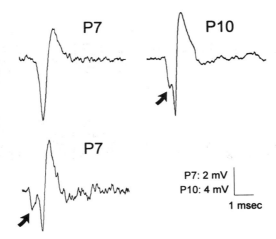

FIGURE 5.9. Spontaneous activity before hearing onset in the gerbil. Extracellularly recorded spontaneous discharges of gerbil ventral cochlear nucleus neurons at P7 and P10 are shown. The arrows indicate "prepotentials" that commonly preceded the spontaneous discharge of the cochlear nucleus neurons. The prepotentials likely reflect spontaneous action potentials that arise within the cochlea and are transmitted centrally along single auditory nerve fibers. (From Rübsamen and Lippe, unpublished observations.)

days before hatching, neural activity in the developing brainstem auditory pathway is dominated by a rhythmic pattern of spontaneous discharge. Second- and third-order neurons discharge spontaneously in synchronous bursts at periodic intervals of 1–5 s. Bursting rate varies systematically as a function of tonotopic location, being faster for progressively higher frequency regions, and increases progressively during development. On E19, just before the embryo enters the airspace and is first exposed to normal intensities of air borne sound, rhythmic bursting gives way to an adultlike steady level of discharge. The rhythmic bursting in the chick embryo appears to be generated in the cochlea; because the auditory nerve also fires rhythmically, brainstem rhythms are abolished after cochlear removal or injection of tetrodotoxin into the perilymph, and auditory ganglion neurons in adult chickens, which normally don't exhibit a rhythmic pattern in their spontaneous firing, discharge spontaneously in a rhythmic manner after the regeneration of hair cells (Lippe 1994; Salvi et al. 1994).

Little is known about the importance of spontaneous activity in either the mature or developing auditory system. The relatively high level of maintained firing in the brainstem auditory pathway of young chickens and late prehatching embryos contributes to the maintenance of neuronal size and the regulation of calcium homeostasis (Tucci and Rubel 1985; Zirpel, Lachica, and Lippe 1995). What influence the intermittent and much lower levels of spontaneous activity that occur in the mammalian auditory

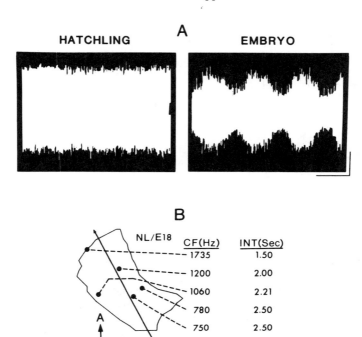

FIGURE 5.10. Rhythmic spontaneous activity in the chick embryo. A: spontaneous multiunit activity in nucleus magnocellularis of a 2-week-old chicken and an 18-day-old embryo. Calibration: 2 sec, 150 μV (Modified from Lippe 1994, Society for Neuroscience.) B: the characteristic frequency (CF) and mean interval between bursts of spontaneous activity (INT) at different recording locations in nucleus laminaris of an 18-day-old embryo are plotted on a horizontal projection of the nucleus. The long arrow overlying the nucleus indicates the orientation of the tonotopic axis in the direction of increasing frequency. Note that the interval between bursts of spontaneous activity decreases (rate of rhythmic bursting increases) progressively for higher frequency locations along the tonotopic axis. A, P, M, and L, anterior, posterior, medial, and lateral, respectively. (Modified from Lippe 1995, Elsevier Science Publishers.)

pathway before hearing onset might have on neural development is not known. In this context, three sets of observations are relevant. First, low-frequency afferent stimulation, or brief tetanic stimulation that resembles the patterned spontaneous bursting exhibited by some immature neurons in vivo, causes prolonged depolarizations in auditory neurons of the gerbil before the onset of hearing (Kotak and Sanes 1995) (Fig. 5.11). This suggests that even the very low level of spontaneous activity that occurs in the auditory pathway before hearing onset may be sufficient to cause long-lasting changes in postsynaptic neurons. Second, manipulations that

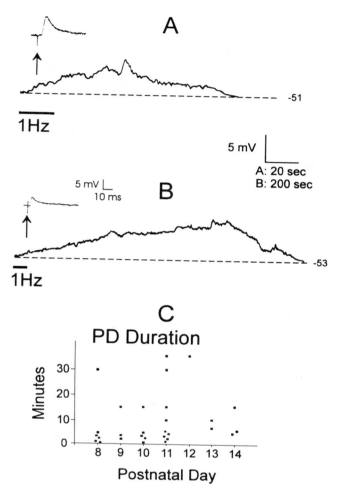

FIGURE 5.11. The influence of low-frequency synaptic transmission on immature lateral superior olive (LSO) neurons in the gerbil. A: whole-cell recording from an 8-day LSO neuron displaying a long-lasting depolarization evoked by a 1-Hz stimulation of its excitatory afferents for 20 s (horizontal bar). The neuron was depolarized for 2.6 minutes. B: whole-cell recording from an 11-day LSO neuron displaying a long-lasting depolarization evoked by a 1-Hz afferent stimulation for approximately 1 minute (horizontal bar). The inset above each trace shows one of the afferent-evoked excitatory postsynaptic potentials. C: the duration of afferent-evoked depolarizations in individual LSO neurons is shown as a function of age. PD, prolonged depolarization. (Modified from Kotak and Sanes 1995, The American Physiological Society.)

perturb afferents at an early age can affect subsequent development processes. Particularly relevant is the finding that early manipulations may cause changes in neural structure and synaptic function even before the onset of hearing. Cochlear removal at P7 causes a reduction in neuronal size in the gerbil cochlea after 2 days of survival, 3 days before hearing onset (Hashisaki and Rubel 1989). Significant neuronal loss occurs when the ablation is performed on P3 (Tierney, Russell, and Moore 1997). Early cochlear removal also alters the normal development of axonal projections in both the gerbil and ferret before hearing onset (Kitzes et al. 1995; Russell and Moore 1995). In addition, treatment with the glycine antagonist strychnine beginning at P3–4 and cochlear removal at P7 affect excitatory and inhibitory transmission in the lateral superior olive when examined at P8–14 (Kotak and Sanes 1996). The shrinkage and loss of neurons that occur when cochlear removal is performed after the onset of hearing in young mammals and birds are activity-dependent processes and result from the elimination of presynaptic activity (Born and Rubel 1985, 1988; Pasic, Moore, and Rubel 1994). Whether the transneuronal effects that have been observed before hearing onset reflect a similar activity-dependent process or are the consequence of an activity-independent trophic mechanism needs to be examined.

One final issue concerns the spatiotemporal pattern of spontaneous activity. As described above, the spontaneous discharge of immature auditory neurons sometimes occurs in an intermittent or synchronously rhythmic pattern. Rhythmic-like spontaneous discharges are not unique to the immature auditory system and occur in other developing sensory pathways as well (Verley and Axelrad 1977; Fitzgerald 1987; Maffei and Galli-Resta 1990; Meister et al. 1991; Wong 1993). How prevalent rhythmic or patterned spontaneous firing might be in the early developing auditory system is not known and remains an important question for future study. The functional significance of the pattern of early spontaneous firing will most probably be better understood once the spatial and temporal integrative properties of immature postsynaptic neurons have been characterized (Sanes 1993; see Sanes and Walsh, Chapter 6). Intermittent synchronous bursts of presynaptic input might be optimal for activating immature auditory neurons and ensuring that activity necessary for growth and development is transmitted along the auditory pathway, particularly during fetal and early neonatal life when the overall level of neural activity in the auditory pathway is very low. The temporal pattern of spontaneous firing could also provide cues for the formation of specific synaptic connections. Retinal ganglion cells in several mammalian species exhibit a bursting pattern of spontaneous synchronous discharge during periods of fetal and neonatal life when photoreceptors are incapable of transmitting visual information. Several lines of evidence implicate this early rhythmic spontaneous activity in the segregation and refinement of axonal projections (Shatz and Stryker 1988; Shatz 1990; Wong 1993; Herrmann and Shatz 1995). It has been speculated that rhythmic spontaneous firing could play a similar instructive role in the developing auditory pathway (Lippe 1995).

8. Development of Cochlear Tonotopy (Table 5.7)

Experimental studies on the development of tonotopy have, in large part, been stimulated by the hypothesis that the spatial representation of sound frequency along the cochlea changes during ontogeny. The idea of a shifting place code was initially proposed to explain what is generally considered to be a paradoxical relationship between the development of responses to tonal stimuli and the anatomical differentiation of the cochlea. With only a few exceptions (Cotanche and Sulik 1984; Lenoir, Puel, and Pujol 1987; Roth and Bruns 1992), the morphological development of the cochlea generally occurs along a basal-to-apical gradient. This appears especially true for the later stages of differentiation. For example, the differentiation and innervation of hair cells, the establishment of intracochlear fluid spaces, the formation of the tectorial membrane and its relationship to the organ of Corti, the thinning of the tympanic cover layer, and actin expression in hair cell stereocilia occur initially in basal or midbasal locations and then spread mainly toward the apex (Pujol and Marty 1970; Rubel 1978; Kraus and Aulbach-Kraus 1981; Romand 1983; Lippe, Ryals, and Rubel 1986; Pirvola, Lehtonen, and Ylikoski 1991; Cole and Robertson 1992). Consequently, at the onset of hearing, the basal portion of the cochlea appears anatomically more mature than the apex. Paradoxically, behavioral and electrophysiological responses to sound are first evoked by low- or mid- to low-frequency sounds, and responses to high frequencies appear later (Crowley and Hepp-Reymond 1966; Rubel 1978; Lippe, Ryals, and Rubel 1986). A similar discrepancy between the spatiotemporal gradient of structural development and responsiveness to sound occurs in the chicken (Saunders, Coles, and Gates 1973; Rebillard and Rubel 1981; Fermin and Cohen 1984). To explain this apparent discrepancy, Rubel (1978) proposed that at the time of hearing onset cochlear transduction mechanisms are only functioning in the basal portion of the cochlea, but they respond to low- or mid- rather than high-frequency sounds. During the course of development, the transduction of low-frequency sounds occurs progressively more apically as the base becomes responsive to higher frequencies (Rubel, Smith, and Miller 1976; Rubel 1978; Rubel, Lippe, and Ryals 1984) (Fig. 5.12).

The body of experimental evidence bearing on the hypothesis of a place code shift has been examined in several previous reviews, each of which provides a comprehensive coverage of the literature up to their respective dates of publication (Lippe, Ryals, and Rubel 1986; Rübsamen 1992; Walsh and Romand 1992; Manley 1996). Sections 8.1 to 8.2.2 describe the main studies that have shaped our thinking about this problem and some of the unresolved issues that still remain. Particular emphasis will be placed on observations made since 1991 and which are not considered in detail in previous reviews.

TABLE 5.7. Development of cochlear tonotopy.

Species (Reference)	Age	Treatment/Measure	Results/Comments
Chicken (1, 2, 3, 4, 5)	E20, P10, P30	Acoustic trauma. Hair cell loss (cell counts)	Position of maximum hair cells loss to 500-, 1,500-, and 3,000-Hz-tone exposure occurs more apically in older birds. Lesion site shifts the equivalent of a 1.5-octave distance along the papilla from E20 to P30; 60% of shift occurs between E20 and P10. Broadband white noise exposure on E20 causes hair cell loss confined to the basal half of the papilla; exposure at P10 and P30 produces loss along entire length of the papilla. In hatchlings, position of cell loss varies as a function of sound frequency for short but not for tall hair cells.
Chicken (6)	P3, P15	Nucleus magnocellularis unit responses	Tonotopic organization is stable. Surface mapping study. Recording sites were not verified histologically. Normalization procedure would make any developmental shift in tonotopic organization difficult to detect.
Chicken (5, 7, 8, 9)	E17, E19, P0, adult	Unit cluster responses in nucleus magnocellularis and nucleus laminaris	CF at any given location in mid- and high-frequency regions increases 1.0 octave between E17 and 2–3 weeks posthatch; 70% of shift occurs between E17 and P0. Insufficient data to determine whether tonotopy shifts in low-frequency region.
Chicken (10)	P1, P10, P30	Acoustic trauma. Hair cell damage (SEM)	Position of damage to 525-, 1,500-, and 3,000-Hz pure-tone exposure does not shift developmentally. Position of damage varies as a function of both the frequency and intensity of sound stimulation. At lower intensities, damage is confined to a strip region along the neural edge of papilla; at higher intensities, a second lesion site occurs inferior and distal to the strip damage.
Chicken (11)	E18, E20, P1, P10, P20, P30	Acoustic trauma. Hair cell damage (SEM), VIII nerve CAP	Lesion site to 1.5-kHz-tone exposure occurs more apically in birds exposed at E20 than at P1; position of the lesion site is stable for exposures between P1 and P30. The frequency of maximum threshold elevation after exposure to a 1.5-kHz tone decreases progressively from 3.9 kHz (exposure on E18) to 2.7 kHz (exposure on P1). Shifting ends at P1 when the frequency of maximum threshold shift stabilizes at ~2.5 kHz.

Chicken (12–14)	E19, P2, P21, adult (3–4 weeks)	Spiral ganglion unit responses. Intracellular labeling of electrophysiologically characterized auditory nerve fibers	E19 to P21: frequency-place map stable for CFs < 1 kHz; relatively few data points for CFs > 1 kHz; highest CF at E19 = 1,623 Hz; middle ear fluids removed at E19. Adult: CF at given cochlear location is higher in adults than in younger birds.
Chicken (15)	P0, P7, P27	Tone-induced 2-DG utilization. Field L	Location of 2-DG frequency contours to pure-tone stimulation shifts developmentally; change equivalent to an ~0.5 octave shift at low (0.2 to 0.3-kHz) and high (1.3 to 5.57-kHz) frequencies and less at intermediate (0.4 to 1.0-kHz) frequencies. Differences in body temperature do not account for developmental shifts.
Gerbil (16–18)	14–30 DAB	CM. summating potential	Best frequency at midbasal cochlear turn (20 kHz) increases 1.5–2.0 octaves between P14 and P30; best frequency at second turn location (2.5 kHz) doesn't change. Potential influence of developmental changes in middle ear transmission is eliminated by differential recording technique.
Gerbil (19)	14–52 DAB	Spiral ganglion unit responses	Best frequency of spinal ganglion neurons innervating midbasal cochlear location (20 kHz) increases 1.5 octaves from 14 to 17 DAB, 1.0 octave from 14 to 15 DAB, and an additional 0.5 octave by 17 DAB when mature CF is achieved. Potential influence of developmental changes in middle ear transmission is eliminated by differential recording technique. Developmental increase in CF is correlated with acquisition of sharply tuned tip on tuning curve and 80-dB increase in sensitivity at CF.
Gerbil (20)	12–18 and 60–120 DAB	Spiral ganglion unit responses. Intracellular labeling of electrophysiologically characterized auditory nerve fibers	Frequency-place map at 6 to 20-kHz sites increases ~1 octave between 18 and 120 DAB; no shift occurs at 2- and 3-kHz sites. Conclusion is based on 7 successfully labeled fibers at E18. Sharpness of tuning at 14–15 DAB was similar to that in adults for the frequency range of ~0.4–30 kHz.
Gerbil (21)	14–192 DAB	Lateral superior olivary unit responses	Best frequency of neurons in mid- and low-frequency locations increases ~1.0–1.5 octaves between 13 DAB and 12–16 weeks postnatal; midfrequency region reaches adult tonotopy earlier (16 DAB) than high-frequency locations; no tonotopic shift occurs in low-frequency locations (<2.5 kHz).

(Continued)

TABLE 5.7. (Continued)

Species (Reference)	Age	Treatment/Measure	Results/Comments
Geril (22)	14–90 DAB	Tone-induced 2-DG utilization. Dorsal cochlear nucleus	At 14 DAB, 12-kHz stimulation produces no elevation in 2-DG uptake; 0.75- and 3.0-kHz stimulation cause increased uptake in locations that in adults respond to higher frequencies; tonotopic shift is equivalent to ~2.5 octaves; adultlike representation at 3.0 kHz is achieved by 18–20 DAB.
Gerbil (23)	12–102 DAB	Tone-evoked DPOAEs	Percentage of animals responding to all frequencies (1.3–13.0 kHz) increases from 0% at 12 DAB to 100% at P18–19. Responses to high frequencies (13.0 kHz) first appear at 13–14 DAB; low-frequency responses (1.3 kHz) appear at 18–19 DAB. High-frequency responses mature faster than low-frequency responses. Input-output functions at any given frequency change from monotonic to nonmonotonic during development.
Gerbil (24, 25)	15–60 DAB	Tone-evoked DPOAEs. Furosemide injection	At 14 DAB, active cochlear emissions for midfrequencies (4–8 kHz) are at adult levels but no responses to 1 or >24 kHz. Active emissions for 10–48 kHz reach adult levels by 29 DAB. Passive emission thresholds improve developmentally for all frequencies but mainly at high frequencies. Cochlear amplifier gain at mid- but not at low or high frequencies is adultlike when the endocochlear potential is at an still immature level. Data support hypothesis that the place code shift results from a developmental change in the passive base cutoff frequency.
Guinea pig (26)	0 DAB to adult	CM	CF at basal turn location does not change developmentally.
Rat (27)	13–22 and 36–37 DAB	Cochlear nucleus single- and multiunit responses. Extracellular labeling of spiral ganglion neurons innervating recording site	Frequency-place map similar at 36–37 DAB in adults at frequencies examined (>12.5 kHz). Composite place-frequency map in 13–22 DAB animals exhibits an ~0.5-octave decrease in frequency for frequency locations examined (4.25–32 kHz); apical 25% of the cochlea (<4 kHz) was not examined. Few data points in younger animals.

Rat (28)	10 DAB to adult	Tone-induced c-Fos immunoreactivity. Brain stem auditory nuclei	Sound-induced C-Fos immunoreactivity is first seen at P14 to both low- and high-frequency sounds; tonotopy of frequency-specific C-Fos bands in the cochlear nucleus, superior olive, medial nucleus of the trapezoid body, and inferior colliculus does not change developmentally.
Rat (29)	15, 18, and 21 DAB	Classical conditioning	After conditioning on 15 DAB to an 8-kHz tone, the peak of the generalization gradient on 18 DAB shifts upward ~0.5 octave to 12 kHz; upward shift is frequency (does not occur after conditioning to a 4-kHz tone) and age dependent (does not take place when conditioning occurs on 18 DAB and generalization is tested on 21 DAB).
Mouse (30)	9–20 DAB	IC single- and multiunit responses	Frequency representation in most dorsal IC (<5 kHz) is stable; representation in more ventral locations (mid- and high frequencies) increase by 1.0–3.4 octaves from 10 to 20 DAB. Sites tuned to <16 kHz acquire adult tuning by 16 DAB; higher frequency sites continue to increase tuning until at least 20 DAB. Size of IC is constant during period of tonotopic change.
Cat (31)	5–40 DAB, adult	Auditory nerve fiber responses	Developmental changes in tuning-curve shape support hypothesis that ontogenetic changes in tonotopic organization reflect both basal-apical and medial-lateral gradients of cochlear differentiation.
Cat (32)	2–28 DAB	IC unit responses	Unit tuning broad and irregular before 6–7 DAB. Orderly dorsoventral tonotopic organization in kittens 11 DAB and older. Frequency representation at low-frequency sites does not change; representation at high-frequency sites appears to increase developmentally. Interpretation of apparent shift in tonotopy complicated by collicular growth.
Cat (33)	4–45 DAB	AVCN unit responses	Data suggest that the basilar membrane in kittens 6 DAB and older supports a traveling wave along its entire length; in kittens younger than 6 DAB, the length of the apical segment of the traveling wave is increased.

(Continued)

TABLE 5.7. (*Continued*)

Species (Reference)	Age	Treatment/Measure	Results/Comments
Cat (34)	5–35 DAB	Tone-induced 2-DG utilization. IC	Location of 2-DG frequency contours to mid- and high-frequency stimulation (2.0–30 kHz) shifts between 10 and 35 DAB; location where 0.5 kHz is represented does not change. Adultlike frequency representation achieved first for lower frequencies.
Cat (35)	2–90 DAB	Tone-induced 2-DG utilization. IC	Size of IC increases. Responsiveness to 2 kHz starts earlier (6 DAB) than to 15 kHz (11 DAB). Two-kilohertz representation in the central IC shifts ~ 1.5 octaves between 6 and 90 DAB; 2-kHz representation in the caudal IC does not shift. Fifteen-kilohertz representation shifts <1 octave between 11 and 90 DAB.
Horseshoe and hipposiderid bats (36, 37)	1–210 DAB	IC single- and multiunit responses	Size of IC does not change postnatally. Frequency place map in low-frequency region is adultlike at or soon after birth. Tonotopic map in high-frequency region increases one-third to one-fourth octave during development. The shift in tonotopy coincides temporally with an increase in sensitivity and tuning sharpness in foveal units. Units in low-frequency sites acquire adult sharp tuning before units in high-frequency regions. In horseshoe bats, the range of frequencies represented in the foveal region matches the frequency composition of echolocation pulses at all ages; no corresponding match is observed in the hipposiderid bat.
Horseshoe bat (38)	8–23 DAB	PVCN unit responses. Extracellular labeling of spiral ganglion neurons innervating recording site	Seventy-eight-kilohertz site in spiral ganglion is tuned to lower frequencies (30–73 kHz) in younger animals.

Phyllostomid bat (39)	1–50 DAB	IC single- and multiunit responses	Size of IC does not change postnatally. Upper frequency of acoustic responsiveness increases from 76 to 110 kHz between 1 and 50 DAB; reflects progressive conversion of nonresponsive regions to acoustically activated areas along the low- to high-frequency axis. Units in low-frequency sites are acoustically responsive and acquire adult tuning sharpness before units in high-frequency regions. Developmental change in the frequency-place map is not uniform throughout the IC: no change in the lowest frequency region (<20 kHz), small decrease in midfrequency sites, and 0.2-octave increase in high-frequency sites (>100 kHz).
Duck (40)	E19 to adult	Cochlear nuclei (nucleus angularis and nucleus magnocellularis) unit responses	Orderly tonotopic organization at all ages. Acoustic responsiveness of anterior n. magnocellularis appears suddenly on E22–23. Nonquantitative mapping procedure.
Mammals, birds (41, 42, 43, 44)			Reviews.

SEM, scanning electron microscopy; 2-DG, 2-deoxyglucose; IC, inferior colliculus. 1, Rubel and Ryals 1983; 2, Ryals and Rubel 1985b; 4, Rubel, Lippe, and Ryals 1984; 5, Rebillard, Ryals, and Rubel 1982; 6, Cohen and Saunders 1994; 7, Lippe and Rubel 1985; 9, Lippe 1987; 10, Cotanche, Saunders, and Tilney 1987; 11, Cousillas and Rebillard 1985; 12, Manley, Brix, and Kaiser 1987; 13, Jones and Jones 1995b; 14, Chen, Salvi, and Shero 1994; 15, Heil and Scheich 1992; 16, Harris and Dallos 1984; 17, Yancey and Dallos 1985; 18, Arjmand, Harris, and Dallos 1988; 19, Echteler, Arjmand, and Dallos 1989; 20, Müller 1996; 21, Sanes, Merickel, and Rubel 1989; 22, Ryan and Woolf 1988; 23, Norton, Bargones, and Rubel 1991; 24, Mills, Norton, and Rubel 1994; 25, Mills and Rubel 1996; 26, Bobbin et al. 1991; 27, Müller 1991; 28, Friauf 1992; 29, Hyson and Rudy 1987; 30, Romand and Ehert 1990; 31, Romand 1987; 32, Aitkin and Moore 1975; 33, Brugge, Javel, and Kitzes 1978; 34, Webster and Martin 1991; 35, Ehret and Romand 1994; 36, Rübsamen and Schäfer 1990a; 37, Rübsamen, Neuweiler, and Marimuthu 1989; 38, Vater and Rübsamen 1989; 39, Sterbing, Schmidt, and Rübsamen 1994; 40, Konishi 1973; 41, Lippe, Ryals, and Rubel 1986; 42, Walsh and Romand 1992; 43, Rübsamen 1992; 44, Manley 1996.

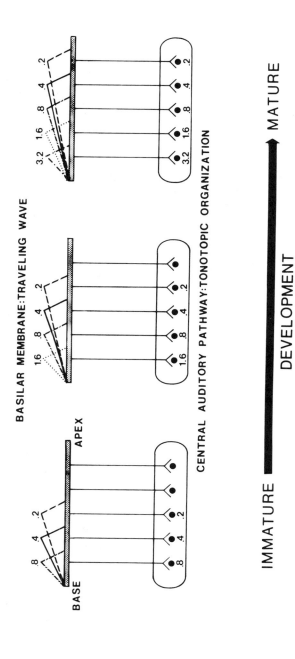

FIGURE 5.12. Model of the developmental shift in the place code originally proposed by Rubel and colleagues. (Modified from Lippe and Rubel 1985, Wiley-Liss, Inc., a subsidiary of Wiley and Sons, Inc.)

8.1 Birds

The hypothesis of a place code shift was first tested in chickens (Figs. 5.13 and 5.14). Rubel and Ryals (1983) reported that the location of maximum hair cell loss to high-intensity pure-tone exposure occurred more apically in older birds. Between E20 and 30 days after hatching (P30), the lesion sites to 500, 1,500 and 3,000 Hz tones shifted apically the equivalent of ~1.5 octaves along the basilar papilla (avian cochlea), with 60% of the shift taking place between late embryogenesis (E20) and 10 days after hatching (Rubel and Ryals 1983; Ryals and Rubel 1985b). Cousillas and Rebillard (1985) also observed a shift in late prehatching embryos comparable in

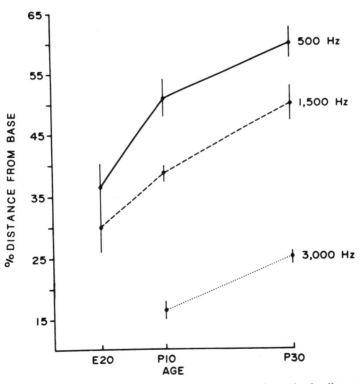

FIGURE 5.13. Development of the frequency-place map along the basilar papilla in the chicken. Maps were obtained by measuring the location along the papilla of maximum hair cell loss after high-intensity pure-tone exposure at embryonic day 20 (E20) P10, and P30. The location of maximal damage to a given frequency shifted progressively more apically in older animals. Values are means ± SE. (From Rubel and Ryals 1983, American Association for the Advancement of Science.)

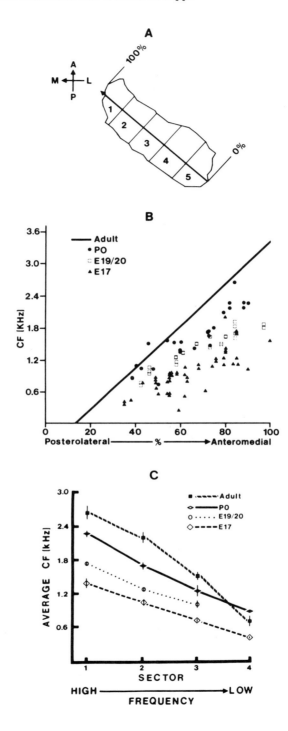

magnitude to that observed by Ryals and Rubel (1985b) and reported that the shift of damage to lower frequency locations on the papilla was associated with a decrease in the frequency of maximum threshold elevation, as would be predicted from the hypothesis of a changing place code. However, their results, as well as the acoustic trauma study of Cotanche, Saunders, and Tilney (1987), suggested that the frequency-place map remains stable after P2. Complementary unit mapping studies showed a change in central tonotopy during embryonic and early postnatal development. The best frequency of neurons at a given location within the mid- and high-frequency regions of the nucleus magnocellularis and nudeus laminaris, second- and third-order auditory nuclei, increased ~1 octave between E17 and 1–2 weeks posthatch (Lippe and Rubel 1983, 1985; Lippe 1987) (Fig. 5.14). Approximately 70% of the shift took place between E17 and P1, and the remainder occurred sometime within the first 2 weeks after hatching. The commonalities and discrepancies in these different lines of evidence suggested that primarily during embryonic development and, to a lesser degree, during at least the first several days after hatching the frequency-place map along the mid- and high-frequency regions of the basilar papilla shifts toward higher values. It was not clear from these studies whether frequency representation in the low frequency apical portion of the papilla also changes and for how long after hatching frequency representation continues to shift.

The interpretation of peripheral tonotopic maps obtained by acoustic trauma and the comparison of cochlear maps across ages is potentially complicated by several factors. These include the ability of the basilar papilla to replace lost hair cells by regeneration (Corwin and Cotanche 1988; Ryals and Rubel 1988; Lippe, Westbrook, and Ryals 1991), the difficulty of unambiguously defining the "center of damage" in the broad lesions frequently caused by acoustic trauma (Cotanche, Saunders, and Tilney 1987), and differences between short and tall hair cells in terms of the frequency specificity of damage to pure-tone overstimulation. The frequency-

←──

FIGURE 5.14. Development of the frequency-place map in nucleus magnocellularis of the chicken obtained by single- and multiunit recording. A: outline drawing of nucleus magnocellularis as seen from the dorsal surface of the brain. The orientation of the tonotopic axis is indicated by the arrow, which points in the direction of increasing frequency. The nucleus is divided into 5 sectors along the tonotopic axis. B: scatterplots showing relationship between characteristic frequency (CF) and percent distance along the tonotopic axis at E17, E19/20, and P0. The solid line shows the best fitting linear regression function for data from adult chickens (2–4 weeks posthatch). C: the mean CF observed in the 5 sectors along the tonotopic axis is compared across ages. The most posterior-lateral sector (5) did not respond to the range of frequencies used in these studies. Values are means ± SE. (From Lippe 1987, Elsevier Science Publishers.)

place maps in acoustic trauma studies are based on lesions of short hair cells, which receive relatively little afferent innervation. In contrast, the location of damage among tall hair cells, which receive almost all of the auditory nerve fibers and provide frequency-specific input to the brain (Manley et al. 1989; Fisher 1992), does not vary with sound frequency (Rebillard, Ryals, and Rubel 1982; Ryals and Rubel 1985a). Because the functional relationship between tall and short hair cells is not known, the implication of these differing patterns of damage for the frequency-specific response of tall hair cells at threshold levels of sound is not known.

Two additional factors that could influence peripheral tonotopic maps, whether obtained by acoustic trauma or unit recording, include developmental changes in the middle ear transfer function and the physical size of the papilla. Because the basilar papilla continues to increase in length until at least 5 weeks after hatching (Ryals, Creech, and Rubel 1984; Tilney et al. 1986), developmental comparisons are routinely done between frequency maps normalized for cochlear length, with the assumptions that growth is comparable in all regions and that hair cells maintain their same percentile positions along the papilla during development. Measurements of hair cell number and density in hatchlings of increasing developmental age are consistent with these assumptions, although the assumption of positional stability has yet to be directly tested (Tilney et al. 1986; Katayama and Corwin 1989).

Maturational changes in the middle ear, particularly the drainage of middle ear fluids during late embryonic development, has often been cited as a variable that could influence tonotopic maps by affecting the intensity and spectrum of sounds reaching the inner ear. The structural and functional development of the middle ear in chickens and other vertebrate species have been reviewed previously (Relkin, Saunders, and Konkle 1979; Saunders, Kaltenbach, and Relkin 1983; Saunders et al. 1986; Woolf and Ryan 1988; Cohen, Bacon, and Saunders 1992; Cohen, Hernandez, and Saunders 1992; Cohen, Rubin, and Saunders 1992; Cohen et al. 1993; Saunders, Doan, and Cohen 1993; Zimmer, Rosin, and Saunders 1994). To date, middle ear function in avian embryos has not been directly measured. Consequently, our understanding of the influence of middle ear conductance on the development of frequency organization during embryogenesis, and on the maturation of neural responses in general, is largely inferential. Information comes from two sources: comparisons between developmental changes in middle ear structure and cochlear and neural responses as well as experimental studies in which middle ear fluids have been drained (e.g., Saunders, Coles, and Gates 1973; Rebillard and Rubel 1981; Jones and Jones 1995a). Clearance of middle ear fluids during embryonic development contributes to an overall improvement in auditory thresholds (Saunders, Coles, and Gates 1973). However, the question of whether middle ear fluid or other middle ear immaturities influence tonotopic maps by, for example, preferentially attenuating the transmission of high-frequency sounds is controversial. The CF threshold in late prehatching embryos is not systematically elevated for progressively higher frequency units, there is no evidence of an increase in the high-frequency slope of high CF neurons, and the upper limit of high-frequency responsiveness is comparable in embryos with fluid-filled middle ears and embryos in which middle ear fluids have been drained (Lippe 1987; Jones and Jones 1995a,b). These data suggest that the embryonic middle ear does not act as a low-pass filter. On the other hand, ontogenetic changes in auditory sensitivity suggesting that the attenuation of sound is greater for high frequencies (Fig. 1 in Rebillard and Rubel 1981) or low frequencies (Fig. 3 in Saunders, Coles, and Gates 1973) or comparable across frequencies (Fig. 2 in Saunders, Coles, and Gates 1973) can also be cited. Thus, in the absence of direct measurements of middle ear admittance, the affect of middle ear transmission on frequency representation in chick embryos remains unclear.

In neonatal and adult chickens, in which middle ear function has been directly measured, the transmission of vibrational energy from the tympanic membrane to the fluids of the inner ear continues to improve beyond 70 days after hatching (Saunders et al. 1986; Cohen, Rubin, and Saunders 1992). Because the developmental increase in middle ear transmission is comparable for frequencies up to 5 kHz, it is highly unlikely that the tonotopic shifts observed in posthatch chickens are attributable to frequency-dependent improvements in middle ear conductance. However, the possibility that maturational changes in sound transmission, although not frequency dependent, might nevertheless contribute to an apparent shift in tonotopy also needs to be considered. Using a damage paradigm similar to that employed earlier by Ryals and Rubel (1985b), Cotanche, Saunders, and Tilney (1987) reported that they failed to find any evidence for a place code shift in posthatch chickens. In addition, they reported that the location of hair cell damage is a function of the intensity and not only of the frequency of sound exposure. At progressively increasing sound intensities, the region of damage enlarged to include a somewhat more distal site of hair cell loss in addition to the proximally located narrow "strip"-shaped lesion produced at lower exposure levels. This observation raises the possibility that increases in the effective intensities of sound reaching the cochlea, due to improvements in middle ear transmission during development, might have contributed to the positional shifts in hair cell damage reported in earlier sound trauma studies. Although the extent to which this occurred is not known, it seems unlikely that developmental changes in the location of sound-induced damage are attributable solely or primarily to increases in sensitivity. The "strip" region of damage appears to consist primarily of tall hair cells, whereas the positional shift reported earlier by Ryals and Rubel (1985b) was restricted to short hair cells. In addition, if a correction is made for developmental changes in sensitivity, the pattern of hair cell damage shown in the summary figures published by Cotanche, Saunders, and Tilney (1987) appears, at least in part, to be consistent with the hypothesis of a changing place code: hair cell damage extended more proximally in younger birds. However, the lack of quantitative data by which to assess the variability of the damage zone makes direct comparisons with the data from Ryals and Rubel (1985b) impossible.

More accurate frequency-place maps can be obtained by labeling cochlear ganglion neurons and correlating the location of the labeled fiber along the cochlea with the unit's CF. Using this procedure, Manley, Brix, and Kaiser (1987) compared frequency-place maps in P2 and P21 chickens, and Jones and Jones (1995b) measured the tonotopic organization at E19 and compared this to the frequency map for P2 chicks reported by Manley, Brix, and Kaiser (1987). These studies reported that no significant difference in the cochlear map of tall hair cells is seen among the three groups of birds, at least up to the highest CFs for which fibers were labeled at the respective ages (E19, 1,692 Hz; P2, 2,380 Hz; P21, 2,000 Hz) (Fig. 5.15A). On the basis of these observations, Jones and Jones (1995a,b) suggested that improvements in middle ear admittance alone or in combination with functional maturation of an immature base, rather than a shifting tonotopic map, account for the low- to high-frequency gradient of functional maturation in the chicken.

Several points are significant in considering the results and interpretation of these fiber-labeling studies. First, the data mainly cover low frequencies. Thus, although the findings generally support the conclusion that the cochlear frequency map is stable for frequencies lower than ~1,100 Hz after E19, it is less clear if frequency organization at higher frequencies is also invariant. Contributing to this uncertainty is the fact that the cochlear map measured by Jones and Jones (1995b) at E19 is highly variable and that many of the data points appear to be shifted toward lower frequencies. Additional recordings at higher frequency locations would help resolve this question. Second, hearing onset in the chicken occurs on E11–12 (Saunders, Coles, and Gates 1973). By E19, the youngest age at which peripheral frequency organization has been mapped by fiber labeling, auditory function is adultlike in many respects (Jones and Jones 1995a). It would be of interest to know whether the cochlear map shifts during the earlier embryonic period when the major changes in auditory function are occurring and when shifts in central tonotopy have been reported (Lippe and Rubel 1985; Lippe 1987). Third, the cochlear frequency map in 1-year-old chickens, which is also derived from the labeling of physiologically characterized spinal ganglion neurons, differs substantially from the maps in late prehatching embryos and younger posthatch chickens (Chen, Salvi, and Shero 1994). The CFs in the most basal 60% of the length along the papilla are significantly higher in adult chickens than in younger birds (Fig. 5.15B).

The reason for the discrepancy between the electrophysiological maps in adult chickens and younger chicks is not known. The difference could be due to age-related changes in the cochlear map. However, it seems highly

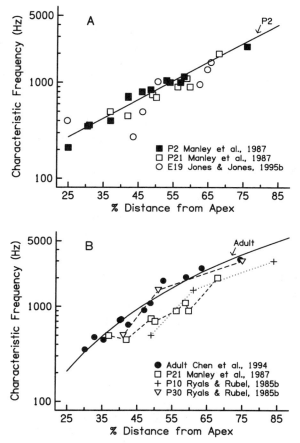

FIGURE 5.15. Development of the cochlear frequency-place map in the chicken. Maps were obtained by labeling single auditory ganglion neurons and correlating the location of the labeled afferent fiber along the basilar papilla with the neuron's CF. A: the CFs of single labeled neurons in 19-day-old-embryos (E19) and 2- and 21-day-old chicks (P2 and P21) are plotted as a function of percent distance from the apical tip of the basilar papilla. The solid line is the best fitting regression function for the P2 data (regression function obtained from Fig. 2 of Jones and Jones 1995b). B: the frequency-place map in adult chickens (posthatch days 105–140) is compared with the map of P21 chickens. The maps derived from noise trauma studies in P10 and P30 chickens are also shown. The solid line is the best fitting regression function for the adult data.

improbable that tonotopy would continue to shift up to 1 year of age, particularly in view of the evidence that frequency representation is stable between P2 and P21. Differences in cochlear temperature might also contribute. The body temperature of the young chicks in the study by

Manley, Brix, and Kaiser (1987) was maintained at 39°C, whereas the adult chickens in the study by Chen, Salvi, and Shero (1994) were maintained at 41.5°C. The CF of auditory nerve fibers in the pigeon (Schermuly and Klinke 1985) as well as in several other nonmammalian species (Moffat and Capranica 1976; Eatock and Manley 1981; Smolders and Klinke 1984; Stiebler and Narins 1990; van Dijk, Lewis, and Wit 1990) shifts to lower frequencies with decreasing head temperature. A comparable effect of temperature on frequency tuning has not been observed in the relatively few studies in mammals (Smolders and Klinke 1978; Gummer and Klinke 1983; Ohlemiller and Siegel 1994), with the exception of the mustached bat (Huffman and Henson 1991, 1993). The mechanism underlying the temperature dependence in nonmammalian vertebrates is not known but could involve the electrical resonant properties of the hair cells, the stiffness of the hair cell stereocilia, or contractile processes modulating basilar membrane stiffness (Fuchs, Nagai, and Evans 1988; Drenckhahn et al. 1991; Cotanche, Henson, and Henson 1992). If frequency tuning in developing chickens is also temperature sensitive, then differences in body temperature could have contributed to the discrepancy in cochlear maps between adult and young birds.

Temperature could conceivably also be an important variable in cochlear maps obtained by sound trauma in awake animals. Until P4, awake hatchling chickens are not capable of thermoregulating, whereas between P4 and P25, chickens maintain a constant body temperature but at levels lower than those in the adult. During this later period, body temperature increases gradually and reaches the adult value of 41.5°C by 1 month after hatching (Kaiser 1992). Because chickens within this age range are homeothermic at progressively higher temperatures, it is possible that temperature dependencies in tuning would be reflected as age-related shifts in the location of hair cell damage, even when environmental temperature is held constant across age and exposure conditions.

To date, the effect of temperature on the development of cochlear frequency maps has not been examined directly. The only observation bearing on this question is the report that putative changes in the chicken's cochlear map, as indicated by developmental shifts in the location of frequency-specific 2-deoxyglucose contours in the auditory forebrain, are not an artifact of temperature (Heil and Scheich 1992). The possible effect of temperature on frequency representation in the developing bird remains an important question for future experimental study.

The final point to consider is that the stability of the cochlear map in late prehatching embryos and young chicks is at variance with the shift of frequency representation in the avian cochlear nucleus. The onset of responses to sound in the chick occurs at E11–12 (Saunders, Coles, and Gates 1973). Lippe and Rubel (1985) and Lippe (1987) reported that the best frequencies of unit clusters in mid- and high-frequency regions of second- and third-order auditory nuclei increase approximately one octave between

E17 and 1–2 weeks after hatching (Fig. 5.14). A large portion of the frequency shift takes place before E19, a period when the development of the cochlear map has not been studied by the labeling of physiologically characterized afferent fibers. However, much of the shift also occurs between E19 and sometime within the first week or two after hatching when fiber labeling studies report no evidence for a change in peripheral tonotopy. To date, there is no conclusive explanation for the difference between these two sets of findings. A recent unit mapping study reported no change in tonotopy in the chicken cochlear nucleus between P3 and P15 (Cohen and Saunders 1994). However, this study was not designed to examine the development of tonotopy, and the method used to compare maps across ages would not allow developmental changes in frequency organization to be reliably detected. Presumably, temperature would not have been a factor in the original cochlear nucleus mapping studies because embryos were maintained at the same temperature. A possible contribution of middle ear immaturities cannot be excluded. However, any effect of middle ear transmission would also be reflected as a change in peripheral tonotopy. Alternatively, the discrepancy between the cochlear and cochlear nucleus mapping studies might reflect the more limited high-frequency sampling in the former studies.

8.1.1 Alternative Models

Currently, there are two main alternative hypotheses for the development of the cochlear frequency map in the chicken (Fig. 5.16). According to the first interpretation, frequency representation does not change developmentally. Responses are evoked initially by low frequencies, and the low-frequency responses occur at their correct (adult) apical location on the basilar papilla. The remainder of the papilla is unresponsive to sound at this early stage, possibly because of immaturities in the basilar papilla itself and/or the middle-ear transfer function. Responses to progressively higher frequencies appear gradually during development at their correct locations along the papilla. The alternative hypothesis proposes that frequency representation shifts developmentally but that the shift is restricted to regions along the papilla that, in the adult, code for medium- and high-frequency sounds. According to this interpretation, low frequencies are always represented at their correct apical location, as appears to be true in mammals (see Section 8.2). Responses to midrange frequency sounds occur progressively more apically as the base becomes responsive to high frequencies.

8.2 Mammals

In a mammal that is precocial with respect to hearing, the guinea pig, tonotopy in the basal cochlea does not change postnatally (Bobbin et al. 1991). In contrast, a large body of data supports the hypothesis that

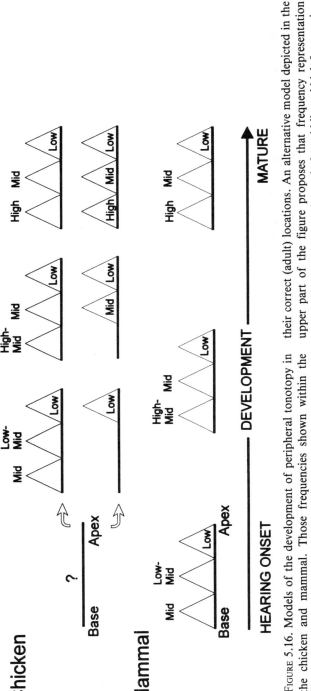

FIGURE 5.16. Models of the development of peripheral tonotopy in the chicken and mammal. Those frequencies shown within the triangular areas representing basilar membrane deflection do not undergo a place code change during development. Frequencies shown above the triangular areas shift their location apically in older animals. Chicken: according to the first model depicted in the middle portion of the figure, frequency representation does not change during development. This interpretation suggests that the limited range of high-frequency responsiveness that is observed during middle and late stages of embryogensis occurs because the middle and basal portions of the papilla are unresponsive to sound during this period of development. The absence of responses is attributed to immaturities in the basilar papilla and/or middle ear rather than to a shifting tonotopic map. When responses to middle and high frequencies first appear during development, they occur at their correct (adult) locations. An alternative model depicted in the upper part of the figure proposes that frequency representation shifts during development but only for middle and high frequencies. The representation of low frequencies at the apex is stable throughout development. Note that the development of tonotopic organization during the first 4–5 days after hearing onset has not yet been studied. Mammal: according to the revised model depicted in the lower portion of the figure, frequency representation in the cochlear apex is invariant from the onset of hearing. The middle and basal portions of the cochlea undergo a place code shift as first proposed by Lippe and Rubel (1983) and Rubel and Ryals (1983). One qualification of this revised model is that the development of frequency representation in approximately the most basal 15–20% portion of the cochlea has not yet been studied.

frequency representation along the cochlea of altricial mammals shifts during the early development of hearing. The first evidence for an age-related change of the place code in mammals came from developmental studies of the gerbil cochlea. Dallos and colleagues reported that the cutoff frequency of the CM and summating potential in the midbasal turn (15 kHz location) increases ~1.5–2.0 octaves between the onset of sound-evoked activity on P12 and the third postnatal week when frequency representation becomes adultlike. In contrast, the cutoff frequency of the CM at a second turn location (~2.5 kHz) remains stable during development (Harris and Dallos 1984; Yancey and Dallos 1985; Arjmand, Harris, and Dallos 1988) (Fig. 5.17). Because the summating potential and CM are believed to originate primarily from OHCs (Dallos and Cheatham 1976; Dallos 1981; Dallos, Santo-Sacchi, and Flock 1982; Dallos 1985), developmental changes in these potentials do not provide direct support for a place code shift in the output of the cochlea, which is mediated by the IHCs. Direct evidence was

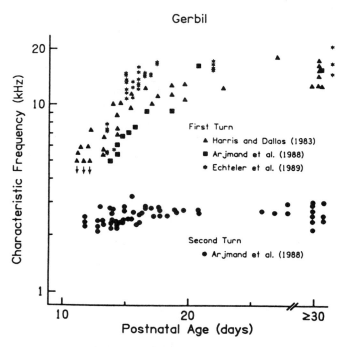

FIGURE 5.17. Developmental change in the cochlear frequency-place map. The CFs of CM potentials measured in the first and second turns of the gerbil cochlea are plotted as a function of postnatal age. Also shown are the CFs of spiral ganglion neurons that were recorded in the first turn of the cochlea. Note that the tonotopic map in the second turn is stable, whereas the frequency represented at a basal turn location increases 1.5–2.0 octaves between ~P12 and P17–20. (From Walsh and Romand 1992, Elsevier Science Publishers.)

provided by the finding that the CFs of spiral ganglion neurons at a constant basal cochlear location increase up to 1.5 octaves between the second and third postnatal weeks (Echteler, Arjmand, and Dallos 1989) (Fig. 5.17). Developmental increases in frequency representation ranging from 1.0 to 1.5 octaves have also been shown for high-frequency regions in the spiral ganglion of the rat and horseshoe bat (Vater and Rübsamen 1989; Müller 1991). In contrast, low-frequency neurons in the spiral ganglion of the gerbil exhibit little or no frequency shift (Müller 1996), which confirms prior findings that tonotopy in the cochlear apex is stable (Arjmand, Harris, and Dallos 1988).

Corroborating evidence for a peripheral place code shift also comes from ontogenetic studies of frequency representation in the central auditory nuclei (Rübsamen 1992) (Fig. 5.18). These studies have the advantage that electrophysiological mapping of central frequency representation does not involve direct intrusion into the cochlea, thus minimizing the possibility that

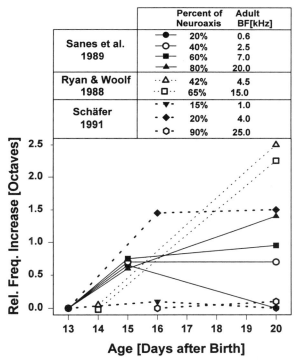

FIGURE 5.18. Development of central frequency maps. Developmental changes in the frequency represented at given locations along the neuroaxis in the dorsal cochlear nucleus (Ryan and Woolf 1988; 2-deoxyglucose technique), lateral superior olive (Sanes, Merickel, and Rubel 1989; unit mapping), and inferior colliculus (Schäfer 1991; unit mapping) of the neonatal gerbil. The increase in frequency is shown relative (Rel. Freq. Increase) to the respective best frequency (BF) values found in the youngest animals tested. (Modified from Rübsamen 1992, Springer-Verlag.)

putative changes in peripheral frequency representation might result from the disruption of highly vulnerable cochlear processes. The interpretation of such studies is complicated, however, by the possibility that developmental changes in central frequency maps could reflect changes in higher auditory centers as well as shifts in peripheral tonotopy. In this context, damage studies in adult mammals suggest that functional topography in the brain may be regulated, in part, by central synaptic mechanisms that select a subset of inputs from a larger number of inputs that are available anatomically (Kaas, Merzenich, and Killackey 1983; Wall et al. 1986; Robertson and Irvine 1989; Roe et al. 1990).

Frequency mapping with the 2-deoxyglucose technique has shown that frequency representation in the cochlear nucleus of the gerbil and the inferior colliculus of the kitten changes during development (Ryan and Woolf 1988; Webster and Martin 1991; Ehret and Romand 1994). Single- and multiunit recording studies report developmental shifts in tonotopy in the mid- and high-frequency regions of the lateral superior olivary nucleus in the gerbil and in the inferior colliculus of the cat, gerbil, house mouse, and several different species of bats (Aitkin and Moore 1975; Sanes, Merickel, and Rubel 1989; Romand and Ehret 1990; Rübsamen 1992) (Fig. 5.18). The magnitude of the reported shifts in central tonotopy varies considerably, ranging from a minimum of ~ 0.25 octaves in the midbrain auditory nucleus of the horseshoe bat to > 4.7 octaves in the murine inferior colliculus. The details of these different studies have been reviewed previously (Rübsamen 1992) and are summarized in Table 5.7.

One intriguing prediction of the hypothesis of a shifting place code is that an animal's perception of its acoustic environment should change during the development of hearing. In particular, a given frequency tone experienced early in development should be functionally equivalent to a higher frequency tone experienced at a later age. In what, to date, is still the only direct test of this prediction, 15-day-old rat pups were classically conditioned to an 8-kHz tone and subsequently tested for generalization to 4-, 8-, and 24-kHz tones (Hyson and Rudy 1987). When tested for generalization shortly after training, the maximum response occurred at the 8-kHz training stimulus. However, 3 days later, on P18, the peak of the generalization gradient had shifted upward 0.5 octave to 12 kHz. The ontogenetic shift in the peak of the generalization gradient appeared to be age specific because no shift occurred in rats trained on P18 and tested on P21. The 0.5-octave shift observed behaviorally is comparable in magnitude to the ontogenetic shift in the rat's cochlear map, although the change in peripheral tonotopy takes place at a somewhat later age (Müller 1991).

The potential behavioral significance of a shifting place code has also been examined in the echolocating horseshoe bat (Rübsamen 1987; Rübsamen and Schäfer 1990a,b). The adult horseshoe bat emits echolocating pulses with pure-tone components closely matched to a narrow frequency band that is greatly overrepresented in the bat's auditory system. During the development of hearing, the specialized "foveal" portion of the inferior colliculus becomes tuned to progressively higher frequency sounds. During this same period, the echolocating pulses change in a parallel fashion such that their frequency content always coincides with the foveal frequencies. The production of echolocating pulses appears to be under auditory feedback control because deafening young bats alters the structure and frequency composition of echolocating pulses. These observations have led to the suggestion that maturational shifts in foveal tuning are directly responsible for ontogenetic changes in the frequency content of the echolocation signal (Rübsamen 1992).

Two points are relevant in considering the collective evidence that supports the hypothesis of a place code shift. First, several variables that complicate the comparison of frequency maps across age in chickens are not factors in mammals. With the exception of the mustached bat (Huffman and Henson 1991, 1993), frequency tuning in adult mammals is not temperature sensitive, although the effect of temperature on frequency tuning in neonatal mammals has yet to be examined (Smolders and Klinke 1978; Gummer and Klinke 1983; Ohlemiller and Siegel 1994). In addition, developmental shifts in peripheral tonotopy are observed in the gerbil even when the potential influence of developmental changes in middle ear conductance has been eliminated by differential recording (Harris and Dallos 1984; Yancey and Dallos 1985; Arjmand, Harris, and Dallos 1988). Furthermore, both central and peripheral frequency maps continue to change in the gerbil beyond 16 days after birth, by which time the middle ear is adultlike in most respects and the cochlea has attained its mature length (Finck, Schneck, and Hartman 1972; Woolf and Ryan 1985, 1988; Harris, Rotche, and Freedom 1990).

Second, it is uniformly reported that tonotopic organization in the mid- and high-frequency regions of the cochlea and central auditory nuclei changes during development. In contrast, tonotopy in the cochlear apex and its central projection sites appears to be developmentally stable (e.g., Arjmand, Harris, and Dallos 1988; Sanes, Mernickel, and Rubel 1989; Rübsamen and Schäfer 1990a; Müller 1996) (Figs. 5.17 and 5.18). The main exceptions come from 2-deoxyglucose studies that reported developmental shifts of low-frequency representation in the cochlear nucleus of the gerbil and the inferior colliculus of the kitten (Ryan and Woolf 1988; Ehret and Romand 1994), although the low-frequency shift in kittens is controversial (Webster and Martin 1991).

8.2.1 Mechanism

Two explanations for the place code shift have been proposed (Lippe and Rubel 1985; Walsh and Romand 1992). The first attributes shifts in frequency representation to maturational changes in the passive mechanical properties of the cochlear partition. The ratio of partition stiffness to mass, a value that decreases from base to apex, determines the place-specific resonant frequencies along the basilar membrane and how far along the membrane the traveling wave will propagate. According to this proposal, a reduction in the mass loading and/or increase in the stiffness of the basilar membrane during development causes the resonance at a given location along the basilar membrane to shift to progressively higher frequencies in older animals. Consequently, the maximum deflection of the traveling wave to a given frequency input would shift apically during development.

The second explanation, initially proposed by Romand (1987), attributes the shift in frequency organization to maturational changes in OHC-mediated active processes. According to this model, early in development, IHCs are functional while the OHC system is immature and makes little or no active contribution to cochlear mechanics. The low-frequency responses at the onset of hearing reflect those of a passive basilar membrane. With the development of the OHC system, the active processes amplify basilar membrane motion, increase tuning sharpness, and cause an upward shift in the frequency of tuning for given cochlear locations. In this model, the stability of tonotopy at low frequencies might be accounted for by evidence suggesting that the active processes make relatively less contribution at the cochlear apex (Sewell 1984a; Cooper and Yates 1994).

Models that account for the place code shift by developmental changes in the passive mechanical properties of the basilar membrane or active nonlinear processes or a combination of both factors have been proposed (Romand 1987; Norton, Bargones, and Rubel 1991; Walsh and Romand 1992; He, Evans, and Dallos 1994; Mills and Rubel 1996). Evidence consistent with each model can be cited. The cochlear partition undergoes a number of structural changes during the early development of hearing, including the thinning of the tympanic cover layer, an increase in the thickness of basilar membrane fiber bands, and a change in the angle of the pillar cells (e.g., Kraus and Aulbach-Kraus 1981; Roth and Bruns 1992; Echteler 1995; Schweitzer et al. 1996). These changes in cellular architecture and geometry are consistent with a reduction in the mass loading and increase in stiffness of the basilar membrane that, in turn, should lead to an ontogenetic increase in place-specific resonant frequencies (Olson and Mountain 1991, 1994). To date, however, these predictions have not been confirmed by direct measurements of the stiffness and mechanical tuning of the basilar membrane in developing animals.

Evidence that OHC-mediated active processes contribute to the place code shift comes primarily from two sources. First, damage experiments in adult animals show that the tuning curve "tip" in high-frequency units depends on the presence and normal functioning of OHCs. Destruction of OHCs by acoustic trauma and ototoxic drugs, or otherwise interfering with OHC-mediated active mechanisms, can cause the CF of primary afferent fibers innervating the cochlear base and the mechanical tuning of the basilar membrane to shift downward by 0.5–1.0 octaves depending on the extent and exact site of damage. In addition, CF thresholds are elevated and sharpness of tuning is decreased (Dallos and Harris 1978; Robertson and Johnstone 1979b; Robertson et al. 1980; Cody and Johnstone 1981; Sellick, Patuzzi, and Johnstone 1982; Leonard and Khanna 1984; Liberman and Dodds 1984b; Sewell 1984a; Johnstone, Patuzzi, and Yates 1986; Ruggero and Rich 1991). In many respects, the tuning curves of auditory nerve fibers in immature animals resemble those in adult animals with damaged OHCs (Walsh and McGee 1990). The results of damage studies in adult animals must, of course, be treated cautiously in drawing conclusions about the contribution of OHC-mediated active processes to the normal development of frequency representation in immature animals. However, these data do suggest that active mechanisms could potentially account for up to a 0.5-octave shift in tonotopy during development. On the other hand, a 0.5-octave change does not account for the full extent of the ontogenetic cochlear shift, which averages ~1.5 octaves (e.g., Harris and Dallos 1984; Arjmand, Harris, and Dallos 1988; Yancey and Dallos 1985; Echteler, Arjmand, and Dallos 1989; Müller 1996).

Second, as originally proposed by Romand (1979, 1987) and more recently confirmed by Echteler, Arjmand, and Dallos (1989), the "tail" and "tip" portions of high-frequency tuning curves appear to follow a different chronology of development. Shortly after the onset of hearing, the tuning curves of spiral ganglion neurons innervating the cochlear base in the gerbil resemble the "tail" portions of frequency-threshold curves of mature, high-frequency units: neurons are relatively insensitive and respond broadly over a range of frequencies well below their adult CF. Over the next 3 days, the high-frequency "tip" of the tuning curve is added and refined: the tip becomes progressively more sensitive and sharper and its CF shifts upward ~1.5 octaves to reach its correct (mature) value on P17 (Figs. 5.17 and 5.19). To the extent that the addition of the tip and the increase in CF during development are due to the active process, these findings implicate active mechanisms in the place code shift. The relative contribution of the active and passive processes to the ontogenetic shift in CF could be assessed by examining the development of ganglion cell tuning curves with the active mechanism inactivated, e.g., by furosemide injection (Ruggero and Rich 1991).

The issue of whether the place code shift in the basal cochlea is due to changes in passive basilar membrane mechanics or the maturation of active processes or both factors has recently been examined by comparing tone-evoked DPOAEs before and after an injection of furosemide in gerbils between P14 and adulthood (Mills, Norton, and Rubel 1994; Mills and Rubel 1996). As described in Section 5, the DPOAE consists of two parts: a low-intensity vulnerable component, which is a primary manifestation of the active cochlear process, and a higher intensity nonvulnerable component, which reflects the passive behavior of the cochlea and the periphery. Lowering of the ECP by a systemic injection of furosemide induces a selective and transient inactivation of the active process, thereby allowing

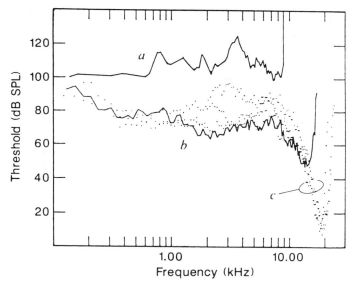

FIGURE 5.19. Development of tuning curves. Tuning curves obtained from single gerbil spiral ganglion neurons at a constant basal cochlear location at 14 (*a*), 15 (*b*), and 17, 22, and 53 DAB (*c*) are shown. Note the addition of the high-frequency "tip" during development, which might partially account for the maturational change of tonotopy in high-frequency units. (From Echteler, Arjmand, and Dallos 1989, Macmillan Magazines Limited.)

one to estimate the gain of the cochlear amplifier and to characterize the passive behavior of the cochlea. The results show that at frequencies < 2 kHz, a frequency range for which the place code is stable, the gain of the cochlear amplifier increases during development. There is no active gain at 14 days after birth (DAB), and gain reaches adult levels by 23 DAB. In contrast, the cochlear amplifier gain for midfrequencies (4–8 kHz) is already mature by 14 DAB, and the gain at 16 kHz reaches its adult level just a few days later on 16 DAB (Fig. 5.6A,B). These observations suggest that the active process in the midbasal cochlea is adultlike at or soon after the onset of hearing and thus cannot account for the place code shift, at least at midrange frequencies. Data on the development of electromotility in OHCs (He, Evans, and Dallos 1994) (Figs. 5.2 and 5.3) and on the ontogeny of tuning properties of ventral cochlear nucleus neurons in the gerbil (Woolf and Ryan 1985) also support the idea that the active mechanism is adultlike soon after the onset of hearing. On the other hand, developmental changes in the shape of spiral ganglion tuning curves suggest that the active mechanism itself, or factors that allow for its functional expression, continue to mature during the first week after hearing onset (Echteler, Arjmand, and Dallos 1989) (Fig. 5.19).

The amplitude of active emissions at frequencies >20 kHz does increase during development. However, Mills and Rubel (1996) attribute this to developmental changes in the passive properties of the basilar membrane rather than to the maturation of active processes. This interpretation derives from the assumption that the passive base cutoff frequency is progressively lower in younger animals. This assumption is consistent with the measurements of basal cutoff frequencies of cochlear potentials in developing gerbils (Harris and Dallos 1984; Arjmand, Harris, and Dallos 1988; Yancey and Dallos 1985) (Fig. 5.17) and the finding that the threshold stimulus level needed to evoke a passive emission at higher frequencies decreases markedly during development (Mills, Norton, and Rubel 1994; Mills and Rubel 1996) (Fig. 5.6B). Mills and his colleagues argue that input signals greater than the low base cutoff frequency would not effectively propagate along the basilar membrane in immature animals and therefore not activate the cochlear amplifier in the basal cochlea, which they suggest is present and fully functional at or soon after the onset of hearing (Mills, Norton, and Rubel 1994; Mills and Rubel 1996). Thus, according to this model, the proposed increase in the passive base cutoff frequency rather than maturational changes in the active process accounts for the place code shift.

Two points should be noted in considering these observations. First, although the data suggest that developmental changes in the active process do not account for the place code shift at midfrequencies, this does not in itself eliminate the possibility that the active mechanism might contribute to an ontogenetic shift in tonotopy. At 14–15 DAB, when the cochlear amplifier gain is adultlike, the cochlear partition may not have reached its adult morphological state (Munyer and Schulte 1995). Structural immaturities could restrict the expression of even a fully adult cochlear amplifier and limit its contribution to frequency organization. Second, developmental changes in the passive base cutoff frequency have not been measured directly but are inferred on the basis of frequency-dependent improvements in passive thresholds. It will be important to test this assumption by direct measurement of basilar membrane motion in developing animals.

8.2.2 Revised Model

Although the data strongly support the hypothesis of a place code shift in altricial mammals, certain aspects of the hypothesis as originally proposed appear to require revision. The original model stated that all cochlear locations become responsive to progressively higher frequency sounds during development. However, data generally support the idea that ontogenetic changes in tonotopy are restricted to mid- and high-frequency regions of the cochlea and the central auditory nuclei, whereas tonotopy in the apex and low-frequency regions in the brain does not change (e.g., Arjmand, Harris, and Dallos 1988; Sanes, Merickel, and Rubel 1989;

Rübsamen 1992; Müller 1996). In addition, it appears that apical sites respond to their correct (adult) low-frequency sounds from the very beginning or shortly after the onset of hearing. If the apical frequency map is stable from the inception of hearing, then the expanding range of frequencies to which animals become responsive during development will be represented on a basilar membrane of fixed length. This, in turn, would suggest that the length of the traveling-wave envelope narrows progressively during development, which might contribute to an increase in the sharpness of tuning of the basilar membrane's passive response.

An additional consideration is the complexity of the model. The original model proposed that functional development occurs along a basal-to-apical gradient, corresponding to the spatiotemporal gradient of cochlear differentiation. However, because the base is changing its frequency response, the development of its adultlike characteristics may be delayed. The data concerning this are controversial (Rübsamen 1992; Walsh and Romand 1992). For example, DPOAEs to high-frequency tones occur earlier and mature faster than responses to low frequencies (Henley et al. 1989; Norton, Bargones, and Rubel 1991; Mills, Norton, and Rubel 1994). In contrast, the basal part of the cochlea and its central projection sites acquire adultlike tonotopy progressively later than lower frequency regions. (Arjmand, Harris, and Dallos 1988; Rübsamen, Neuweiler, and Marimuthu 1989; Sanes, Merickel, and Rubel 1989; Rübsamen and Schäfter 1990a; Rübsamen 1992). Recordings from the auditory nerve in kittens generally show that mid- and high-frequency fibers acquire adultlike tuning sharpness before low-frequency fibers (Dolan, Teas, and Walton 1985; Romand 1987; Walsh and McGee 1987). However, regions in the inferior colliculus that receive input from more apical regions of the cochlea are the first to develop sharply tuned responses in the kitten and several species of bats (Rübsamen 1992). Further complicating this issue is the fact that developmental studies have generally not looked at the actual high-frequency responses of most animals. For example, although adult gerbils exhibit cochlear, neural, and behavioral responses up to 60 kHz or greater (Brown 1973a,b; Ryan 1976), developmental studies, with only a few exceptions (e.g., Mills and Rubel 1996; Rübsamen 1992), have generally been limited to 20 kHz.

A revised model of the place code shift hypothesis for altricial mammals, based on the sum of evidence from developmental studies of central and peripheral frequency maps, is shown in Figure 5.16. According to this interpretation, the entire length of the basilar membrane is capable of supporting a traveling wave at or very soon after the onset of hearing. Frequency representation in the cochlear apex is developmentally stable. From the very onset of hearing, the apex responds to its correct (adult) frequency, although the sensitivity and sharpness of tuning are reduced. In contrast, the more basal regions of the cochlea (mid- and high-frequency regions) undergo a shift in frequency organization such that each location becomes responsive to progressively higher frequencies in older animals.

Shifts in the cochlear map result largely from maturational changes in the mechanical properties of the cochlear partition. It is likely that the active mechanism also contributes to the shift in frequency organization. Whether this is due to maturational changes in the active process itself, structural changes within the cochlea that allow for the expression of a functionally mature active mechanism, or both factors remains to be determined.

9. Summary

From the large number of studies conducted in altricial mammals over the past 20 years, we now have a good understanding of the time course of development of various measures of cochlear function (CM, ECP, and CAP). Active cochlear mechanisms are in place and functional from the very onset of hearing.

To date, there have been no studies of the biophysical properties, including the types and distribution of ion channels, of hair cells in developing mammals, although information of this sort is available for immature hair cells in the chick (Fuchs and Sokolowski 1990). In addition, there are no developmental studies looking at the mechanical response of the basilar membrane (as reported by Ruggero and Rich [1991] for adult chinchillas). Such studies, as well as intracellular in vivo recordings from OHCs in immature animals, although extremely difficult to perform, would provide important information about the developmental dynamics of active cochlear processes.

Extremely low levels of maintained (spontaneous) activity occur in the auditory system before the onset of hearing. This maintained activity could potentially influence neural development in two ways. First, it could serve a trophic function. Second, it might provide instructive cues during the formation or refinement of synaptic connections. The importance of spontaneous activity before the onset of sensory function has been a major topic of research in the visual system and needs to be further examined in the auditory pathway.

Presently, there is no single comprehensive model for the development of tonotopy in the peripheral auditory system of birds and mammals. The question of whether a place code shift occurs at all in chickens is still a matter of controversy. This is somewhat ironic because it was the early evidence supporting a tonotopic shift in chickens that directly led to the test of the place code shift hypothesis in mammals, where it has been substantially confirmed. Whether the contradictory findings in chickens are attributable to temperature effects on frequency tuning, the limited amount of data we currently have regarding the development of frequency organization in the basal part of the papilla, or other factors remains to be determined.

The available data in mammals indicate that the representation of low frequencies at the cochlear apex is developmentally stable, whereas the tonotopy at more basal (mid- and high-frequency) locations shifts. The extent to which developmental changes in the passive mechanical properties of the basilar membrane and active cochlear processes contribute to this shift needs to be further examined. In this context, direct measurements of cochlear partition stiffness (e.g., Olson and Mountain 1991, 1994) and basilar membrane tuning in immature animals should be especially informative.

Acknowledgments. The authors were supported by National Institute on Deafness and Other Communication Disorders Grant DC-00774 to W. R. Lippe and by Deutsche Forschungsgemeinschaft Ru 390/2-1 to R. Rübsamen during the preparation of this chapter.

References

Adrian ED (1931) The microphonic action of the cochlea: an interpretation of Wever and Bray's experiments. J Physiol (Lond) 71:28–29.

Aitkin LM, Moore DR (1975) Inferior colliculus. II. Development of tuning characteristics and tonotopic organization in central nucleus of neonatal cat. J Neurophysiol 38:1208–1216.

Anastasio TJ, Correia MJ, Perachio AA (1985) Spontaneous and driven responses of semicircular canal primary afferents in the unanesthetized pigeon. J Neurophysiol 54:335–344.

Änggard L (1965) An electrophysiological study of the development of cochlear function in the rabbit. Acta Otolaryngol Suppl 203:1–64.

Anniko M (1985) Histochemical, microchemical (microprobe), and organ culture approaches to the study of auditory development. Acta Otolaryngol Suppl 421:10–18.

Anniko M, Nordemar H (1980) Embryogenesis of the inner ear. IV. Postnatal maturation of the secretory epithelia of the inner ear in correlation with the elemental composition in the endolymphatic space. Arch Otorhinolaryngol 229:281–288.

Anniko M, Wroblewski R (1986) Ionic environment of cochlear hair cells. Hear Res 22:279–293.

Anniko M, Wroblewski R, Wersäll J (1979) Development of endolymph during maturation of the mammalian inner ear. A preliminary report. Arch Otorhinolaryngol 225:161–163.

Arjmand E, Harris D, Dallos P (1988) Developmental changes in frequency mapping of the gerbil cochlea: comparison of two cochlea locations. Hear Res 32:93–97.

Ashmore JF (1987) A fast motile response in guinea pig outer hair cells: the cellular basis of the cochlear amplifier. J Physiol (Lond) 388:323–347.

Bargones JY, Burns EM (1988) Suppression tuning curves for spontaneous otoacoustic emissions in infants and adults. J Acoust Soc Am 83:1809–1816.

Bobbin RP, Fallon M, Li L, Berlin CI (1991) Guinea pigs show post-natal stability in frequency mapping at the basal turn. Hear Res 51:231–234.

Bock GR, Steel KP (1983) Inner ear pathology in the deafness mutant mouse. Acta Otolaryngol 96:39–47.

Bock GR, Webster WR (1974) Spontaneous activity of single units in the inferior colliculus of anesthetized and unanesthetized cats. Brain Res 76:150–154.

Bonfils P, Dumont A, Marie P, Francois M, Narcy P (1990) Evoked otoacoustic emissions in newborn hearing screening. Laryngoscope 100:186–190.

Born DE, Rubel EW (1985) Afferent influences on brain stem auditory nuclei of the chicken: neuron number and size following cochlea removal. J Comp Neurol 231:435–445.

Born DE, Rubel EW (1988) Afferent influences on brain stem auditory nuclei of the chicken: presynaptic action potentials regulate protein synthesis in nucleus magnocellularis neurons. J Neurosci 8:901–919.

Born DE, Durham D, Rubel EW (1991) Afferent influences on brainstem auditory nuclei of the chick: nucleus magnocellularis neuronal activity following cochlea removal. Brain Res 557:37–47.

Bosher SK, Warren RL (1971) A study of the electrochemistry and osmotic relationship of the cochlear fluids in the neonatal rat at the time of development of the endocochlear potential. J Physiol (Lond) 212:739–761.

Brown AM (1973a) High frequency peaks in the cochlear microphonic response of rodents. J Comp Physiol 83:377–392.

Brown AM (1973b) High levels of responsiveness from the inferior colliculus of rodents at ultrasonic frequencies. J Comp Physiol 83:393–406.

Brown AM, McDowell B, Forge A (1989) Acoustic distortion products can be used to monitor the effects of chronic gentamicin treatment. Hear Res 42:143–156.

Brownell WE (1990) Outer hair cell electromotility and otoacoustic emission. Ear Hear 11:82–92.

Brownell WE, Manis PB, Ritz LA (1979) Ipsilateral inhibitory responses in the cat lateral superior olive. Brain Res 177:189–193.

Brownell WE, Bader CR, Bertrand D, Ribaupierre Y (1985) Evoked mechanical responses of isolated cochlear outer hair cells. Science 227:194–196.

Brugge JF (1986) Development of the auditory nerve. In: Aslin RN (ed) Advances In Neural and Behavioral Development. Vol. 2. NJ: Ablex Publishing, pp. 73–94.

Brugge JF, O'Connor TA (1984) Postnatal functional development of the dorsal and posteroventral cochlear nuclei of the cat. J Acoust Soc Am 75:1548–1562.

Brugge JF, Javel E, Kitzes L (1978) Signs of functional maturation of peripheral auditory system in discharge patterns of neurons in anteroventral cochlear nucleus. J Neurophysiol 41:1557–1579.

Burns EM, Hoberg DL, Arehart K, Campbell SL (1992) Prevalence of spontaneous otoacoustic emissions in neonates. J Acoust Soc Am 91:1571–1575.

Brusilow SW (1976) Propanolol antagonism to the effect of furosemide on the composition of endolymph in guinea pigs. Can J Physiol Pharmacol 54:42–48.

Charpak S, DuBois-Dauphin M, Raggenbass M, Dreifuss JJ (1989) Vasopressin excites neurones located in the dorsal cochlear nucleus of the guinea-pig brainstem. Brain Res 483:164–169.

Chen L, Salvi R, Shero M (1994) Cochlear frequency-place map in adult chickens: intracellular biocytin labeling. Hear Res 81:130–136.

Chuang SW, Gerber SE, Thornton AR (1993) Evoked otoacoustic emissions in preterm infants. Int J Pediatr Otorhinolaryngol 26:39–45.

Church MW, Williams HL, Holloway JA (1984) Postnatal development of the brainstem auditory evoked potential and far-field cochlear microphonic in non-sedated rat pups. Dev Brain Res 14:23–31.

Cody AR, Johnstone BM (1981) Acoustic trauma: single neuron basis for the "half-octave shift." J Acoust Soc Am 70:707–711.

Cody AR, Russell J (1987) The responses of hair cells in the basal turn of the guinea-pig cochlea to tones. J Physiol (Lond) 383:551–556.

Cohen YE, Saunders JC (1994) The effect of acoustic overexposure on the tonotopic organization of the nucleus magnocellularis. Hear Res 81:11–21.

Cohen YE, Bacon CK, Saunders JC (1992) Middle ear development. III. Morphometric changes in the conducting apparatus of the Mongolian gerbil. Hear Res 62:187–193.

Cohen YE, Hernandez HN, Saunders JC (1992) Middle-ear development: II. Morphometric changes in the conducting apparatus of the chick. J Morphol 212:265–267.

Cohen YE, Rubin DM, Saunders JC (1992) Middle ear development: I. Extra-stapedius response in the neonatal chick. Hear Res 58:1–8.

Cohen YE, Doan DE, Rubin DM, Saunders JC (1993) Middle-ear development. V: Development of umbo sensitivity in the gerbil. Am J Otolaryngol 14:191–198.

Cole KS, Robertson D (1992) Early efferent innervation of the developing rat cochlea studied with a carbocyanine dye. Brain Res 575:223–230.

Cooper NP, Yates GK (1994) Nonlinear input-output functions derived from the responses of guinea-pig cochlear nerve fibers: variations with characteristic frequency. Hear Res 78:269–285.

Corwin JT, Cotanche DA (1988) Regeneration of sensory hair cells after acoustic trauma. Science 240:1772–1774.

Cotanche DA, Sulik KK (1984) The development of stereociliary bundles in the cochlear duct of chick embryos. Dev Brain Res 16:181–193.

Cotanche DA, Saunders JC, Tilney LG (1987) Hair cell damage produced by acoustic trauma in the chick cochlea. Hear Res 25:267–286.

Cotanche DA, Henson MM, Henson OW Jr (1992) Contractile proteins in the hyaline cells of the chicken cochlea. J Comp Neurol 324:353–364.

Cousillas H, Rebillard G (1985) Age-dependent effects of a pure tone trauma in the chick basilar papilla: evidence for a development of the tonotopic organization. Hear Res 19:217–226.

Crawford AC, Fettiplace RR (1980) The frequency selectivity of auditory nerve fibers and hair cells in the cochlea of the turtle. J Physiol (Lond) 306:79–125.

Crowley DE, Hepp-Reymond MC (1966) Development of cochlear function in the ear of infant rat. J Comp Physiol Psychol 62:427–432.

Dallos P (1973) Cochlear potentials and cochlear mechanics. In: A. Møller (ed) Basic Mechanisms in Hearing. New York: Academic Press, pp. 335–376.

Dallos P (1981) Cochlear physiology. Annu Rev Psychol 32:153–190.

Dallos P (1985) Response characteristics of mammalian cochlear hair cells. J Neurosci 5:1591–1608.

Dallos P (1992) The active cochlea. J Neurosci 12:4575–4585.

Dallos P, Cheatham MA (1976) Production of cochlear potentials by inner and outer hair cells. J Acoust Soc Am 60:510–512.

Dallos P, Harris D (1978) Properties of auditory nerve response in absence of outer hair cells. J Neurophysiol 41:365–383.

Dallos P, Santo-Sacchi I, Flock Å (1982) Intracellular recordings of cochlear outer hair cells. Science 218:582–584.

Davis H (1983) An active process in cochlear mechanics. Hear Res 9:79–90.

Dieler R, Shehata-Dieler WE, Brownell WE (1991) Concomitant salycilate-induced alterations of outer hair cell subsurface cisternae and electromotility. J Neurocytol 20:637–653.

Dolan DF, Teas DC, Walton JP (1985) Postnatal development of physiological responses in auditory nerve fibers. J Acoust Soc Am 78:544–554.

Drenckhahn D, Merte C, von Düring M, Smolders J, Klinke R (1991) Actin, myosin and alpha-actinin containing filament bundles in hyaline cells of the caiman cochlea. Hear Res 54:29–38.

Durham D, Rubel EW, Steel KP (1989) Cochlear ablation in deafness mutant mice: 2-deoxyglucose analysis suggests no spontaneous activity of cochlear origin. Hear Res 43:39–46.

Eatock RA, Manley GA (1981) Auditory nerve fibre activity in the Tokay gecko. II. Temperature effect on tuning. J Comp Physiol 142:219–226.

Echteler SM (1995) Structural correlates of frequency-place map development. Abstr Assoc Res Otolaryngol, pp. 111.

Echteler SM, Arjmand E, Dallos P (1989) Developmental alterations in the frequency map of the mammalian cochlea. Nature (Lond) 341:147–149.

Eggermont JJ (1991) Maturational aspects of periodicity coding in cat primary auditory cortex. Hear Res 57:45–56.

Ehret G, Romand R (1994) Development of tonotopy in the inferior colliculus II: 2-DG measurements in the kitten. Eur J Neurosci 6:1589–1595.

Evans BN, Dallos P (1993) Stereocilia displacement induced somatic motility of outer hair cells. Proc Natl Acad Sci USA 90:8347–8351.

Evans EF, Klinke R (1982) The effect of intracochlear and systemic furosemide on the properties of single cochlear nerve fibers. J Physiol (Lond) 131:409–428.

Fermin CD, Cohen GM (1984) Developmental gradients in the embryonic chick's basilar papilla. Acta Otolaryngol 97:39–51.

Fernández C, Hinojosa R (1974) Postnatal development of endocochlear potential and stria vascularis in the cat. Acta Otolaryngol 78:173–186.

Finck A, Schneck CD, Hartman AF (1972) Development of cochlea function in the neonate mongolian gerbil (*Meriones unguiculatus*). J Comp Physiol Psychol 78:375–380.

Fisher FP (1992) Quantitative analysis of the innervation of the chicken basilar papilla. Hear Res 61:167–178.

Fitzgerald M (1987) Spontaneous and evoked activity of fetal primary afferents in vivo. Nature (Lond) 326:603–605.

Friauf E (1992) Tonotopic order in the adult and developing auditory system of the rat as shown by c-fos immunocytochemistry. Eur J Neurosci 4:798–812.

Fuchs PA, Sokolowski HA (1990) The acquisition during development of Ca-activated potassium currents by cochlear hair cells of the chick. Proc R Soc Lond B Biol Sci 241:122–126.

Fuchs PA, Nagai T, Evans MG (1988) Electrical tuning in hair cells isolated from the chick cochlea. J Neurosci 8:2460–2467.

Furukawa T, Matsuura S (1978) Adaptive rundown of excitatory post-synaptic potentials at synapses between hair cells and eighth nerve fibres in the goldfish. J Physiol (Lond) 276:193–209.

Geisler CD (1986) A model of the effect of outer hair cell motility on cochlear vibrations. Hear Res 24:125–132.

Gummer AW, Klinke R (1983) Influence of temperature on tuning of primary-like units in the guinea pig cochlear nucleus. Hear Res 12:367–380.

Gummer AW, Mark RF (1994) Patterned neural activity in brain stem auditory areas of a prehearing mammal, the tammar wallaby (*Macropus eugenii*). Neuro-Report 5:685–688.

Harris DM, Dallos P (1984) Ontogenetic changes in frequency mapping of a mammalian ear. Science 225:741–743.

Harris DM, Rotche R, Freedom T (1990) Postnatal growth of cochlear spiral in mongolian gerbil. Hear Res 50:1–6.

Harvey D, Steel KP (1992) The development and interpretation of the summating potential response. Hear Res 61:137–146.

Hashisaki G, Rubel EW (1989) Effects of unilateral cochlea removal on anteroventral cochlear nucleus neurons in developing gerbils. J Comp Neurol 283:465–473.

He DZ, Evans BN, Dallos P (1994) First appearance and development of electromotility in neonatal gerbil outer hair cells. Hear Res 78:77–90.

Heil P, Scheich H (1992) Postnatal shift of tonotopic organization in the chick auditory cortex analogue. NeuroReport 3:381–384.

Henley CM, Rybak LP (1995) Ototoxicity in developing mammals. Brain Res Rev 20:68–90.

Henley CM, Owings MH, Stagner BB, Martin GK, Lonsbury-Martin BL (1989) Postnatal development of 2f1-f2 otoacoustic emission in pigmented rat. Hear Res 43:141–148.

Herrmann K, Shatz CJ (1995) Blockade of action potential activity alters initial arborization of thalamic axons within cortical layer 4. Proc Natl Acad Sci USA 92:11244–11248.

Hirsch JA, Oertel D (1988) Intrinsic properties of neurones in the dorsal cochlear nucleus of mice, in vitro. J Physiol (Lond) 396:535–548.

Holley M (1991) High-frequency force generation in outer hair cells from the mammalian ear. Bioessays 13:115–120.

Holley MC, Kalinec F, Kachar B (1992) Structure of the cortical cytoskeleton in mammalian outer hair cells. J Cell Sci 102:569–580.

Hubbard AE, Mountain DC (1983) Alternating current delivered into the scala media alters sound pressure at the eardrum. Science 222:510–512.

Hudspeth AJ (1989) How the ear's work works. Nature (Lond) 341:397–404.

Hudspeth AJ, Corey DP (1977) Sensitivity, polarity and conductance change in the response of vertebrate hair cells to controlled mechanical stimuli. Proc Natl Acad Sci USA 74:2407–2411.

Huffman RF, Henson OW Jr (1991) Cochlear and CNS tonotopy: normal physiological shifts in the mustached bat. Hear Res 56:79–85.

Huffman RF, Henson OW Jr (1993) Labile cochlear tuning in the mustached bat. II. Concomitant shifts in neural tuning. J Comp Physiol A Sens Neural Behav Physiol 171:735–748.

Hyson R, Rudy JW (1987) Ontogenetic change in the analysis of sound frequency in the infant rat. Dev Psychobiol 20:189–207.

Jaramillo F, Hudspeth AJ (1991) Localization of the hair cell's transduction channels at the hair bundle's top by iontophoretic application of a channel blocker. Neuron 7:409–420.

Johnstone BM, Patuzzi R, Yates GK (1986) Basilar membrane measurements and the traveling wave. Hear Res 22:147–153.

Jones SM, Jones TA (1995a) Neural timing characteristics of auditory primary afferents in the chicken embryo. Hear Res 82:139–148.

Jones SM, Jones TA (1995b) The tonotopic map in the embryonic chicken cochlea.

Hear Res 82:149–157.

Kaas JHM, Merzenich MM, Killackey HP (1983) The reorganization of somato-sensory cortex following peripheral nerve damage in adult and developing mammals. Annu Rev Neurosci 6:325–356.

Kaiser A (1992) The ontogeny of homeothermic regulation in post-hatching chicks: its influence on the development of hearing. Comp Biochem Physiol A Comp Physiol 103:105–111.

Kaltenbach JA, Falzarano PR (1994) Postnatal development of the hamster cochlea. I. Growth of hair cells and the organ of Corti. J Comp Neurol 340:87–97.

Kaltenbach JA, Falzarano PR, Simpson TH (1994) Postnatal development of the hamster cochlea. II. Growth and differentiation of stereocilia bundles. J Comp Neurol 350:187–198.

Kandler K, Fraiuf E (1995) Development of glycinergic and glutamatergic synaptic transmission in the auditory brainstem of perinatal rats. J Neurosci 15:6890–6904.

Katayama A, Corwin JT (1989) Cell production in the chicken cochlea. J Comp Neurol 281:129–135.

Kemp DT (1978) Stimulated acoustic emissions from within the human auditory system. J Acoust Soc Am 5:1386–1391.

Kettner RE, Feng JZ, Brugge JF (1985) Postnatal development of the phase-locked response to low frequency tones of auditory nerve fibers in the cat. J Neurosci 5:275–283.

Kiang NY-S, Watanabe T, Thomas EC, Clark LF (1965) Discharge Patterns of Single Fibers in the Cat's Auditory Nerve. Cambridge, MA: MIT Press.

Kitzes LM, Kageyama GH, Semple MN, Kil J (1995) Development of ectopic projections from the ventral cochlear nucleus to the superior olivary complex induced by neonatal ablation of the contralateral cochlea. J Comp Neurol 353:341–363.

Koerber KC, Pfeiffer WB, Kiang NY-S (1966) Spontaneous spike discharges from single units in the cochlear nucleus after destruction of the cochlea. Exp Neurol 16:119–130.

Komune S, Nakagawa T, Hisashi K, Kimitsuki T, Uemura T (1993) Mechanism of lack of development of negative endocochlear potential in the guinea pig with hair cell loss. Hear Res 70:197–204.

Konishi M (1973) Development of auditory neuronal responses in avian embryos. Proc Nat Acad Sci USA 70:1795–1798.

Kotak VC, Sanes DH (1995) Synaptically evoked prolonged depolarizations in the developing auditory system. J Neurophysiol 74:1611–1620.

Kotak VC, Sanes DH (1996) Developmental influence of glycinergic transmission: regulation of NMDA receptor-mediated EPSPs. J Neurosci 16:1836–1843.

Kraus H-J, Aulbach-Kraus K (1981) Morphological changes in the cochlea of the mouse after the onset of hearing. Hear Res 4:89–102.

Kuijpers W (1974) Na-K-ATPase activity in the cochlea of the rat during develop-ment. Acta Otolaryngol 78:341–344.

Laukli E, Mair IW (1981) Development of surface-recorded cochlear and early neural potentials in the cat. Arch Otorhinolaryngol 233:1–12.

Lenoir M, Puel J-L (1987) Development of 2f1-f2 otoacoustic emission in the rat. Hear Res 29:265–271.

Lenoir M, Puel J-L, Pujol R (1987) Stereocilia and tectorial membrane development in the rat cochlea. A SEM study. Anat Embryol 175:477–487.

Leonard DG, Khanna SM (1984) Histological evaluation of damage in cat cochleas used for measurement of basilar membrane mechanics. J Acoust Soc Am 75:515–527.

Lewis RS, Hudspeth AJ (1983) Voltage and ion-dependent conductances in solitary vertebrate hair cells. Nature (Lond) 304:538–541.

Liberman MC (1978) Auditory-nerve response from cats raised in a low-noise chamber. J Acoust Soc Am 63:442–455.

Liberman MC (1980a) Morphological differences among radial afferent fibers in the cat cochlea: an electron-microscopic study of serial sections. Hear Res 3:45–63.

Liberman MC (1980b) Efferent synapses in the inner hair cell area of the cat cochlea: an electron-microscopic study of serial sections. Hear Res 3:189–204.

Liberman MC (1982) Single-neuron labeling in the cat auditory nerve. Science 216:1239–1241.

Liberman MC, Dodds LW (1984a) Single-neuron labeling and chronic cochlear pathology. II. Stereocilia damage and alteration of spontaneous discharge rates. Hear Res 16:43–53.

Liberman MC, Dodds LW (1984b) Single-neuron labeling and chronic cochlear pathology. III. Stereocilia damage and alteration of threshold tuning curves. Hear Res 16:55–74.

Liberman MC, Oliver ME (1984) Morphometry of intracellularly labeled neurons of the auditory nerve: correlations with functional properties. J Comp Neurol 223:163–176.

Lippe WR (1987) Shift of tonotopic organization in brain stem auditory nuclei of the chicken during late embryonic development. Hear Res 25:205–208.

Lippe WR (1994) Rhythmic spontaneous activity in the developing avian auditory system. J Neurosci 14:1486–1495.

Lippe WR (1995) Relationship between frequency of spontaneous bursting and tonotopic position in the developing avian auditory system. Brain Res 703:205–213.

Lippe WR, Rubel EW (1983) Development of the place principle: tonotopic organization. Science 219:514–516.

Lippe WR, Rubel EW (1985) Ontogeny of tonotopic organization of brain stem auditory nuclei in the chicken: implications for development of the place principle. J Comp Neurol 237:273–289.

Lippe WR, Ryals BM, Rubel EW (1986) Development of the place principle. In: Aslin RN (ed) Advances In Neural and Behavioral Development. Vol. 2. Norwood, NJ: Ablex Publishing, pp. 155–203.

Lippe WR, Westbrook EW, Ryals BM (1991) Hair cell regeneration in the chicken cochlea following aminoglycoside toxicity. Hear Res 56:203–210.

Long GR, Tubis A (1988) Investigations into the nature of the association between threshold microstructure and otoacoustic emissions. Hear Res 36:125–139.

Maffei L, Galli-Resta L (1990) Correlation in the discharges of neighboring rat retinal ganglion cells during prenatal life. Proc Natl Acad Sci USA 87:2861–2864.

Mair IW, Elverland HH, Laukli HH (1978) Development of early auditory evoked responses in the cat. Audiology 17:469–488.

Manley GA (1996) Ontogeny of frequency mapping in the peripheral auditory system of birds and mammals: a critical review. Aud Neurosci 3:199–214.

Manley GA, Robertson D (1976) Analysis of spontaneous activity of auditory neurones in the spiral ganglion of the guinea-pig cochlea. J Physiol (Lond) 258:323–336.

Manley GA, Brix J, Kaiser A (1987) Developmental stability of the tonotopic organization of the chick's basilar papilla. Science 237:665–666.

Manley GA, Gleich O, Kaiser A, Brix J (1989) Functional differentiation of sensory cells in the avian auditory periphery. J Comp Physiol A Sens Neural Behav Physiol 164:289–296.

Manley GA, Kaiser A, Brix J, Gleich O (1991) Activity patterns of primary auditory-nerve fibres in chickens: development of fundamental properties. Hear Res 57:1–15.

Marty R, Thomas J (1963) Réponse électro-corticale à la stimulation du nerf cochléaire chez la chat nouveau-né. J Physiol (Paris) 55:165–166.

McGuirt JP, Schmiedt RA, Schulte BA (1995) Development of cochlear potentials in the neonatal gerbil. Hear Res 84:52–60.

Meister M, Wong ROL, Baylor DA, Shatz CJ (1994) Synchronous bursts of action potentials in ganglion cells of the developing mammalian retina. Science 252:939–943.

Mills DM, Rubel EW (1996) Development of the cochlear amplifier. J Acoust Soc Am 100:1–15.

Mills DM, Norton SJ, Rubel EW (1993) Vulnerability and adaptation of distortion product otoacoustic emissions to endocochlear potential variation. J Acoust Soc Am 94:2108–2122.

Mills DM, Norton SJ, Rubel EW (1994) Development of active and passive mechanics in the mammalian cochlea. Aud Neurosci 1:77–99.

Moffat AJM, Capranica RR (1976) Effects of temperature on the response properties of auditory nerve fibers in the American toad (*Bufo americanus*). J Acoust Soc Am 60:S80.

Moore DR (1981) Development of the cat peripheral auditory system: input-output functions of cochlear potentials. Brain Res 219:29–44.

Morlet T, Collet L, Salle B, Morgon A (1993) Functional maturation of cochlear active mechanisms and of the medial olivocochlear system in humans. Acta Otolaryngol 113:271–277.

Müller M (1991) Developmental changes of frequency representation in the rat cochlea. Hear Res 56:1–7.

Müller M (1996) The cochlear place-frequency map of the adult and developing mongolian gerbil. Hear Res 94:148–156.

Munyer PD, Schulte BA (1995) Developmental expression of proteoglycans in the tectorial and basilar membrane of the gerbil cochlea. Hear Res 85:85–94.

Mysliveček J (1983) Development of the auditory evoked responses in the auditory cortex in mammals. In: Romand R (ed) Development of Auditory and Vestibular Systems. New York: Academic Press, pp. 167–209.

Nakazawa K, Spicer SS, Schulte BA (1995) Postnatal expression of the facilitated glucose transporter, GLU5, in gerbil outer hair cells. Hear Res 82:93–99.

Norton SJ (1992) Cochlear function and otoacoustic emissions. Semin Hear 13:1–14.

Norton SJ, Widen JE (1990) Evoked otoacoustic emissions in normal-hearing infants and children: emerging data and issues. Ear Hear 11:121–127.

Norton SJ, Bargones JY, Rubel EW (1991) Development of otoacoustic emissions in gerbil: evidence for micromechanical changes underlying development of the place code. Hear Res 51:73–92.

Ohlemiller KK, Siegel JH (1994) Cochlear basal and apical differences reflected in the effects of cooling on responses of single auditory nerve fibers. Hear Res

80:174–190.

Olson ES, Mountain DC (1991) In vivo measurement of basilar membrane stiffness. J Acoust Soc Am 89:1262–1275.

Olson ES, Mountain DC (1994) Mapping the cochlear partition's stiffness to its cellular architecture. J Acoust Soc Am 95:395–400.

Pasic TR, Moore DR, Rubel EW (1994) Effect of altered neuronal activity on cell size in the medial nucleus of the trapezoid body and ventral cochlear nucleus of the gerbil. J Comp Neurol 348:111–120.

Pickles JO (1988) An Introduction to the Physiology of Hearing. London: Academic Press.

Pirvola U, Lehtonen E, Ylikoski J (1991) Spatiotemporal development of cochlear innervation and hair cell differentiation in the rat. Hear Res 52:345–355.

Probst R, Lonsbury-Martin BL, Martin GK (1991) A review of otoacoustic emissions. J Acoust Soc Am 89:2027–2067.

Probst R, Harris FP, Hauser R (1993) Clinical monitoring using otoacoustic emissions. Br J Audiol 27:85–90.

Pujol R, Lavigne-Rebillard M (1992) Development of neurosensory structures in the human cochlea. Acta Otolaryngol 112:259–264.

Pujol R, Marty R (1968) Structural and physiological relationships of the maturing auditory system. In: Vilek L, Trojan S (eds) Ontogenesis of the Brain. Prague, Czechoslovakia: Charles University Press, pp. 377–385.

Pujol R, Marty R (1970) Postnatal maturation in the cochlea of the cat. J Comp Neurol 139:115–126.

Pujol R, Lavigne-Rebillard M, Uziel A (1991) Development of the human cochlea. Acta Otolaryngol Suppl 482:7–12.

Pujol R, Zajic G, Dulon D, Raphael Y, Altschuler RA, Schacht J (1991) First appearance and development of motile properties in outer hair cells isolated from the guinea pig cochlea. Hear Res 57:129–141.

Rebillard G, Rubel EW (1981) Electrophysiological study of the maturation of auditory responses from the inner ear of the chick. Brain Res 229:15–23.

Rebillard G, Ryals BM, Rubel EW (1982) Relationship between hair cell loss on the chick basilar papilla and threshold shift after acoustic overstimulation. Hear Res 8:77–81.

Relkin EM, Saunders JC, Konkle DF (1979) The development of middle-ear admittance in the hamster. J Acoust Soc Am 66:133–139.

Robertson D, Irvine DRF (1989) Plasticity of frequency organization in auditory cortex of guinea pigs with partial unilateral deafness. J Comp Neurol 282:456–471.

Robertson D, Johnstone BM (1979a) Effects of divalent cations on spontaneous and evoked activity of single mammalian auditory neurones. Pflügers Arch 380:7–12.

Robertson D, Johnstone BM (1979b) Aberrant tonotopic organization in the inner ear damaged by kanamycin. J Acoust Soc Am 66:466–469.

Robertson D, Cody AR, Bredberg G, Johnstone BM (1980) Response properties of spiral ganglion neurons in cochleas damaged by direct mechanical trauma. J Acoust Soc Am 67:1295–1303.

Roe AW, Pallas SL, Jong-On H, Sur M (1990) A map of visual space induced in primary auditory cortex. Science 250:818–820.

Romand R (1971) Maturation des potentiels cochléaires dans la périod périnatale chez le chat et chez le cobaye. J Physiol (Paris) 63:763–782.

Romand R (1979) Development of auditory nerve activity in kittens. Brain Res

173:554-556.

Romand R (1983) Development of the cochlea. In: Romand R (ed) Development of Auditory and Vestibular Systems. New York: Academic Press, pp. 47-88.

Romand R (1984) Functional properties of auditory nerve fibers during postnatal development in the kitten. Exp Brain Res 56:395-402.

Romand R (1987) Tonotopic evolution during development. Hear Res 28:117-123.

Romand R, Dauzat M (1981) Spontaneous activity and signal processing in the cochlear nerve of kittens. J Acoust Soc Am 69:S52.

Romand R, Ehret G (1990) Development of tonotopy in the inferior colliculus. I. Electrophysiological mapping in house mice. Dev Brain Res 54:221-234.

Romand R, Despers G, Giry N (1987) Factors affecting the onset of inner ear function. Hear Res 28:1-7.

Roth B, Bruns S (1992) Postnatal development of the rat organ of Corti I. General morphology, basilar membrane, tectorial membrane and border cells. Anat Embryol 185:559-569.

Rubel EW (1978) Ontogeny of structure and function in the vertebrate auditory system. In: Jacobson M (ed) Handbook of Sensory Physiology. Vol. IX. Development of Sensory Systems. New York: Springer-Verlag, pp. 135-237.

Rubel EW, Ryals BM (1983) Development of the place principle: acoustic trauma. Science 219:512-514.

Rubel EW, Smith DJ, Miller JC (1976) Organization and development of brain stem auditory nuclei of the chicken: ontogeny of n. magnocellularis and n. laminaris. J Comp Neurol 166:469-489.

Rubel EW, Lippe WR, Ryals BM (1984) Development of the place principle. Ann Otol Rhinol Laryngol 93:609-615.

Rubel EW, Born DE, Deitch JS, Durham D (1985) Recent advances toward understanding auditory system development. In: Berlin C (ed) Hearing Science. San Diego, CA: College Hill Press, pp. 110-157.

Rübsamen R (1987) Ontogenesis of the echolocation system in rufous horseshoe bat, *Rhinolophus rouxi* (audition and vocalization in early postnatal development). J Comp Physiol A Sens Neural Behav Physiol 161:899-913.

Rübsamen R (1992) Postnatal development of central auditory frequency maps. J Comp Physiol A Sens Neural Behav Physiol 170:129-143.

Rübsamen R, Schäfer M (1990a) Ontogenesis of auditory fovea representation in the inferior colliculus of the Sri Lankan rufous horseshoe bat, *Rhinolophus rouxi*. J Comp Physiol A Sens Neural Behav Physiol 167:757-769.

Rübsamen R, Schäfer M (1990b) Audiovocal interactions during development? Vocalisation in deafened young horseshoe bats vs. audition in vocalisation-impaired bats. J Comp Physiol A Sens Neural Behav Physiol 167:771-784.

Rübsamen R, Neuweiler G, Marimuthu G (1989) Ontogenesis of tonotopy in inferior colliculus of a hipposiderid bat reveals postnatal shift in frequency-place code. J Comp Physiol A Sens Neural Behav Physiol 165:755-769.

Rübsamen R, Mills DM, Rubel EW (1995) Effects of furosemide on distortion product otoacoustic emissions and on neuronal responses in the anteroventral cochlear nucleus. J Neurophysiol 74:1628-1638.

Ruggero MA, Rich NC (1991) Furosemide alters organ of Corti mechanics: evidence for feedback of outer hair cells upon the basilar membrane. J Neurosci 11:1057-1067.

Russell FA, Moore DR (1995) Afferent reorganization within the superior olivary complex of the gerbil: development and induction by neonatal, unilateral cochlear removal. J Comp Neurol 352:607–625.

Ryals BM, Rubel EW (1985a) Differential susceptibility of avian hair cells to acoustic trauma. Hear Res 19:73–84.

Ryals BM, Rubel EW (1985b) Ontogenic changes in the position of hair cell loss after acoustic overstimulation in avian basilar papilla. Hear Res 19:135–142.

Ryals BM, Rubel EW (1988) Hair cell regeneration after acoustic trauma in adult *Coturnix* quail. Science 240:1774–1776.

Ryals BM, Creech HB, Rubel EW (1984) Postnatal changes in the size of the avian cochlear duct. Acta Otolaryngol 98:93–97.

Ryan AF (1976) Hearing sensitivity of the mongolian gerbil, *Meriones unguiculatis*. J Acoust Soc Am 59:1222–1226.

Ryan AF, Woolf NK (1983) Energy dispersive x-ray analysis of inner ear fluids and tissue during the ontogeny of cochlear function. Scanning Electron Microsc 1:201–207.

Ryan AF, Woolf NK (1988) Development of tonotopic representation in the Mongolian gerbil: a 2-deoxyglucose study. Dev Brain Res 41:61–70.

Rybak LP, Morizono T (1982) Effect of furosemide upon endolymph potassium concentration. Hear Res 7:223–231.

Rybak LP, Weberg A, Whitworth C (1991) Development of the stria vascularis in the rat. ORL (Basel) 53:72–77.

Rybak LP, Whitworth C, Scott V, Weberg A (1991) Ototoxicity of furosemide during development. Laryngoscope 101:1167–1174.

Rybak LP, Whitworth C, Scott V (1992) Development of endocochlear potential and compound action potential in the rat. Hear Res 59:189–194.

Salvi RJ, Saunders SS, Hashino E, Chen L (1994) Discharge patterns of chicken cochlear ganglion neurons following kanamycin-induced hair cell loss and regeneration. J Comp Physiol A Sens Neural Behav Physiol 174:351–369.

Sanes DH (1993) The development of synaptic function and integration in the central auditory system. J Neurosci 13:2627–2637.

Sanes DH, Merickel M, Rubel EW (1989) Evidence for an alternation of the tonotopic map in the gerbil cochlea during development. J Comp Neurol 279:436–445.

Saunders JC, Coles RG, Gates GR (1973) The development of auditory evoked responses in the cochlea and cochlear nuclei of the chick. Brain Res 63:59–74.

Saunders JC, Kaltenbach JA, Relkin EM (1983) The structural and functional development of the outer and middle ear. In: Romand R (ed) Development of Auditory and Vestibular Systems. New York: Academic Press, pp. 3–25.

Saunders JC, Relkin EM, Rosowski JJ, Bahl C (1986) Changes in middle-ear input admittance during postnatal auditory development in chicks. Hear Res 24:227–235.

Saunders JC, Doan DE, Cohen YE (1993) The contribution of middle-ear sound conduction to auditory development. Comp Biochem Physiol A Comp Physiol 106:7–13.

Schacht J, Zenner HP (1987) Evidence that phosphoinositides mediate motility in cochlear outer hair cells. Hear Res 31:155–161.

Schäfer M (1991) Vergleichende Untersuchung zur Ontogenese von Frequenzkarten bei Säugetieren mit unterschiedlich spezialisiertem Hörsystem. Dissertation Fakultät für Biologie, Ruhr-Universität Bochum.

Schermuly L, Klinke R (1985) Change of characteristic frequencies of pigeon primary auditory afferents with temperature. J Comp Physiol A Sens Neural

Behav Physiol 156:209–211.

Schmiedt RA (1989) Spontaneous rates, thresholds and tuning of auditory-nerve fibers in the gerbil: comparisons to cat data. Hear Res 42:23–35.

Schweitzer L, Lutz C, Hobbs M, Weaver SP (1996) Anatomical correlates of the passive properties underlying the developmental shift in the frequency map of the mammalian cochlea. Hear Res 97:84–94.

Sellick R, Patuzzi R, Johnstone BM (1982) Measurement of basilar membrane motion in the guinea pig using the Mössbauer technique. J Acoust Soc Am 72:131–141.

Sewell WF (1984a) The effects of furosemide on the endocochlear potential and auditory nerve fiber tuning curves in cats. Hear Res 14:305–314.

Sewell WF (1984b) Furosemide selectively reduces one component in rate-level functions from auditory nerve fibers. Hear Res 15:69–72.

Shatz CJ (1990) Impulse activity and the patterning of connections during CNS development. Neuron 5:745–756.

Shatz CJ, Stryker MP (1988) Prenatal tetrodotoxin infusion blocks segregation of retinogeniculate afferents. Science 242:87–89.

Shehata WE, Brownell WE, Dieler R (1991) Effects of salycilate on shape, electromotility and membrane characteristics of isolated outer hair cells from guinea pig cochlea. Acta Otolaryngol 111:707–718.

Shnerson A, Pujol R (1981) Age related changes in the C57BL/6J mouse cochlea. I. Physiological findings. Brain Res 254:65–75.

Shnerson A, Willott JF (1979) Development of inferior colliculus response properties in C57BL/6J mouse pubs. Exp Brain Res 37:373–385.

Shotwell SL, Jakobs R, Hudspeth AJ (1981) Directional sensitivity of individual vertebrate hair cells to controlled deflection of their hair bundles. Ann NY Acad Sci 374:1–10.

Slepecky N, Ulfendahl M (1992) Actin-binding and microtubule-associated proteins in the organ of Corti. Hear Res 57:201–215.

Slepecky N, Ulfendahl M, Flock Å (1988) Shortening and elongation of isolated outer hair cells in response to application of potassium gluconate, acetylcholine and cationized ferritin. Hear Res 34:119–126.

Smolders JWT, Klinke R (1978) Effect of temperature on tuning properties of primary auditory fibres in caiman and cat. Pflügers Arch 373 Suppl:R84.

Smolders JWT, Klinke R (1984) Effects of temperature on the properties of primary auditory fibres of the spectacled caiman, *Caiman crocodilus* (L.). J Comp Physiol 155:19–30.

Steel KP, Barkway C (1989) Another role for melanocytes: their importance for normal stria vascularis development in the mammalian inner ear. Development 107:453–463.

Sterbing JS, Schmidt U, Rübsamen R (1994) The postnatal development of frequency-place code and tuning characteristics in the auditory midbrain of the phyllostomid bat, *Carollia perspicillata*. Hear Res 76:133–146.

Stiebler IB, Narins PM (1990) Temperature-dependence of auditory nerve response properties in the frog. Hear Res 46:63–82.

Tasaki I, Davis H, Legouix JP (1952) The space-time pattern of the cochlear microphonics (guinea pig), as recorded by differential electrodes. J Acoust Soc Am 24:502–519.

Tierney TS, Russell FA, Moore DR (1997) Susceptibility of developing cochlear nucleus neurons to deafferentiation-induced death abruptly ends just before the onset of hearing. J Comp Neurol 378:295–306.

Tilney LG, Tilney M, Saunders JC, DeRosier DJ (1986) Actin filaments, stereocilia, and hair cells of the bird cochlea. III. The development and differentiation of hair cells and stereocilia. Dev Biol 116:100–118.

Tokimoto T, Osako S, Matsuura S (1977) Development of auditory evoked cortical and brain stem responses during the early postnatal period in the cat. Osaka City Med J 23:141–153.

Tucci DL, Rubel EW (1985) Afferent influences on brain stem auditory nuclei of the chicken: effects of conductive and sensorineural hearing loss on n. magnocellularis. J Comp Neurol 238:371–381.

Uziel A, Romand R, Marot M (1981) Development of cochlear potentials in rats. Audiology 20:89–100.

van Dijk P, Lewis ER, Wit HP (1990) Temperature effects on auditory nerve fiber response in the American bullfrog. Hear Res 44:231–240.

Vater M, Rübsamen R (1989) Postnatal development of the cochlea in horseshoe bats. In: Wilson JP, Kemp DT (eds) Cochlear Mechanisms. New York: Plenum, pp. 217–225.

Verley R, Axelrad H (1977) Functional maturation of rat trigeminal nerve. Neurosci Lett 5:133–139.

Wall JT, Kaas JH, Sur M, Nelson RJ, Felleman DJ, Merzenich MM (1986) Functional reorganization in somatosensory cortical areas 3b and 1 of adult monkeys after median nerve repair: possible relationships to sensory recovery in humans. J Neurosci 6:218–233.

Waller HJ, Godfrey DA (1994) Functional characteristics of spontaneously active neurons in rat dorsal cochlear nucleus in vitro. J Neurophysiol 71:467–478.

Walsh EJ, McGee J (1986) The development of function in the auditory periphery of cats. In: Altschuler RA, Bobbin RP, Hoffman DW (eds) Neurobiology of Hearing: The Cochlea. New York: Raven Press, pp. 247–269.

Walsh EJ, McGee J (1987) Postnatal development of auditory nerve and cochlear nucleus neuronal responses in kittens. Hear Res 28:97–116.

Walsh EJ, McGee J (1990) Frequency selectivity in the auditory periphery: similarities between damaged and developing ears. Am J Otolaryngol 11:23–32.

Walsh EJ, Romand R (1992) Functional development of the cochlea and cochlear nerve. In: Romand R (ed) Development of Auditory and Vestibular Systems 2. Amsterdam: Elsevier, pp. 161–219.

Weaver SP, Schweitzer L (1994) Development of gerbil outer hair cells after the onset of cochlear function: an ultrastructural study. Hear Res 72:44–52.

Weaver SP, Hoffpauir J, Schweitzer L (1994) Distribution of actin in developing outer hair cells in the gerbil. Hear Res 72:181–188.

Webster WR, Martin RL (1991) The development of frequency representation in the inferior colliculus of the kitten. Hear Res 55:70–80.

Webster WR, Aitkin LM (1975) Central auditory processing. In: Gazzaniga MS, Blakemore C (eds) Handbook of Psychobiology. New York: Academic Press, pp. 325–364.

Whitehead ML, Lonsbury-Martin BL, Martin GK, McCoy J (1996) Otoacoustic emissions: animal models and clinical observations. In: Van De Water TR, Popper AN, Fay RR (eds) Clinical Aspects of Hearing. New York: Springer-Verlag, pp. 199–257.

Wilson JP (1980) Evidence for a cochlear origin for acoustic re-emissions, threshold fine-structure and tonal tinnitus. Hear Res 2:233–252.

Wong ROL (1993) The role of spatio-temporal firing patterns in neuronal development of sensory systems. Curr Opin Neurobiol 3:595–601.

Woolf NK, Ryan AF (1984) Development of auditory function in the cochlea of the Mongolian gerbil. Hear Res 13:277–283.

Woolf NK, Ryan AF (1985) Ontogeny of neural discharge patterns in the ventral cochlear nucleus of the Mongolian gerbil. Dev Brain Res 17:131–147.

Woolf NK, Ryan AF (1988) Contributions of the middle ear to the development of function in the cochlea. Hear Res 35:131–143.

Woolf NK, Ryan AF, Harris JP (1986) Development of mammalian endocochlear potential: normal ontogeny and effects of anoxia. Am J Physiol (Regulatory Integrative Comp Physiol 19) 250:R493–R498.

Wu SH, Oertel D (1987) Maturation of synapses and electrical properties of cells in the cochlear nucleus. Hear Res 30:99–110.

Yancey D, Dallos P (1985) Ontogenetic changes in cochlear characteristic frequency at a basal turn location as reflected in the summating potential. Hear Res 18:189–195.

Young ED, Brownell WE (1976) Responses to tones and noise of single cells in dorsal cochlear nucleus of unanesthetized cats. J Neurophysiol 39:282–300.

Yao X, ten-Cate WJF, Curtis LM, Rarey KE (1994) Expression of Na^+, K^+-ATPase $\alpha 1$ subunit mRNA in the developing rat cochlea. Hear Res 80:31–37.

Zajic G, Anniko M, Schacht J (1983) Cellular localization of adenylate cyclase in the developing and mature inner ear of the mouse. Hear Res 10:249–261.

Zenner HP (1986a) K^+-induced motility and depolarization of cochlear hair cells: direct evidence for a new pathophysiological mechanism in Ménière's disease. Arch Otorhinolaryngol 243:108–111.

Zenner HP (1986b) Motile responses in outer hair cells. Hear Res 22:83–90.

Zenner HP, Zimmermann R, Gitter AH (1988) Active movements of the cuticular plate induce sensory hair motion in mammalian outer hair cells. Hear Res 34:233–240.

Zidanic M, Brownell WE (1990) Fine structure of the intracochlear potential field. I. The silent current. Biophys J 57:1253–1268.

Zimmer WM, Rosin DF, Saunders JC (1994) Middle-ear development. VI: Structural maturation of the rat conducting apparatus. Anat Rec 239:475–484.

Zirpel L, Lachica EA, Lippe WR (1995) Deafferentation increases the intracellular calcium of cochlear nucleus neurons in the embryonic chick. J Neurophysiol 74:1355–1357.

6
The Development of Central Auditory Processing

Dan H. Sanes and Edward J. Walsh

1. Introduction

The range of acoustic features that are encoded and interpreted by the mature central auditory system has become a rapidly expanding topic of research (Bregman 1990; Handel 1990; Yost 1991). However, our understanding of the developmental processes that underlie auditory perception remains rudimentary. This chapter focuses on developmental mechanisms that contribute to central auditory processing of acoustic stimuli. The term *processing* will refer to the neural mechanisms by which the central auditory system represents acoustic information. In principle, this representation may take the form of a single neuron's discharge pattern or the activity pattern of an entire central auditory structure.

One challenge to studying the processing of simple acoustic stimuli is to understand the relative contributions of the peripheral and central auditory systems. For example, developmental alterations in the cochlea's tonotopic map are mirrored in the central nervous system (Rübsamen 1992; Lippe and Norton, Chapter 5; 1992). Because it is practically impossible to isolate the cochlea's contribution to a sound-evoked response, few developmental studies provide an unconditional assay of central auditory processing. Even binaural coding properties that derive from central synaptic integration may exhibit developmental changes that are of cochlear origin. For example, if the rate of cochlear development is not synchronized at the two ears, then a seemingly immature binaural coding property may actually reflect cochlear asymmetry.

A general constraint on the study of information processing in the developing central nervous system (CNS) concerns the magnitude of neural response. For single neurons, it has commonly been found that the most effective sound stimulus evokes fewer discharges in neonatal animals compared to adults. If information (i.e., number of action potentials per unit time) is diminished, then one must consider whether the neuronal computations that underlie perception in developing animals require longer stimulus durations, are performed in a unique manner, or are inadequate to support an accurate perception. A related conceptual problem is our inability to assign specific auditory perceptions to discrete populations of neurons (however, see Riquimaroux, Gaioni, and Suga 1992). Therefore, when one finds that only a subset of neurons are functionally mature at a certain age, it remains unclear whether or not they are able to support mature psychophysical attributes. It is important to understand that these sorts of questions will only be addressed through a rigorous correlation of neurophysiological data with the psychophysical attributes of immature animals (Werner and Gray, Chapter 2).

In this chapter, we will begin by focusing on the cellular data that directly illustrate the onset of central auditory function and the maturation of ionic and synaptic properties that underlie stimulus processing. The discussion then moves to sound-evoked neural responses in neonates and how they differ from those found in the adult nervous system. We will then examine the development of binaural processing, including the formation of spatial maps, to illustrate how central auditory computations are assembled. Finally, we will consider whether activity levels within the developing auditory system regulate its subsequent functional maturation.

2. The Onset of Central Auditory Synaptic Function

The onset of auditory function may be detected in two fundamental ways: first, *as the time when central synapses are capable of neurotransmission* and second, *as the time when sound first elicits a response from various central auditory nuclei.* As described below, the onset of synaptic transmission appears to occur immediately after innervation but is not necessarily associated with the onset of response to airborne sound.

It is known from previous in vivo electrophysiological studies that sound-evoked responses are not generally obtained from the gerbil cochlea, the cochlear nucleus, or the lateral superior olive before postnatal day 12 (Finck, Schneck, and Hartman 1972; Woolf and Ryan 1984, 1985; Sanes and Rubel 1988a). However, when the stapes is mechanically activated with a piezoelectric driver, a cochlear microphonic is elicited by postnatal day 10 (Woolf and Ryan 1988). Intracellular recordings obtained from the lateral superior olive (LSO) demonstrate that both excitatory and inhibitory synaptic transmission can be evoked with direct electrical stimulation of afferents by postnatal day 1 (Fig. 6.1; Sanes 1993). These results suggest that cochlear activity may be processed in some form by the auditory CNS as soon as it becomes electrically active.

Depending on the mouse strain examined, sound-evoked behavioral responses (i.e., a reflexive movement of the pinnae called the Preyer reflex; Preyer 1908) and neural responses are first observed between postnatal days 8–12 (Alford and Ruben 1963; Mikaelian and Ruben 1965; Shnerson and Willott 1979; Saunders, Dolgin, and Lowry 1980). With use of the

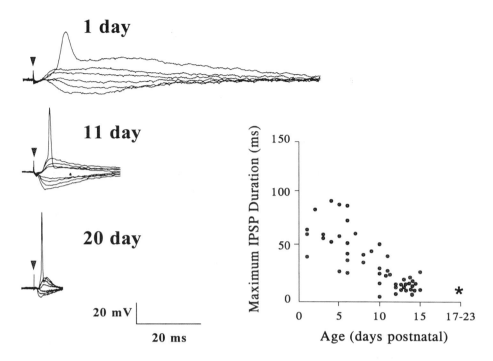

FIGURE 6.1. The development of postsynaptic potential duration in the gerbil lateral superior olive. *Left*: intracellular recordings were obtained in a brain slice preparation, and a family of excitatory and inhibitory synaptic responses are shown for single neurons at 3 different ages. Electric stimuli were applied to pathways from the cochlear nucleus and medial nucleus of the trapezoid body. A stimulus artifact occurs before each set of postsynaptic potentials (arrowheads). *Right*: a scatterplot of maximum inhibitory postsynaptic potential (IPSP) durations are shown along with the mean value (*) from animals aged 17–23 days postnatal (Sanes 1990). There was a dramatic decrease in PSP duration from 1 to 12 days. (From Sanes 1993.)

brain-slice preparation, it was shown that electrical stimulation of auditory nerve fibers elicits synaptic transmission in the cochlear nucleus (CN) at postnatal day 4, the earliest age examined (Wu and Oertel 1987). In the mouse, spiral ganglion cells penetrate the brain by day 12 of gestation (Sher 1971), although the precise time of CN innervation has not been determined. Taken together, the analyses of gerbil and mouse development suggest that synaptic transmission precedes the earliest time at which the central auditory system processes external acoustic stimuli.

Although the onset of synaptic function appears to occur well before sound-evoked neural activity in some mammals, this is not true for all species. In the chicken, auditory nerve fibers innervate the nucleus magnocellularis at embryonic day 10 (Jhaveri and Morest 1982a,b), and synaptically-evoked action potentials are first elicited with electrical stimulation at embryonic day 11 (Jackson, Hackett, and Rubel 1982). Auditory-evoked potentials recorded within the nucleus magnocellularis also indicate that transmission occurs at about embryonic day 11, although it is possible that these compound responses were due to an auditory nerve volley (Saunders, Coles, and Gates 1973). Therefore, it is clear that central synaptic transmission in the chicken is activated by sound stimuli very soon after auditory terminal synaptogenesis (Cant, Chapter 7).

At present, these are the only species for which central synaptic transmission has been assessed *independently*, that is, with direct electrical stimulation of the afferent pathway before the first sound-evoked responses. One may tentatively conclude that synaptic function, albeit grossly immature, begins almost immediately after the arrival of afferent terminals at their central auditory target. Neurites growing in tissue culture secrete neurotransmitter from growth cones and form efficacious synapses on muscle cells within minutes of making an initial contact (Young and Poo 1983; Xie and Poo 1986). Thus the relative maturity of the cochlea appears to dictate the age at which central auditory processing of acoustic cues commences for a given species. In the chicken, sound-evoked activity is processed almost immediately after ingrowth, whereas in the rodent this process is delayed for more than a week due to a protracted period of inner ear development.

Despite the rapid development of synaptic transmission in the central auditory system, it seems likely that synaptic efficacy remains immature for some time. In the developing gerbil, it has been shown that sound-evoked 2-deoxyglucose uptake is delayed by a few days in the superior olivary complex as compared with the ventral cochlear nucleus (Ryan, Woolf, and Sharp 1982). One interpretation of these results is that excitatory synaptic transmission from the CN is present but is subthreshold and cannot elicit postsynaptic action potentials. Thus the number of active terminals in ascending auditory structures would be severely limited. A second line of evidence for immature synaptic function comes from microionophoretic studies in the CN. In neonatal kittens, acoustically unresponsive neurons can be transformed into acoustically responsive neurons after the application of glutamate (Walsh, McGee, and Fitzakerley 1993; see Section 4.2). These results support the concept that synaptic transmission in the CN is present but may be subthreshold for some neurons.

3. The Maturation of Cellular Properties

The most direct cellular criteria for assessing mature central auditory function are the ionic and synaptic properties of individual neurons. These properties can be examined independently of cochlear processing when assessed with an in vitro preparation. At the molecular level, both neurotransmitter receptors and ion channels are known to be supplanted or exchanged during the course of development in several neuronal systems (Spitzer 1991). At the synaptic level, there may be alterations in the strength, number, and location of terminals, all of which may influence the processing of acoustic stimuli.

3.1 Membrane Properties

Although there are presently no developmental studies within the auditory CNS examining developmental transformations of voltage-gated ion channels, it is possible to examine the membrane properties that derive from these membrane proteins. The intrinsic membrane properties of mouse CN neurons (Wu and Oertel 1987) and gerbil and rat LSO neurons (Sanes 1993; Kandler and Friauf 1995) have been examined during development. In general, current-voltage relationships are similar in neurons from neonatal and more mature animals, although membrane time constant (τ) input resistance (R_{in}) vary with age. In the mouse CN, R_{in} decreases and τ increases with age (Wu and Oertel 1987). The situation is somewhat different in the gerbil and rat LSO in that both R_{in} and τ decrease with age (Sanes 1993; Kandler and Friauf 1995). One interpretation of the latter result is that there is a decrease in membrane resistance (R_m) but no change in membrane capacitance (C_m). Because $R_{in} \propto R_m$ and $\tau = R_m C_m$, there would be a decrease in both values during development. A decrease in R_m could result from the addition of a resting conductance (e.g., membrane channel) in older animals. A shorter membrane time constant permits synaptic potentials to rise and decay more rapidly, a property that may be critical for accurate temporal processing in adult animals.

In all three rodents examined, action potential duration is greater in immature animals. The maximum rate of discharge elicited by depolarizing current injection was shown to increase commensurately in the rat LSO (Kandler and Friauf 1995). This could result from different Na^+- or K^+-channel kinetics. However, several neurons in the gerbil LSO produce action potentials with a prolonged plateau, a phenotype that is often indicative of a Ca^{2+} conductance as shown in other developing systems (Spitzer 1981). In the mouse CN, current-evoked action potentials become significantly broader and smaller at rapid discharge rates (Wu and Oertel 1987). In the gerbil LSO, current injection initially evokes action potentials, which occasionally regress into a series of membrane oscillations in some neurons. These data suggest that action potential duration may limit the ability of central auditory neurons to respond to rapidly changing stimuli. The decrement in action potential amplitude indicates that the sodium extrusion mechanism may also be immature, thus limiting the number of action potentials that can be produced in response to a stimulus. The speculative nature of our comments highlights the need for direct analyses of ionic channels in the developing auditory CNS.

3.2 Synaptic Properties

We have noted above that central auditory synaptic transmission occurs at or before the time when cochlear transduction of airborne sound begins. However, the developmental state of central auditory synapses may be quite rudimentary when they first process sound stimuli, and we should consider the rate at which different central properties develop.

By using intracellular recordings from the LSO in the gerbil brain-slice preparation, it was found that the mean duration of excitatory and inhibitory postsynaptic potentials (EPSPs and IPSPs, respectively) decrease dramatically during the first two postnatal weeks (Fig. 6.1). When several electrical stimuli are delivered in short succession (e.g., 7 pulses delivered at 100 Hz), the temporal summation of EPSPs and IPSPs results in prolonged membrane depolarizations and hyperpolarizations, respectively, in neonatal animals. This synaptic property would be expected to limit the ability of developing animals to perform time-critical computations, such as inter-aural time difference discrimination, and temporal integration tasks, such as gap detection (Werner and Gray, Chapter 2). In neonates, fatigue is usually a major factor affecting information processing (Jackson, Hackett, and Rubel 1982; Sanes and Constantine-Paton 1985a; Wu and Oertel 1987; Eggermont 1991), and as such, the temporal summation of synaptic potentials observed in neonatal animals may serve to mitigate against this process. That is, some evoked responses in neurons from young animals are due to the temporal summation of multiple subthreshold events (Sanes 1993).

In the mouse CN, synaptic fatigue is more obvious during the first 2 postnatal weeks. The maximum stimulus rate that is able to elicit a train of action potentials increases from 140 to 500/s (Wu and Oertel 1987). In addition, the minimum synaptic latency is more variable in young animals and decreases with age. At a more qualitative level, it appears that identical electrical stimuli evoke diverse synaptic responses from stellate cells at 7–9 days postnatal but not at day 23 (see Fig. 3 in Wu and Oertel 1987).

Two unusual synaptic response properties are observed during the first postnatal week in the gerbil LSO (Sanes 1993). First, delivery of a short train of stimuli (100 Hz for 70 ms) to the ipsilateral excitatory pathway could elicit a very long-lasting depolarization, as recorded intracellularly. Because the decay time of this depolarization (over 1 s) far outlasts that observed with a single stimulus, it is possible that synaptic transmission recruits a metabotropic response. In fact, a recent study has demonstrated that synaptically evoked depolarization can last for up to *30 minutes* and that such prolonged depolarizations are also evoked by *trans* (1S, 3R)-1-amino-1,3-cyclopentanedicarboxylic acid (*t*-ACPD), a metabotropic gluta-mate-receptor agonist (Kotak and Sanes 1995). Second, delivery of one or several stimulus pulses to the medial nucleus of the trapezoid body (MNTB) often leads to a large hyperpolarization (i.e., IPSP) followed by a rebound depolarization or action potential. Hyperpolarizing current injection can reverse the polarity of the stimulus-evoked IPSP and obliterate the rebound action potential, suggesting that the synaptic hyperpolarization elicits the rebound depolarization, possibly by causing low-voltage-activated calcium channels to open. This may be significant with regard to the developmental influence of inhibitory transmission on the maturation of pre- and post-synaptic processes (Sanes and Chokshi 1992; Sanes et al. 1992; Sanes and Takacs 1993; Kotak and Sanes 1996).

3.3 Synaptic Elimination

In several neuronal systems, it has been found that the amount of afferent convergence onto a postsynaptic neuron decreases during development due to normal synaptic elimination. Inappropriate axonal projections were recognized early in the century (Ramon y Cajal 1908, 1919; Tello 1917) and have now been described at the developing neuromuscular junction (Redfern 1970; Betz, Caldwell, and Ribchester 1979), cerebellum (Mariani and Changeux 1981), autonomic ganglia (Lichtman 1977), cricket ganglia (Chiba, Shepherd, and Murphey 1988), spinal cord (Conradi and Ronnevi 1975), cortex (Cragg 1975; Rakic 1976; Hubel et al. 1977; Innocenti Fiore, and Caminiti 1977; Ivy and Killackey 1982; Rakic et al. 1986), and thalamus (Rakic 1976; Sretevan and Shatz 1986).

This process has also been observed in the central auditory system with an in vitro preparation. By recording intracellular potentials from the chick nucleus magnocellularis while delivering incremental electrical stimuli to the auditory nerve, Jackson and Parks (1982) demonstrated that the number of increments in EPSP size (presumably reflecting the recruitment of an increasing number of afferents) decreases four afferents per cell at embryonic day 13 to two afferents per cell at posthatch day 4 (Fig. 6.2). In accord with this result, it was found that the branch points and terminal swellings of individual eighth nerve fibers decrease over the same time course (Jackson and Parks 1982).

Similar observations have been made with respect to the inhibitory projection from the MNTB to the LSO in the gerbil. By examining the arborization pattern of single horseradish peroxidase (HRP)-labeled MNTB arbors, it was found that their terminal boutons in the LSO become more restricted by ~30% during the first 3 postnatal weeks (Sanes and Siverls 1991). A subsequent electrophysiological analysis demonstrated that the number of quantized increases in IPSP size within LSO neurons also decrease during the first 3 postnatal weeks (Sanes 1993). Thus, synaptic contacts appear to be inappropriately localized during the initial period of development, and some fraction of these contacts are eliminated to yield a mature array of connections (Cant, Chapter 7). The relationship between this process and acoustic stimulus processing remains to be determined.

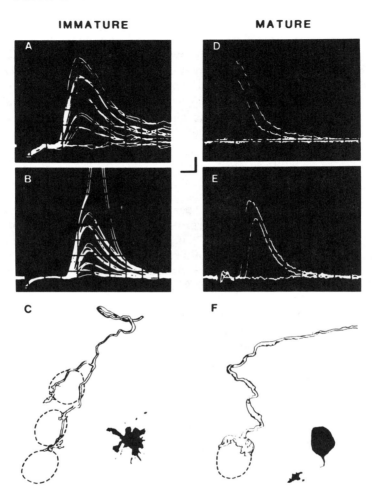

FIGURE 6.2. The reduction of synaptic contacts on chicken cochlear nucleus neurons during development. Intracellular recordings were obtained in the brain chunk preparation, and the number of eighth nerve-evoked quantal increases in excitatory postsynaptic potentials was found to be greater in immature (A and B) than in mature (D and E) animals. There is also an ontogenetic decrease in the number of branched auditory nerve fibers (C) to yield fibers that make single contacts (F) in mature animals. (From Jackson and Parks 1982.)

4. Transmitter Systems

Although remarkably little is known about the development of neurotransmitter systems within the central auditory system, it is widely recognized that many synapses located throughout the auditory neuraxis are formed and are functional, although not necessarily adultlike, before the age that responses to airborne sounds can be elicited. Strong evidence from several laboratories indicates that low-molecular-weight neurotransmitters, most likely amino acids or related compounds, mediate the most frequently studied aspects of neurotransmission within and among brain stem auditory nuclei (Godfrey et al. 1977; Wenthold 1978, 1979, 1980; Altshuler et al. 1981; Potashner 1983; Adams and Mugnaini 1987). The roles that cholinergic, peptidergic and adrenergic receptors, as well as other transmitter or transmitter-like substances, play in the development of central processing are unknown and delineate an underrepresented area of auditory neuroscience. In the following sections, we consider what is known about developmental changes in the availability and actions of both excitatory and inhibitory amino acid neurotransmitters, as well as other neurotransmitter systems when possible.

4.1 Developmental Alterations of Glutamatergic Systems

Although the widespread expression of glutamate-mediated neurotransmission in the auditory brain stem of mammals has been extensively documented, its development has received less attention. Nonetheless, it is clear that central glutamatergic auditory synapses undergo significant developmental regulation. Zhou and Parks (1992) were the first investigators to provide strong evidence that the expression of glutamate-receptor subtypes on auditory neurons is age dependent (Fig. 6.3). Their findings on second-order neurons in chicks indicate that N-methyl-D-aspartate (NMDA) receptors are expressed early in development, but their functional influence declines during later stages of embryonic life. Additionally, Zhou and Parks (1992) found that the functional expression of non-NMDA receptors increases throughout early development, with α-amino-3-hydroxy-5-methyl-4-isoxazole propionic acid (AMPA) exerting the dominant influence on *N. magnocellularis* neuron function in younger embryos and kainic acid (KA) potency being higher in older chick embryos.

4.2 Postsynaptic Considerations

Recently, it has been shown that the expression of glutamate-receptor subunits on auditory neurons changes during development (Hunter and Wenthold 1993). In particular, the GluR1 subunit is expressed at relatively high levels in auditory brain stem nuclei during the first postnatal week in rats and declines in the adult, whereas GluR2-4 subunits are expressed at high levels throughout development and into adulthood. Because glutamate receptors that do not contain the GluR2 subunit are more permeable to Ca^{2+} (Wisden and Seeburg 1993) and because inward Ca^{2+} currents have a profound influence on several aspects of neuronal development (Spitzer

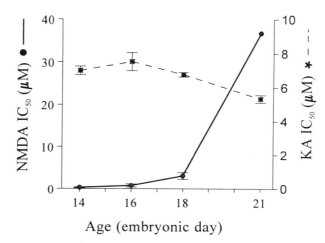

Age (embryonic day)

FIGURE 6.3. A developmental alteration in the functional characteristics of glutama-tergic transmission in the chicken cochlear nucleus. The concentration of glutamate-receptor agonists (N-methyl-D-aspartate [NMDA] and kainic acid [KA]) that was necessary to diminish the auditory nerve-evoked response by 50% (IC_{50}) is shown during development. The IC_{50} of NMDA increases, indicating that this receptor subtype declines with age. The IC_{50} of KA decreases, indicating that this receptor increases in number or efficacy with age. (Data from Zhou and Parks 1992.)

1991), these results suggest the possibility that homomeric GluR1 AMPA receptors act like NMDA receptors in their capacity to mediate Ca^{2+}-induced intraneuronal actions, including metabotropism, however, without conditional (i.e., voltage-dependent) restraints.

Differential expression of receptor subtypes, even within a class (e.g., AMPA receptors), may produce operationally diverse neuronal populations during development. For example, "flip" and "flop" versions of GluR subunits have been described as variants of AMPA-receptor-channel complexes. The relative expression of these versions changes during development and appears to influence glutamate- and/or AMPA-induced cellular currents (Sommer et al. 1990; reviewed by Seeburg 1993). In addition, the regulatory importance of voltage-dependent Ca^{2+} flux triggered by activation of the NMDA-receptor-ionophore complex on cellular development is widely recognized (Cline 1991). However, recent findings suggest that non-NMDA ionotropic glutamate receptors conduct Ca^{2+} currents differentially depending on subunit composition as well. Although there is no evidence that such receptor channels display voltage dependence, the observation that complexes exhibiting relatively high Ca^{2+} conductance are preferentially expressed early in development (Hunter and Wenthold 1993) suggests that activity-driven ionotropic actions may promote cellular differentiation or fuel metabotropic actions that are important during development. Thus alterations in receptor and channel isoforms during development could be important to a comprehensive understanding of developmental processes affecting stimulus processing.

Results from the in vivo microionophoresis studies of Walsh, McGee, and Fitzakerley (1993) indicate that glutamate receptors and their ionophores are functional before the onset of hearing. Exogenously applied glutamate or NMDA raise acoustically evoked discharge rates of most (65 and 80%, respectively) CN neurons studied in kittens during the first postnatal month. However, functional differences among the various classes of glutamate receptors have not been easily distinguished in vivo. Results from CN recordings reveal two clear categories of immature neurons, neither of which can be activated by sound alone. In one set of neurons, the combined action of glutamate and sound produce a clear, synapse-mediated response, verifying the functional integrity of auditory nerve connections with the cell in question (Fig. 6.4A). Neurons belonging to the other category of "unresponsive" cells manifest no sign of synapse-mediated activation in that rate enhancements resulting from exposure to glutamate are unrelated to acoustic stimulation (Fig. 6.4B). The source of "presynaptic" immaturity that limits neuronal responsivity in this context is unknown. However, based on what is known about the time course of synaptogenesis in the brain, it is possible that either sound-evoked EPSPs are subthreshold or that evoked IPSPs are large enough to suppress all activity. There are two further possibilities: that a significant number of auditory nerve fiber synapses have not yet formed on neurons comprising this neuronal subset or that the synapses can not be activated by the cochlea.

Through the microionophoresis of glutamate and other neuroactive agents, it is also possible to alter the temporal discharge properties of auditory neurons during sound stimulation. When glutamate is microionophoresed adjacent to a cell during acoustic stimulation, some neurons are converted from a phasic (i.e., onset) to a tonic, sustained response (Walsh, McGee, and Fitzakerley 1993). Thus the exogenous administration of glutamate reveals the presence of acoustically driven inputs that are not ordinarily observed. Under conditions of combined acoustic and glutamate stimulation, the temporal responses (i.e., poststimulus time histograms) of neurons that were unresponsive to acoustic stimulation alone are essentially identical to responses produced by immature CN neurons that do respond to sound. That is, in the presence of glutamate, these neurons exhibit low-frequency discharge bursts of roughly 10 Hz in response to combined stimulation (see Section 5.6), have acoustic thresholds of at least 120 dB sound pressure level (SPL), and exhibit frequency-dependent and poststimulus fatigue characteristics that are similar to normally responsive neurons at the same age.

Certain aspects of glutamate neurotransmission remain immature for some time into postnatal life. Chief among these is the diminished dynamic

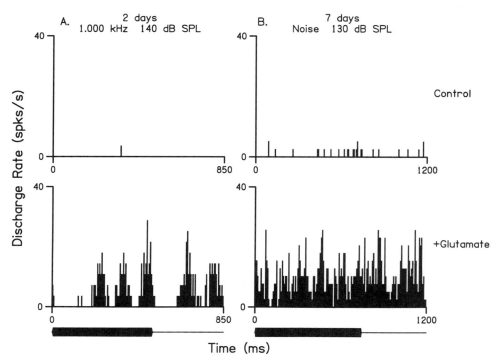

FIGURE 6.4. Examples of peristimulus time histograms representing control (sound alone; *top*) and experimental conditions (sound plus glutamate; *bottom*) for two neurons from the cochlear nucleus of perinatal kittens. Neither neuron responded to sound presented at the highest level available. A: a clear synapse-mediated response was produced under conditions of combined stimulation. B: glutamate elevated overall discharge rate but there was no evidence of sound-evoked rate increase. SPL, sound pressure level; spks, spikes.

range over which CN neurons operate in immature animals. This is evident in both the intensity-dependent character of input-output relationships as well as in the range of currents over which spike rates change under microionophoresis conditions (Fig. 6.5). It is also the case that maximal discharge rates evoked by saturating doses of glutamate tend to be lower in neurons recorded from young kittens compared to maximal discharge rates produced by acoustic stimulation alone in older animals. It is not yet clear how a dynamic range of drug dosage relates to a neuron's coding properties. However, it is clear that any coding properties that require a large dynamic range would be compromised in immature animals as a result of such response range compression.

Recently, it has been shown that glutamate may influence the consolidation of CN synapses during development (Walsh, McGee, and Fitzakerley 1993). Although the number of observations is small, in a distinct subpo-

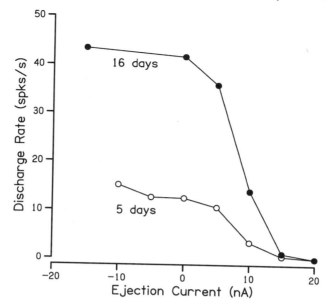

FIGURE 6.5. Examples of dose-dependent effects of γ-aminobutyuric acid (GABA) on discharge rates produced in response to characteristic frequency tone bursts of cochlear nucleus neurons from 5- and 16-postnatal-day-old kittens. The range of discharge rates increased significantly between 5 and 19 postnatal days.

pulation of acoustically unresponsive, immature CN neurons, exposure to glutamate promotes, and in some cases extends, the acoustic responsivity of postsynaptic neurons. As mentioned above, most neurons in the acoustic part of the statoacoustic nerve and the CN complex of cats are unresponsive to acoustic stimulation during the early days of postnatal life. In some of these neurons, exposure to microionophoretically applied glutamate transforms acoustically unresponsive CN neurons to acoustically responsive neurons after termination of glutamate ionophoresis (Fig. 6.6). Neurons in this group continue to respond to acoustic stimulation for as long as they can be recorded from. A more common outcome of glutamate exposure is an improvement of acoustic thresholds (Fig. 6.7). These preliminary observations suggest that the earliest actions of glutamate may be to promote further synaptic maturation (i.e., a trophic influence), possibly through a metabotropic receptor.

4.3 Presynaptic Considerations

Results from the in vitro and in vivo pharmacological studies reflect the relative potencies of selective ligands during electrical or acoustic stimulation, providing insights into neurotransmitter-receptor kinetics. It is equally

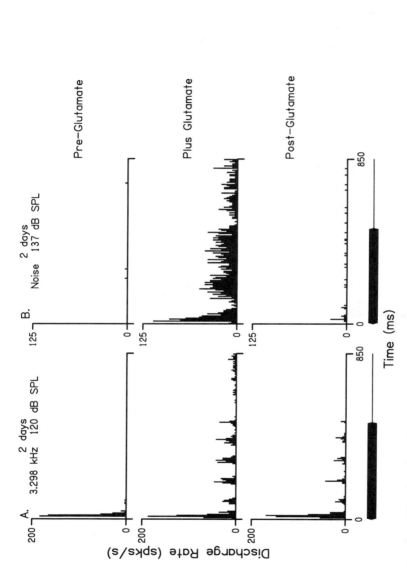

FIGURE 6.6. Examples of peristimulus time histograms representing control (pre-glutamate, sound alone; *top*), experimental (sound plus glutamate; *middle*) and recovery (post-glutamate, sound alone; *bottom*) in the neonatal kitten cochlear nucleus. A: the neuron responded phasically to sound but was converted to a tonic, albeit rhythmic, responder as a consequence of glutamate administration. B: the neuron did not respond to sound but did produce synapse-mediated activity when sound was combined with exogenous glutamate. Both neurons showed signs of continued sound-induced activity in the absence of glutamate after exposure to it.

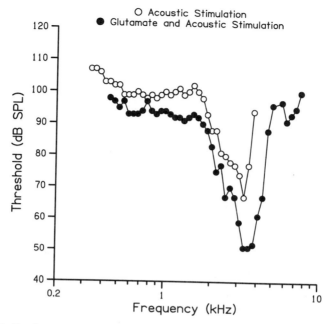

FIGURE 6.7. Tuning curves from kitten cochlear nucleus neuron obtained during simultaneous glutamate and acoustic stimulation are shown along with the corresponding tuning curve obtained with acoustic stimulation alone. Thresholds, both at characteristic frequency (CF) and off-CF, were substantially improved as a consequence of glutamate exposure.

important to understand whether or not a neurotransmitter of interest is present within specific areas of the central auditory system and the amount available for release at the synapse. To our knowledge, neurochemical estimates of glutamate concentrations during development have only been made for the rat inferior colliculus with high-performance liquid chromatography (McGee and Walsh, unpublished observations). In that study, glutamate concentrations increased monotonically during the period that inner ear transduction is acquired in the rat (i.e., birth through 21 postnatal days), after which concentrations plateaued at a value significantly higher than that measured at birth (Fig. 6.8). These results are consistent with the escalation of glutamatergic-antagonist potency in the chick (Zhou and Parks 1992), suggesting that a similar developmental trend may occur among diverse mammalian species and among auditory nuclei throughout the vertebrate brain stem. Although efforts are being made to identify that portion of the total glutamate pool actually available for neurotransmission during development (McGee, personal communication), early estimates of overall glutamate concentrations throughout development support the

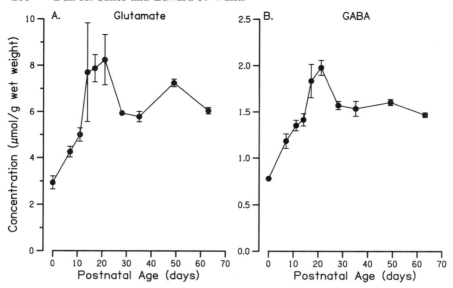

FIGURE 6.8. Developmental changes in the concentrations of glutamate (A) and GABA (B) within the inferior colliculus of rats are shown. For these amino acids, concentrations increase during the early postnatal period and reach a maximum at the end of the third week. They then decline and attain adultlike values after the first postnatal month.

notion that the maturational state of presynaptic synapse elements does limit the acquisition of normal, adultlike synapse function.

4.4 Inhibitory Neurotransmission in the Auditory CNS

As with excitatory amino acid neurotransmitters, McGee and Walsh (unpublished observations) have shown that γ-aminobutyric acid (GABA) concentrations in the inferior colliculus increase monotonically during the period that auditory function is acquired and slightly decrease and saturate thereafter (Fig. 6.8). Glycine concentrations, on the other hand, either follow monotonic age curves or show no concentration change when considered in wet weight terms. On the basis of these measurements, one may tentatively conclude that GABAergic systems change more dramatically during the period that CNS systems become functionally organized than do glycinergic systems.

Immunocytochemical results are consistent with a developmental increase in the expression of GABA and glycine during the postnatal period (Snead, Altschuler, and Wenthold 1988; Code, Burd, and Rubel 1989; Schweitzer, Cecil, and Walsh 1993). It is likely that inhibitory and excitatory transmitter systems emerge at roughly the same developmental age in at least some auditory nuclei (Schweitzer and Cant 1984; Code, Burd, and Rubel 1989;

Schweitzer and Cecil 1992). For example, the GABA-synthesizing enzyme glutamic acid decarboxylase (GAD) first appears in the dorsal CN around the time that auditory nerve fibers and descending inhibitory inputs arrive, and a second wave of GAD expression occurs several days later, presumably in association with the formation of internuncial connections (Schweitzer, Cecil, and Walsh 1993). Interestingly, recent findings made by Schweitzer and colleagues (Riggs et al. 1995) indicate that glycine receptors are evident among neurons of the dorsal CN (DCN) as much as one week earlier than GABA is first expressed in the structure.

Finally, $[^3H]$-strychnine autoradiography in the gerbil LSO indicates that glycine receptors are present at or soon after birth (Sanes and Wooten 1987). However, the relative density of $[^33H]$-strychnine binding sites is modified along the tonotopic axis during the first three postnatal weeks. As with glutamate, the ionophoretic application of GABA in vivo allows one to define a dynamic range (i.e., the range of currents over which discharge rate is modified). As shown in Figure 6.5, the operating characteristics of auditory brain stem neurons are quite immature in neonatal cats. Although exogenously applied GABA reduces the spontaneous and/or acoustically evoked discharge rates in general, it is interesting to note that the action, like that associated with NMDA-receptor actions, appears to be voltage dependent in that the GABA efficacy was directly related to control discharge rates (i.e., efficacy is high when discharge rates are elevated) for CN neurons recorded throughout the period of development (Walsh, McGee, and Fitzakerley 1990, 1995).

5. The Quantity and Precision of Sound-Evoked Discharge

5.1 Maintained Discharge Levels

The *maintained* activity that central auditory neurons produce in the absence of sound stimuli may play a role in both information processing and synaptic maturation, as has been suggested for the visual system (Levick 1973; Shatz 1990). The few studies in which maintained discharge has been considered during postnatal development have shown it to increase dramatically. However, it is conceivable that anesthesia, which is known to suppress maintained activity, has a different efficacy in neonates and adults. Nevertheless, it is now clear that spontaneous discharge of central auditory neurons occurs before the onset of sound-evoked responses (Gummer and Mark 1994; Kotak and Sanes 1995).

The mean spontaneous discharge rate of cat auditory nerve fibers increases over an extended period of time during early postnatal life, and fibers eventually display a broad range of spontaneous rates (Romand 1984;

Walsh and McGee 1987). Primary-like and chopper units in the CN and primary auditory cortical area (AI) neurons exhibit an increase of spontaneous rates over a similar period of time (Eggermont 1992; Fitzakerley, McGee, and Walsh 1991). The percentage of neurons exhibiting spontaneous discharge also increases during early postnatal life in the cochlear nuclei and the inferior colliculus (Romand and Marty 1975; Shnerson and Willott 1979; Brugge and O'Connor 1984; Woolf and Ryan 1985).

Although there is a general increase in spontaneous discharge rates, it appears that this activity is qualitatively different from that found in the adult. For example, neurons in the CN of prehatchling chicks and postnatal wallabys (*Macropus eugenii*) exhibit periodic spontaneous discharge patterns (Lippe 1994; Gummer and Mark 1994). During the first two postnatal weeks, there is an increase in the percentage of neurons in the cat CN that display a less regular, yet maintained, bursting pattern after offset of sound stimulation (Fitzakerley, McGee, and Walsh 1991). In the neonatal mouse, periodic discharge patterns have been correlated with the cardiac cycle (Sanes 1984), and these patterns are not observed in nonauditory areas. It is possible that pulsations in the large basilar spiral artery found in immature vertebrates produce the bursting activity described here. Interestingly, this vessel undergoes a dramatic decrease in diameter during the first 2 postnatal weeks in rodents and cats (Retzius 1884; Wada 1923; Axelsson, Ryan, and Woolf 1986). Although clear, regularly occurring bursts have not been observed in cats, poststimulus bursts have been noted in a small population of CN neurons that respond rhythmically to acoustic stimulation (Walsh and McGee 1987).

5.2 Evoked Discharge Levels

It has generally been found that sound-evoked discharge rates are extremely low in young animals, even when an optimal stimulus configuration is employed. The sensitivity, maximum discharge rate, and operating range (the range of sound levels across which neuronal discharge is found to vary) of auditory neurons are often well correlated. For neurons in the cat VCN, the mean maximum discharge rate and dynamic range appear to reach maturity simultaneously at about postnatal day 20 (Brugge, Javel, and Kitzes 1978; Brugge, Kitzes, and Javel 1981; see Fig. 11 in Walsh and McGee 1987). In the gerbil LSO, there is a positive correlation between mean maximum discharge rate and operating range from 13 to 16 days postnatal (Fig. 6.9A). However, the dynamic range of gerbil ventral CN (VCN) neurons reaches maturity at postnatal day 18, whereas the mean maximal discharge rate continues to increase through day 30 (Fig. 6.9B; Woolf and Ryan 1985).

Therefore, fewer action potentials are available to central auditory nuclei for processing. This observation suggests that an animal's perceptual capacities may be limited by this relatively small information flow through

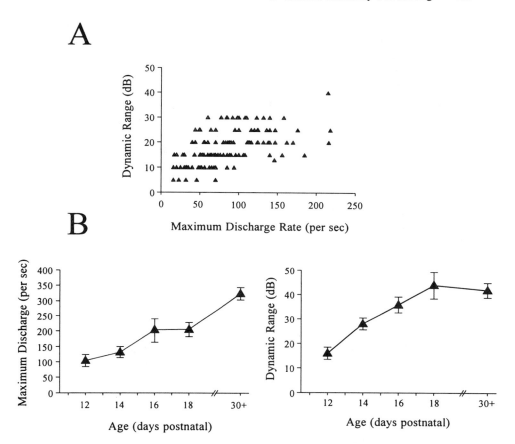

FIGURE 6.9. The development of discharge rate and operating range in the gerbil auditory brain stem. A: there is a positive correlation between dynamic range and maximum discharge rate to 100 spikes/sec for single lateral superior olive (LSO) neurons in 13- to 16-day gerbils. For larger values of maximum discharge there is no correlation (data from Sanes and Rubel 1988). B: the maximum discharge rate of single neurons in the gerbil cochlear nucleus continues to increase through postnatal day 30 (*left*), even though the dynamic range is mature at 18 days postnatal. (Adapted from Woolf and Ryan 1985.)

the CNS (see Werner and Gray, Chapter 2). For example, human infants require sound stimulation for >500 ms to elicit an appropriate head orientation response (Clarkson et al. 1989). This is a much longer duration than that required for adults to perform simple perceptual tasks (Tobias and Zerlin 1959; Ricard and Halfter 1973). Analyses of absolute thresholds and startle responses also suggest that extended temporal integration is required for optimal performance in human infants compared to adults (Watson and Gengel 1969; Blumenthal, Avenando, and Berg 1987; Thorpe and Schneider 1987).

Most analyses of central auditory coding properties consider the cumulative response to many iterations of each sound stimulus. These analyses ignore variation of the neuronal response from trial to trial, such as the failure of a neuron to produce any response at all. In recordings from kitten auditory nerve fibers and CN neurons, Fitzakerley, McGee, and Walsh (1991) demonstrated that the sound-evoked discharge rates to a tone pulse exhibit a greater coefficient of variation during the first three postnatal weeks compared to those found in adults. The variability of sound-evoked response has also been demonstrated for cat inferior colliculus neurons. The discharge rate to a binaural stimulus of equal intensity at each ear [0 dB interaural intensity difference (IID)] was measured three times, and the range of values was divided by the maximum contralaterally evoked discharge rate. This value decreases from 48% at days 11–19 to 27% at days 41–50 (Moore and Irvine 1981). Studies of variability of the neural response may provide a means for comparing neurophysiological and behavioral data concerning response reliability in developing animals (Werner and Gray, Chapter 2).

5.3 Stimulus Following

The low sound-evoked discharge rates exhibited by neurons in developing animals may be curtailed even further by a rapid decrease in response to repeated sound stimuli (Ellingson and Wilcott 1960; Mourek et al. 1967). Stimulus-following capacity has been assessed by measuring the amplitude of click-evoked compound potentials at increasing repetition rates (Saunders, Coles, and Gates 1973; Sanes and Constantine-Paton 1985a). The underlying assumption is that a decreased response amplitude signifies the failure of some neurons to respond, although it may also result from a less synchronized discharge among the elements of the population being recorded from. Recordings obtained from the chick CN demonstrate that stimulus rates as low as 2 Hz lead to a decrement in the compound action potential amplitude at embryonic days 12–18, and a rate of 5 Hz leads to a 50% decrement at embryonic days 12–14 (Saunders, Coles, and Gates 1973). A similar finding was obtained in developing mice (Sanes and Constantine-Paton 1985a), although the following response was found to mature several days earlier in the CN compared with the inferior colliculus.

The ability of single neurons to follow repetitive clicks has been studied at the level of the auditory cortex in kittens (Eggermont 1991). By analyzing the percentage of times that a neuron responded to each click within a train of stimuli, as rate varies between 1 and 32 Hz (i.e., interstimulus interval decreases from 1,000 to 30 ms), it was shown that the following response improves with postnatal age. Figure 6.10 shows the percent response, termed entrainment, of cortical neurons to each of the first nine clicks from an 8-Hz train. There is a dramatic improvement in entrainment, particularly for later occurring pulses, that continues through postnatal day 60.

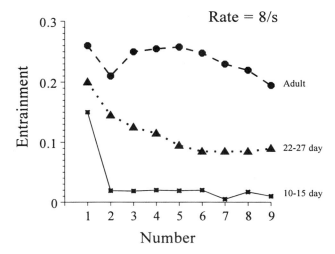

FIGURE 6.10. The development of click-following response by single neurons in the kitten auditory cortex. The degree of entrainment, defined as the percentage of clicks that produced at least one action potential, is plotted against each click in a 8-Hz train. In very young animals, the neurons are only able to respond to the first click in the series (10–15 day). However, in the adult there is almost no change in response efficacy during the click train. (Data from Eggermont 1991.)

Eggermont (1991) noted that the maturation of this response property occurs much later in development than that observed for threshold, latency, and maximum discharge rate at lower areas of the cat auditory pathway. The response of cortical neurons to amplitude-modulated noise matures much more rapidly than that to click stimuli (Eggermont 1993), suggesting that the specific afferent pathways that are recruited by diverse stimuli are differentially susceptible to fatigue.

5.4 Prolonged Discharge Latency

A universal finding in the area of developmental auditory physiology is that response latencies are prolonged in immature animals. This has been measured by using both compound potentials and single neuron responses (Ellingson and Wilcott 1960; Romand 1971; Pujol 1972; Romand, Granier, and Marty 1973; Hecox and Galambos 1974; Romand and Marty 1975; Brugge, Javel, and Kitzes 1978; Woolf and Ryan 1985; Walsh, McGee, and Javel 1986; Sanes and Rubel 1988a). Taken together, these studies demonstrate that neurons in more rostral areas of the auditory pathway undergo a greater and more prolonged shortening of response latency when compared with auditory nerve fibers.

One mechanism by which this could occur is suggested by the development of synaptic potentials in the gerbil LSO. During the first 3 postnatal weeks, the rising slope of EPSPs increases from 5.8 to 10.9 mV/ms and the rising slope of action potentials increases from 24.7 to 102 mV/ms (Sanes 1993). If a similar change in synaptic transmission occurs at every central auditory nucleus, then each set of connections will introduce an additional delay to the signal. A second factor that might be expected to reduce response latency at all levels of the neuraxis is myelination. The addition of myelin lamellae to axons of the auditory nerve and brain stem in cats exhibits an extended period of maturation (Walsh et al. 1985). However, when one compares the time course of myelination with the response latency of auditory nerve fibers, there is practically no correspondence (see Fig. 4 in Walsh and McGee 1986). Although other factors contribute to conduction velocity, such as axon diameter and sodium-channel density, the developmental alterations in response latency appear to be due largely to immature synaptic transmission.

5.5 Precision of Temporal Discharge

The precision with which auditory neurons discharge is most clearly illustrated by their synchronized response to low-frequency sinusoidal stimuli (phase locking). This property is thought to play a fundamental role in frequency analysis and the detection of interaural time differences (Werner and Gray, Chapter 2). Brugge, Javel, and Kitzes (1978) and Kettner et al. (1985) have conducted an extensive developmental analysis of this property in auditory nerve fibers and anteroventral CN neurons. When phase locking is quantified with a metric that establishes the degree to which a neuron discharges at the same phase within a sinusoidal stimulus (Goldberg and Brown 1969), it is found that the maturation of anteroventral CN (AVCN) neurons lags behind that of auditory nerve fibers (Fig. 6.11). Both the degree of phase locking and the maximum frequency at which it occurs are found to increase with age (Brugge, Javel, and Kitzes 1978; Kettner, Feng, and Brugge 1985).

5.6 Rhythmic Discharge and Efferent Control

During the early developmental period, rhythmic or bursting discharge patterns of auditory neurons appear and clearly distinguish immature animals from adults. These rhythmic discharges have been observed both in the absence of acoustic stimulation (i.e., "spontaneous rhythms") or during acoustic stimulation depending on the species studied. Rhythmic "spontaneous" discharge activity has been observed in immature chicks (Lippe 1994), and stimulated rhythms have been observed during the first 2 postnatal weeks in kitten auditory nerve fibers and central neurons (Pujol 1969; Carlier, Abonnenc, and Pujol 1975; Romand and Marty 1975; Brugge and O'Connor 1984; Walsh and McGee 1988). In addition, unambiguous

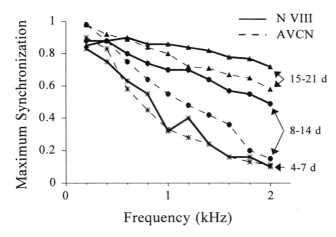

FIGURE 6.11. The development of phase-locked discharge in the kitten cochlear nerve and nucleus. The maximum synchronization of all recorded auditory nerve fibers (N VIII; solid lines) and anteroventral cochlear nucleus (AVCN) neurons (dashed line) during the first three postnatal weeks. At 4–7 days postnatal, the maximum synchronization rate falls off very rapidly with stimulus frequency and is roughly equivalent for auditory nerve and AVCN neurons. At 8–14 days postnatal, the maximum synchronization has shown a greater improvement in the auditory nerve compared with AVCN neurons. By 15–21 days postnatal, the maximum synchronization has improved in both loci and is once again similar to one another. (Data from Kettner, Feng, and Brugge 1985.)

"poststimulation" rhythmic activity after the termination of acoustic stimulation has been observed in a small population of CN neurons in cats; however, this activity remains time locked to the previous stimulation (Walsh and McGee 1988). Although the poststimulus epoch does not reflect true spontaneous discharge activity, such activity occurs after stimulation and thus resembles the spontaneous rhythmic activity observed in chicks. The frequency of the vast majority of acoustically evoked rhythms ranges between 5 and 15 Hz, is intensity dependent, is not correlated with stimulus frequency, and is not associated with the phenomenon of phase locking to the stimulus. The functional relevance, if any, of this phenomenon is unclear. However, one is drawn to consider mechanisms that have been proposed to explain synapse consolidation in other systems as a rationale for the phenomenon (Shatz 1990).

Numerous other explanations for rhythmic responses are possible, including the prospect that it is an epiphenomenon. In this context, it is interesting that individual fibers of the crossed olivocochlear bundle of adult cats respond rhythmically to sounds near the acoustic threshold and that the frequency of the bursts is in the range of that reported for immature auditory neurons throughout the system (Robertson and Gummer 1985; Liberman and Brown 1986). During the neonatal period when rhythmic responses are observed in cats, OCB projection patterns appear to be distinct from adults. It appears that crossed OCB fibers do not form readily identifiable axosomatic contacts with outer hair cells, that fibers of the uncrossed OCB may be delayed in their arrival at the cochlea, and that crossed OCB projections may actually form transient axosomatic contacts with inner hair cells in perinatal mammals (Pujol, Carlier, and Devigne 1978; Lenoir, Shnerson, and Pujol 1980; Simmons et al. 1990). Recently, it has been found that contralateral sound stimulation leads to an enhancement of cochlear output in young kittens more frequently than in adults (Fig. 6.12; Jenkins, Walsh, and McGee 1993). This is in contrast to the suppressive influence of contralateral stimulation observed in adult cats (Warren and Liberman 1989). These physiological results, in conjunction with a transient efferent projection to inner hair cells, leads to the hypothesis that transduction at the inner hair cell level may be modulated directly at a time when outer hair cells lack efferent endings. The view that OCB neurons produce the acoustically evoked rhythmic activity in immature kittens is supported by a study in which the brain stem was sectioned off the midline at the floor of the fourth ventricle in a position corresponding to the location of the olivocochlear bundle (OCB) and temporal response profiles of neurons recorded before and after the lesion were

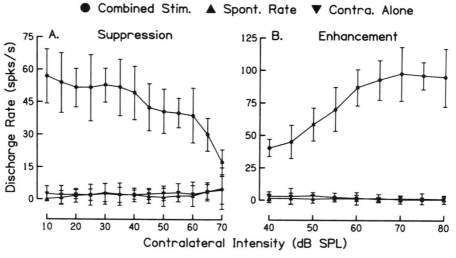

FIGURE 6.12. The intensity-dependent influence of contralateral acoustic stimulation on auditory nerve fiber discharge rates from an adult (A) and a 17 day old kitten (B) are shown. (A) The auditory nerve fiber had a threshold of 32 dB SPL at a CF of 1.14 kHz. (B) The auditory nerve fiber had a threshold of 52 dB SPL at a CF of 12.03 kHz. In young kittens, contralateral sound stimulation produced rate enhancement more frequently than suppression, while control discharge rates were most commonly suppressed in adults.

compared (Walsh, McGee, and Fitzakerley 1995). Response rhythmicity was abolished as an apparent consequence of the section, supporting the hypothesis that the OCB is the source of bursting discharge behavior among immature auditory neurons.

6. The Development of Spatial Processing

A developmental analysis of complex processing has been restricted largely to those coding properties thought to underlie sound localization. These include IID coding, and two-dimensional spatial receptive fields that derive from both monaural and binaural properties. The underlying developmental issues that concern us are twofold: how peripheral immaturities affect binaural processing and whether limitations on synaptic integration lead to immature coding properties.

6.1 Interaural Level Differences

The binaural property for which the greatest ontogenetic information exists is IID processing, although the literature does not always provide a consistent picture. Neurons that are responsive to IIDs are generally excited by sound stimuli delivered to one ear and inhibited by sound stimuli delivered to the other ear (termed excitatory-inhibitory [EI] cells). The development of EI neuron responses to IIDs has been studied in the LSO, inferior colliculus (IC), and AI auditory cortex. Qualitatively, all studies have shown that the discharge rate of EI neurons varies with binaural intensity difference from the earliest age examined (Moore and Irvine 1981; Brugge, Reale, and Wilson 1988; Sanes and Rubel 1988a; Blatchley and Brugge 1990). However, it remains to be seen whether the topographic organization of IID-sensitive neurons, such as those found in the adult cat and bat midbrain (Fuzessery, Wenstrup, and Pollak 1985; Wenstrup, Ross, and Pollak 1985; Wise and Irvine 1985), become more precise during development.

Of the three studies that were performed in kittens, one found that the discharge rate varied irregularly with IID during the first 4 postnatal weeks (Moore and Irvine 1981), whereas two studies from the same research group found that the discharge rate consistently varied with intensity difference (Brugge, Reale, and Wilson 1988; Blatchley and Brugge 1990). Irregular IID functions have also been described in the developing gerbil LSO (Sanes and Rubel 1988a), although they were more conspicuous in the kitten IC.

The major quantitative differences between immature and mature IID functions in the gerbil LSO are decreased dynamic range and resolution (i.e., discharge/ΔdB). In 13- to 16-day animals, the IID dynamic range is 18–19 dB and the resolution is 3.6–3.8 spikes/s/dB (Fig. 6.13). In LSO neurons from adult animals, both of these parameters double in size, and

A

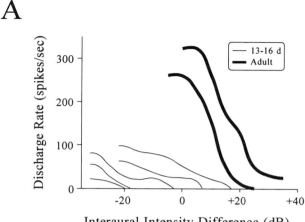

Interaural Intensity Difference (dB)

B

Age	Dynamic Range (dB)	IID Resolution (discharges/dB)	Neurons
Adult	28.2 (10.3)	7.9 (2.7)	42
15-16 day	19.2 (7.0)	3.8 (2.4)	47
13-14 day	18.0 (4.0)	3.6 (1.5)	28

FIGURE 6.13. The development of interaural intensity difference (IID) coding by single neurons in the gerbil LSO. A: representative IID functions are shown for 5 single LSO neurons in 13 to 16-day gerbils and 2 single neurons in adult gerbils. The IID functions from young animals are shallower and are somewhat irregular in shape. Negative values of IID denote relatively greater intensity to the ipsilateral ear. B: parametric analyses of IID functions revealed that dynamic range increased by 50% and resolution (i.e., the number of discharges per dB change in intensity difference) increased over 100% between 13 to 16 days postnatal and the adult. Values are means; nos. in parentheses, SE. (Adapted from Sanes and Rubel 1988.)

this at least partially reflects maturation at more peripheral locations. For example, Woolf and Ryan (1985) showed that the mean dynamic range of CN neurons increases to a mature value of 44 dB by 18 days postnatal in the gerbil.

Appropriate processing of IIDs also requires that the afferents from each ear are well matched in the frequency domain. When characteristic frequency and tuning of the excitatory and inhibitory pathways were compared with one another in the gerbil LSO, it was found that the tonotopic maps were well aligned from the earliest response to sound (Sanes and Rubel 1988a). There was, however, a difference in the IID ranges encoded by developing LSO neurons. In young animals, the discharge rate can be modulated from maximum to minimum when ipsilateral sound levels are greater than contralateral sound levels. Additionally, the average midpoint of the IID function shifts approximately 10 dB during development (Sanes and Rubel 1988a). This parameter has not been quantified in the other studies of intensity difference coding. This effect may be due to a transformation of the transmitter system that mediates contralateral inhibition. Glycine-receptor expression is at first uniform within the LSO but assumes an inhomogeneous distribution pattern during the first 3 postnatal weeks (Sanes and Wooten 1987).

6.2 Maps of Space

Neurons with delimited spatial receptive fields (i.e., azimuth and elevation) are topographically ordered within the auditory midbrain of the barn owl (Knudsen and Konishi 1978; Knudsen 1982), and a similar map of auditory space is found in the deep and intermediate layers of the mammalian superior colliculus (Palmer and King 1982; King and Hutchings 1987). Developmental analyses of spatial maps have now been obtained for both a precocial mammal (guinea pig) and an altricial mammal (ferret). Although the development of IID processing in the auditory brain stem has not been studied in these species, it appears that neurons with spatially tuned receptive fields mature over an extended time course.

Electrophysiological recordings were obtained from multiple units in the guinea pig superior colliculus (SC) during the first 6 months of postnatal development, and the response to short noise bursts from speaker locations along the azimuth were evaluated (Withington-Wray, Binns, and Keating 1990a). These recordings demonstrate that there is a progressive decrease in the range of azimuthal sound locations that elicit a response from an SC recording site, nearing maturity by 32 days postnatal. In adult guinea pigs the auditory receptive fields are discrete and topographically ordered, whereas in the young animal (0-30 days postnatal) the receptive fields are not spatially tuned (Fig. 6.14).

Even when spatially tuned responses begin to emerge in the guinea pig at 16–30 days after birth, they are predominantly responsive to loci at 20° or 90° off the midline. It is possible that these immature receptive fields are partially attributable to the restricted IID ranges that were described for LSO neurons in developing gerbils (Sanes and Rubel 1988a). Although mean absolute response thresholds do not change significantly during the period of auditory space map development, there is a continuous increase of the interaural distance through 64 days postnatal. Finally, the authors noted the limited behavior capability of neonatal guinea pigs to localize sound (Clements and Kelly 1978) prior to 32 days postnatal, and suggest that the topographical organization of the SC may be unnecessary when processing the extremely long duration stimuli (5 secs) provided to young animals

1-11 day

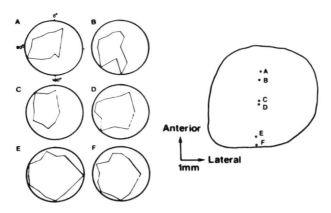

Adult

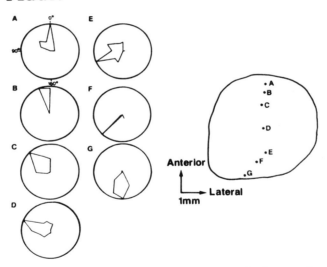

FIGURE 6.14. The maturation of the azimuthal space map in the superior colliculus (SC) of guinea pig. The receptive fields are illustrated on polar plots that show the relative magnitude of response for a recording site, the location of which is shown to the *right*. Representative recordings obtained from 1 to 11-day guinea pigs (*top*) exhibit a very large receptive field and lack of topographic organization along the anteroposterior access of the SC. Representative recordings from an adult animal (*bottom*) exhibit well-defined receptive fields that are topographically organized. Midline stimulation (0°) selectively activate sites in the rostral SC, whereas stimuli located behind the animals head (180°) activate the sites in the caudal SC. (From Withington-Wray, Binns, and Keating 1990.)

(Withington-Wray, Binns, and Keating 1990a). It should be noted that gradual emergence of a functional map in the superior colliculus is analogous to Gray's (1991) behavioral finding that young chickens are able to discriminate between sound stimuli prior to the appearance of a consistent relationship between these same stimuli.

More recently, the maturation of a topographic map in the ferret superior colliculus has been compared to the development of monaural and binaural cues provided by the auditory periphery (King and Carlile 1991; Carlile 1990). It was found that spatially tuned multi-unit responses and a topographic representation of azimuth were attained after 50 days postnatal (King and Carlile 1991). Moreover, adult-like response characteristics are well correlated with the maturation of interaural intensity cues and spectral transformation functions of the external ear (Carlile 1990).

7. The Influence of Processing on Functional Maturation

The development of synaptic circuits that underlie various auditory computations is largely established prior to the onset of response to airborne sound. However, as the prolonged development of spatially tuned neurons suggests, a period of use may be required to establish the mature phenotype. Here, we discuss the evidence for the influence of stimulus processing on the maturation of frequency coding and sound localization.

7.1 Frequency Coding

The range of frequencies to which a single neuron responds is a consequence of its excitatory and inhibitory innervation (Greenwood and Maruyama 1965). Although there have been numerous studies documenting sharper frequency tuning of central auditory neurons during development, it is difficult to dissociate central changes from the concurrent maturation of the cochlea (see Sanes and Rubel 1988b). However, a number of experimental manipulations have led to an alteration in the frequency-coding properties of central auditory neurons. All of these studies were predated by an experiment in which chicks were raised with bilateral ear plugs until posthatch days 3–4 and assessed for frequency discrimination with an habituation generalization paradigm (Kerr, Ostapoff, and Rubel 1979). The results indicated that animals were poorer at frequency discrimination even though the cochlea appeared to be functionally normal, indicating that central auditory processing was modified by the ear plugging.

On the basis of these results, one might postulate that the frequency-response characteristics of individual central neurons would also be influenced by environmental stimuli. For example, single-neuron frequency-tuning curves from the inferior colliculus of mice that were reared in an acoustic environment of repetitive clicks (20/s) were significantly broader than those of normal animals even though absolute thresholds were normal (Sanes and Constantine-Paton 1985b). The authors proposed that a normal period of synaptic rearrangement is dependent on the differential activity patterns of afferents and was prevented from occurring by synchronizing the discharge pattern of the afferent population. An analogous finding has been obtained in neonatally deafened cats that were subsequently stimulated with electrical pulses (30 pulses/s; 1 hour/day; 5 days/week; 12 weeks) delivered to one cochlea with an implanted electrode array (Snyder et al. 1990). Because frequency tuning could no longer be assessed with acoustic stimuli, the authors determined the dorsoventral extent of the inferior colliculus that was activated by one electrode pair of the array that lay along the cochlea's tonotopic axis (i.e., spatial-tuning curves). It was found that the spatial-tuning curves obtained from stimulated animals were significantly broader than those obtained from deafened, unstimulated animals.

The activity-dependent modification of frequency coding can also be demonstrated by rearing neonatal rats in a constant pure-tone environment (4 or 20 kHz) or a frequency-modulated sound (3–5 kHz) for the first 3–5 postnatal weeks (Poon, Chen, and Hwang 1990; Poon and Chen 1992). Both of these manipulations lead to a disproportionate number of inferior colliculus neurons responding to the frequency that was used in the rearing environment. Moreover, there is a concomitant increase in the depth of the inferior colliculus responding to the exposure frequency (e.g., presumed isofrequency contour). A somewhat analogous finding was described by Sanes and Constantine-Paton (1985b) after rearing mice in an environment of repetitive pulses of two tones (11 and 14 kHz). Twin-peaked tuning curves (e.g., those having two best frequencies) with a response within the exposure-frequency range were altered such that the higher frequency peak had a lower than normal threshold. These results suggest that coactivation of the 11- and 14-kHz afferents lead to their selective stabilization, similar to findings from other developing sensory pathways (Shatz 1990).

7.2 Sound Localization

The neuronal-response properties underlying azimuthal sound localization require frequency-specific convergence of afferents from each ear. As spatial-tuning properties become more elaborate in higher areas of the nervous system, there is a convergence of afferents from peripheral binaural and monaural computational circuits. Several studies have sought to determine whether the neuronal response properties thought to underlie sound localization are dependent on experience. The most common manipulation has been to unilaterally attenuate sound transmission during development, thus altering the usual synaptic interaction between afferents from each ear.

Owls that are reared with one ear plugged gradually compensate for the artificially imposed interaural time and intensity differences and are eventually able to localize sounds accurately (Knudsen, Esterly, and Knudsen 1984a). Although sound localization is initially quite poor when the ear plug is removed, animals younger than 40 weeks at the time of ear plug removal can recover a normal accuracy of localization (Knudsen, Knudsen, and Esterly 1984b). These behavioral results strongly suggest that the central auditory system makes use of an activity-mediated mechanism to generate functional specificity. Binaural capabilities after surgical reversal of a unilateral atresia, a condition in which the ear canal is closed (Wilmington, Gray, and Jahrsdoerfer 1994), has now been studied in humans: The fundamental observation is that recovery of time and intensity difference discrimination, as assessed with earphones, is *superior* to the recovery of free-field sound localization. Taken together, these studies suggest that central auditory circuits are differentially dependent on acoustic stimulation during development.

The effect of unilateral ligation of the external auditory meatus (i.e., unilateral sound attenuation) on the development of intensity difference coding has been examined in three laboratories. Although the specific findings of each group are dissimilar, they provide varying degrees of support for the idea that central auditory pathways can be modified during development. In the rat, monaural attenuation commencing at 10 days postnatal and lasting 3–5 months was found to effect IID processing by EI neurons in the inferior colliculus. Both the ipsilateral (e.g., inhibitory) and the contralateral (e.g., excitatory) pathways were less efficacious if subjected to ear canal ligature during development (Silverman and Clopton 1977) but not during adulthood (Clopton and Silverman 1977). Moreover, the alteration in IID functions could not be explained by the magnitude of the chronic threshold shift at the level of the cochlea (Silverman and Clopton 1977). Because binaural click stimuli were used to assess IID coding, it remains possible that an alteration in monaural properties, such as response latency, could account for the poorer efficacy (Clopton and Silverman 1978).

A similar paradigm was employed in developing kittens, and the data indicate that the contralateral pathway was relatively unaffected by a period of attenuation. However, there was a decreased percentage of neurons sensitive to IIDs, from 57% ($N = 21$) in normal adults to 9% ($N = 32$) in ligation-reared animals (Moore and Irvine 1981). Because the results were, again, not correlated with increased thresholds at the manipulated ear, it was suggested that the decreased IID sensitivity resulted from a diminished efficacy of the ipsilateral inhibitory pathway. Finally, the IID coding properties of cortical AI neurons in the cat have also been examined after a period of unilateral attenuation during development. In general, the efficacy of the previously attenuated pathway is decreased, and the residual threshold shift at the manipulated ear can explain much, *but not all,* of the shift in IID functions (Brugge et al. 1985). A number of methodological differences for assessing IID coding exist between the studies cited above and prevent their direct comparison (e.g., tonal vs. click stimuli, control of average binaural level, and selection of sample population). It is notable that unilateral cochlea removal in neonatal gerbils leads to profound changes in the afferent projection pattern to the inferior colliculus, as well as the response profile of excitatory-excitatory (EE) neurons (Nordeen, Killackey, and Kitzes 1983; Kitzes and Semple 1985; Moore and Kitzes 1985), although this manipulation is far more severe than unilateral atresia.

The auditory space map in the superior colliculus (see above) has been the subject of experimental analysis in ferrets, guinea pigs, and barn owls. Because the superior colliculus is a multisensory area, the developmental influence of both audition and vision on the auditory space map have been examined in each species. Knudsen (1982) has demonstrated that the two-dimensional map of auditory space (i.e., azimuth and elevation) is normally aligned with the visuotopic map such that light or sound stimuli from one point in space activates the same site in the superior colliculus. This alignment is largely preserved in animals reared with unilateral ear plugs (i.e., perturbations of the binaural cues) due to a compensatory shift in the auditory space map relative to the stable visuotopic map (Knudsen 1983; Knudsen and Mogdans 1992). However, in some animals, the unilateral attenuation does result in some distortions of the auditory map. For example, the variability of best ILD (i.e., elevation) or interaural time difference (ITD; i.e., azimuth) can be much larger for a given visuotopic location compared with that observed in normal animals (Knudsen 1985; Mogdans and Knudsen 1992). The compensatory changes underlying the shift in ILD tuning probably occur at several loci including the brain stem nucleus that first responds to level difference stimuli (Mogdans and Knudsen 1994). The alignment between maps fails to emerge after bilateral eyelid suture during development (Knudsen 1988; Knudsen, Esterly, and du Lac 1991). Furthermore, when young owls were reared with prismatic spectacles that displaced the visual field by 23–34°, the neural representation of auditory space was found to remain aligned with the *new* visual receptive field (Knudsen and Brainard 1991). That is, the auditory receptive fields at a physical location in the optic tectum were influenced by the pattern of visually driven neural activity.

A similar set of experimental manipulations has been performed on the ferret and the guinea pig. In the ferret, it was found that the auditory and visual maps of azimuth remained aligned in animals reared with unilateral attenuation (King et al. 1988). When the plugs were removed, the best auditory position shifted by ~30°, suggesting that a rather large compensatory change had occurred to allow the maps to remain in alignment. However, there was more variation in the topography of best auditory position for recordings obtained at a given rostrocaudal position in the SC (i.e., a specific visuotopic location) in manipulated animals. When the visual system was manipulated by rotating one eye by 180° and enucleating the other eye, the auditory space map was completely disrupted (King et al. 1988). It has recently been found that blockade of NMDA receptors prevents the maturation of spacial tuning and the alignment of visual and auditory maps (Schnupp et al. 1995).

A novel rearing paradigm, the use of omnidirectional white noise, was employed in developing guinea pigs to minimize location-specific acoustic stimulation. It was found that the azimuthal receptive field sizes failed to undergo their normal decrease in size, and there was no apparent topographic representation of best azimuth along the rostrocaudal axis of SC (Withington-Wray, Binns, and Keating 1990b). An analysis of cochlear microphonic thresholds demonstrated that the results were not due to peripheral damage. Dark-rearing the guinea pigs from birth produced an identical set of results (Withington-Wray, Binns, and Keating 1990c).

Therefore, where experimental manipulations and assay protocols overlap, the findings obtained on auditory space maps in the superior colliculus (i.e., optic tectum) are relatively consistent among species. Both the binaural receptive fields of individual neurons and the alignment of auditory and visual maps of the horizon require coherent activity pathways from both sensory systems during development.

7.3 Cellular Mechanisms

Although the cellular events underlying activity-dependent changes in neuronal function have not been fully elucidated, there are a few studies that are suggestive. For example, it has been found in the chicken CN that protein synthesis is maintained at normal levels by spontaneous glutamatergic transmission but not by postsynaptic action potentials (Born and Rubel 1988; Hyson and Rubel 1989). Therefore, it is clear that neurotransmision informs the postsynaptic cell about more than just the presence of sound during development. In fact, there is substantial literature concerning the influence of excitatory and inhibitory transmission on neuronal morphology in the developing central auditory system (Levi-Montalcini 1949; Coleman and O'Connor 1979; Parks 1979; Webster and Webster 1979; Feng and Rogowski 1980; Conlee and Parks 1981; Parks 1981; Trune 1982; Smith, Gray, and Rubel 1983; Webster 1983; Deitch and Rubel 1984; Born and Rubel 1988; McMullen et al. 1988; Moore and Kowalchuk 1988; Hashisaki and Rubel 1989; Sanes and Chokshi 1992; Sanes et al. 1992; Sanes and Takacs 1993).

Results from denervation studies suggest that receptors and ionic channels depend on synaptic activity during development. It has been demonstrated that primary afferent innervation of the chicken CN is necessary for the normal transition of glutamate-receptor subtypes during development (Zhou and Parks 1993; also see Fig. 3). After denervation, CN neurons become innervated by afferents from the contralateral CN, and these aberrant synapses display an immature sensitivity to non-NMDA-receptor antagonists. A complementary set of results have been obtained in the gerbil LSO. In vivo manipulations were performed to decrease glycinergic transmission in the LSO during early postnatal development, and the strength of inhibitory and excitatory synapses was examined subsequently in a brain slice preparation (Kotak and Sanes 1996). The deprived inhibitory synapses were found to be markedly weaker, and this was attributed in part to a depolarized IPSP equilibrium potential. In contrast, the excitatory afferents, which were not directly manipulated, produced unusually long duration EPSPs. These long-duration EPSPs apparently resulted from the de novo functional expression of NMDA receptors (Kotak and Sanes 1996). In addition, LSO neurons from experimental animals displayed broad rebound depolarizations after membrane hyperpolarization, and these were abolished in the presence of Ni^{2+}. Thus synaptic activity could have a potent effect on the development of stimulus processing by regulating the functional expression of membrane receptors, channels, and pumps.

8. Summary

An optimist might conclude from this survey that the central auditory system provides fertile territory for studies on the development of stimulus processing. Several areas of study, such as the maturation of ionic channels, have yet to receive experimental treatment. However, a pessimist would probably conclude that our level of understanding has advanced too slowly, with a substantial number of normative studies that are yet to be performed. Moreover, many of the topics covered above rely on experimental studies from only one or two laboratory groups. It is clear that any well-performed experiment will constitute a major contribution to the field.

A number of general conclusions can be drawn from the literature covered in this chapter. The fundamental difference between stimulus processing by central auditory neurons in young and adult animals is their response efficacy and reliability. Auditory neurons in young animals generally exhibit longer response latencies, smaller response amplitudes, and a greater variability of response than neurons from adults display to equivalent stimuli. To date, none of these functional limitations has been directly correlated with the restricted perceptual abilities of young animals (see Werner and Gray, Chapter 2).

Afferent connections and synaptic transmission are generally present before the onset of response to airborne sound in mammals. However, the evidence suggests that synaptic connections are subject to anatomic, functional, and biochemical modification. These modifications may underlie certain enhanced sound-evoked responses found in adult neurons, although this awaits experimental investigation.

Auditory experience influences the normal maturation of certain behavioral attributes in birds (Gottlieb 1980a,b, 1982, 1983) and humans (Kuhl et al. 1992). Although the neural bases for these particular forms of behavioral plasticity may reside within associative areas of the forebrain, electrophysiological evidence from brain stem regions in several species does suggest that central auditory synaptic connections *are* also influenced by the amount or temporal pattern of sound during development. An improved understanding of auditory processing in developing animals will most likely result from an elucidation of the intrinsic developmental programs of single neurons as well as their dependence on cellular interactions, including synaptic activity itself.

Acknowledgments. The authors were supported by National Institute on Deafness and Other Communication Disorders Grants DC-00540 to Dan H. Sanes and DC-01007, DC-00982, and DC-00215 to Edward J. Walsh; a National Science Foundation grant to Dan H. Sanes; March of Dimes Grant 1-F493-0236 to Dan H. Sanes; and a Deafness Research Foundation grant to Edward J. Walsh during the preparation of this chapter.

References

Adams JC, Mugnaini E (1987) Patterns of glutamate decarboxylase immunostaining in the feline cochlear nuclear complex studied with silver enhancement and electron microscopy. J Comp Neurol 262:375–401.

Alford BR, Ruben RJ (1963) Physiological, behavioral and anatomical correlates of the development of hearing in the mouse. Ann Otol Rhinol Laryngol 72:237–248.

Altschuler RA, Neises GR, Harmison GG, Wenthold RJ, Fex J (1981) Immunocytochemical localization of aspartate aminotransferase immunoreactivity in cochlear nucleus of the guinea pig. Proc Natl Acad Sci USA 78:6553–6557.

Axelsson A, Ryan A, Woolf N (1986) The early postnatal development of the cochlear vasculature in the gerbil. Acta Otolaryngol 101:75–87.

Betz WJ, Caldwell JH, Ribchester RR (1979) The size of motor units during postnatal development of rat lumbrical muscle. J Physiol 297:463–478.

Blatchley BJ, Brugge JF (1990) Sensitivity to binaural intensity and phase difference cues in kitten inferior colliculus. J Neurophysiol 64:582–597.

Blumenthal TD, Avenando A, Berg WK (1987) The startle response and auditory temporal summation in neonates. J Exp Child Psychol 44:64–79.

Born DE, Rubel EW (1988) Afferent influences on brain stem auditory nuclei of the chicken: presynaptic action potentials regulate protein synthesis in nucleus magnocellularis neurons. J Neurosci 8:901–919.

Bregman AS (1990) Auditory Scene Analysis: The Perceptual Organization of Sound. Cambridge: MIT Press.

Brugge JF, O'Connor TA (1984) Postnatal functional development of the dorsal and posteroventral cohclear nuclei of the cat. J Acoust Soc Am 75:1548–1562.

Brugge JF, Javel E, Kitzes LM (1978) Signs of functional maturation of peripheral auditory system in discharge patterns of neruons in anteroventral cochlear nucleus of kitten. J Neurophysiol 41:1557–1579.

Brugge JF, Kitzes LM, Javel E (1981) Postnatal development of frequency and intensity sensitivity of neurons in the anteroventral cochlear nucleus of kittens. Hear Res 5:217–229.

Brugge JF, Orman SS, Coleman JR, Chan JCK, Phillips DP (1985) Binaural interactions in cortical area AI of cats reared with unilateral atresia of the external ear canal. Hear Res 20:275–287.

Brugge JF, Reale RA, Wilson GF (1988) Sensitivity of auditory cortical neurons of kittens to monaural and binaural high frequency sound. Hear Res 34:127–140.

Carlier E, Abonnenc M, Pujol R (1975) Maturation des responses unitaires a la stimulation tonale dans le nerf cochleaire du chaton. J Physiol Paris 70:129–138.

Carlile S (1991) Postnatal development of the spectral transfer functions and interaural level differences of the auditory periphery of the ferret. Soc Neurosci Abstr 17:232.

Chiba A, Shepherd D, Murphey RK (1988) Synaptic rearrangement during postembryonic development in the cricket. Science 240:901–905.

Clarkson MG, Clifton RK, Swain IU, Perris EE (1989) Stimulus duration and repetition rate influences newborns' head orientation towards sound. Dev Psychobiol 22:683–705.

Clements M, Kelly JB (1978) Auditory spatial responses of young guinea pigs (*Cavia porcellus*) during and after ear blocking. J Comp Physiol Psychol 92:34–44.

Cline HT (1991) Activity-dependent plasticity in the visual systems of frogs and fish. Trends Neurosci 14:104–111.

Clopton BM, Silverman MS (1977) Plasticity of binaural interaction. II. Critical period and changes in midline response. J Neurophysiol 40:1275–1280.

Clopton BM, Silverman MS (1978) Changes in latency and duration of neural responding following developmental auditory deprivation. Exp Brain Res 32:39–47.

Code RA, Burd GD, Rubel EW (1989) Development of GABA immunoreactivity in brainstem auditory nuclei of the chick: ontogeny of gradients in terminal staining. J Comp Neurol 284:504–518.

Coleman JR, O'Connor P (1979) Effects of monaural and binaural sound deprivation on cell development in the anteroventral cochlear nucleus of rats. Exp Neurol 64:553–566.

Conlee JW, Parks TN (1981) Age- and position-dependent effects of monaural acoustic deprivation in nucleus magnocellularis of the chicken. J Comp Neurol 202:373–384.

Conradi S, Ronnevi L-O (1975) Spontaneous elimination of synapses on cat spinal motoneurons after birth: do half of the synapses on the cell bodies disappear? Brain Res 92:505–510.

Cragg BG (1975) The development of synapses in the visual system of the cat. J Comp Neurol 160:147–166.

Deitch JS, Rubel EW (1984) Afferent influences on brain stem auditory nuclei of the chicken: time course and specificity of dendritic atrophy following deafferentation. J Comp Neurol 229:66–79.

Eggermont JJ (1991) Maturaional aspects of periodicity coding in cat primary auditory cortex. Hear Res 57:45–56.

Eggermont JJ (1992) Stimulus induced and spontaneous rhythmic firing of single units in cat primary auditory cortex. Hear Res 61:1–11.

Eggermont JJ (1993) Differential effects of age on click-rate and amplitude modulation-frequency coding in primary auditory cortex of the cat. Hear Res 65:175–192.

Ellingson RJ, Wilcott RC (1960) Development of evoked responses in visual and auditory cortices of kittens. J Neurophysiol 23:363–375.

Feng AS, Rogowski BA (1980) Effects of monaural and binaural occlusion on the morphology of neurons in the medial superior olivary nucleus of the rat. Brain Res 189:530–534.

Finck A, Schneck CD, Hartman AF (1972) Development of cochlear function in the neonate Mongolian gerbil (*Meriones unguiculatus*). J Comp Physiol Psychol 78:375–380.

Fitzakerley JL, McGee J, Walsh EJ (1991) Variability in discharge rate of cochlear nucleus neurons during development. Soc Neurosci Abstr 17:304.

Fuzessery ZM, Wenstrup JJ, Pollak GD (1985) A representation of horizontal sound location in the inferior colliculus of the mustache bat (*Pteronotus p. parnellii*). Hear Res 20:85–89.

Godfrey DA, Carter JA, Berger SJ, Lowry OH, Matschinsky FM (1977) Quantitative histochemical mapping of candidate transmitter amino acids in cat cochlear nucleus. J Histochem Cytochem 25:417–431.

Goldberg JM, Brown PB (1969) Responses of binaural neurons of dog superior olivary complex to dichotic tonal stimuli: some physiological mechanisms of sound localization. J Neurophysiol 32:613–636.

Gottlieb G (1980a) Development of species identification in ducklings. VI. Specific embryonic experience required to maintain species-typical perception in Peking ducklings. J Comp Physiol Psychol 94:579–587.

Gottlieb G (1980b) Development of species identification in ducklings. VII. Highly specific early experience fosters species-specific perception in wood ducklings. J Comp Physiol Psychol 94:1019–1027.

Gottlieb G (1982) Development of species identification in ducklings. IX. The necessity of experiencing normal variations in embryonic auditory stimulation. Dev Psychobiol 15:507–517.

Gottlieb G (1983) Development of species identification in ducklings. X. Perceptual specificity in the wood duck embryo requires sib stimulation for maintenance. Dev Psychobiol 16:323–334.

Gray L (1991) Development of frequency dimension in chickens (*Gallus gallus*). J Comp Psychol 105:85–88.

Greenwood DD, Maruyama N (1965) Excitatory and inhibitory response areas of auditory neurons in the cochlear nucleus. J Neurophysiol 28:863–892.

Gummer AW, Mark RF (1994) Patterned neural activity in brain stem auditory areas of a prehearing mammal, the tammar wallaby (*Macropus eugenii*). Neuro-Report 5:685–688.

Handel S (1990) *Listening*. Cambridge: MIT Press.

Hashisaki GT, Rubel EW (1989) Effects of unilateral cochlea removal on antero-ventral cochlear nucleus neurons in developing gerbils. J Comp Neurol 283:465–473.

Hecox K, Galambos R (1974) Brainstem auditory evoked responses in human infants and adults. Arch Otolaryngol 99:30–33.

Hubel DH, Wiesel TN, LeVay S (1977) Plasticity of ocular dominance columns in monkey striate cortex. Philos Trans R Soc Lond B Biol Sci 278:377–409.

Hunter C, Wenthold RJ (1993) Expression of the AMPA-selective glutamate receptor, GluR1, is limited in principal cell populations of the rat brain stem auditory nuclei. Assoc Res Otolaryngol Abstr 16:124.

Hyson RL, Rubel EW (1989) Transneuronal regulation of protein synthesis in the brain-stem auditory system of the chick requires synaptic activation. J Neurosci 9:2835–2845.

Innocenti GM, Fiore L, Caminiti R (1977) Exuberent projection into the corpus callosum from the visual cortex of newborn cats. Neurosci Lett 4:237–242.

Ivy GO, Killackey HP (1982) Ontogenetic changes in the projections of neocortical neurons. J Neurosci 2:735–743.

Jackson H, Parks TN (1982) Functional synapse elimination in the developing avian cochlear nucleus with simultaneous reduction in cochlear nerve axon branching. J Neurosci 2:1736–1743.

Jackson H, Hackett JT, Rubel EW (1982) Organization and development of brain stem auditory nuclei in the chick: ontogeny of postsynaptic responses. J Comp Neurol 210:80–86.

Jenkins JJ, Walsh EJ, McGee J (1993) Developmental changes of auditory nerve responses to efferent stimulation. Soc Neurosci Abstr 19:534.

Jhaveri S, Morest DK (1982a) Sequential alterations of neuronal architecture in nucleus magnocellularis of the developing chicken: a Golgi study. Neuroscience 7:837–853.

Jhaveri S, Morest DK (1982b) Sequential alterations of neuronal architecture in nucleus magnocellularis of the developing chicken: an electron miscroscope study. Neuroscience 7:855–870.

Kandler K, Friauf E (1995) Development of electrical membrane properties and discharge characteristics of superior olivary complex neurons in fetal and postnatal rats. Eur J Neurosci 7:1773–1790.

Kerr LM, Ostapoff EM, Rubel EW (1979) Influence of acoustic experience on the ontogeny of frequency generalization gradients in the chicken. J Exp Psychol 5:97–115.

Kettner RE, Feng J-Z, Brugge JF (1985) Postnatal development of the phase-locked response to low frequency tones of auditory nerve fibers in the cat. J Neurosci 5:275–283.

King AJ, Carlile S (1991) Maturation of the map of auditory space in the superior colliculus of the ferret. Soc Neurosci Abstr 17:231.

King AJ, Hutchings ME (1987) Spatial response properties of acoustically responsive neurones in the superior colliculus of the ferret: a map of auditory space. J Neurophysiol 57:596–624.

King AJ, Hutchings ME, Moore DR, Blakemore C (1988) Developmental plasticity in the visual and auditory representations in the mammalian superior colliculus. Nature 332:73–76.

Kitzes LM, Semple MN (1985) Single-unit responses in the inferior colliculus: effects of neonatal unilateral cochlear ablation. J Neurophysiol 53:1483–1500.

Knudsen EI (1982) Auditory and visual maps of space in the optic tectum of the owl. J Neurosci 2:1177–1194.

Knudsen EI (1983) Early auditory experience aligns the auditory map of space in the optic tectum of the barn owl. Science 222:939–942.

Knudsen EI (1985) Experience alters the spacial tuning of auditory units in the optic tectum during a sensitive period in the barn owl. J Neurosci 5:3094–3109.

Knudsen EI (1988) Early blindness results in a degraded auditory map of space in the optic tectum of the barn owl. Proc Natl Acad Sci USA 85:6211–6214.

Knudsen EI, Brainard MS (1991) Visual instruction of the neural map of auditory space in the developing optic tectum. Science 253:85–87.

Knudsen EI, Konishi M (1978) A neural map of auditory space in the owl. Science 200:793–795.

Knudsen EI, Mogdans J (1992) Vision-independent adjustment of unit tuning to sound localization cues in response to monaural occlusion in developing owl optic tectum. J Neurosci 12:3485–3493.

Knudsen EI, Esterly SD, Knudsen PF (1984a) Monaural occlusion alters sound localization during a sensitive period in the barn owl. J Neurosci 4:1001–1011.

Knudsen EI, Knudsen PF, Esterly SD (1984b) A critical period for the recovery of sound localization accuracy following monaural occlusion in the barn owl. J Neurosci 4:1012–1020.

Knudsen EI, Esterly SD, du Lac S (1991) Stretched and upside-down maps of auditory space in the optic tectum of blind-reared owls; acoustic basis and behavioral correlates. J Neurosci 11:1727–1747.

Kotak VC, Sanes DH (1995) Synaptically-evoked prolonged depolarizations in the developing central auditory system. J Neurophysiol 74:1611–1620.

Kotak VC, Sanes DH (1996) Developmental influence of glycinergic transmission: regulation of NMDA receptor-mediated EPSPs. J Neurosci 16:1836–1843.

Kuhl PK, Williams KA, Lacerda F, Stevens KN, Lindblom B (1992) Linguisitic experience alters phonetic perception in infants by 6 months of age. Science 255:606–608.

Lenoir M, Shnerson A, Pujol R (1980). Cochlear receptor development in the rat with emphasis on synaptogenesis. Anat Embryol 160:253–262.

Levi-Montalcini R (1949) The development of the acoustico-vestibular centers in the chick embryo in the absence of the afferent root fibers and of descending fiber tracts. J Comp Neurol 91:209–241.

Levick WR (1973) Maintained discharge in the visual system and its role for information processing. In: Jung R (ed) Handbook of Sensory Physiology. Vol 7. New York: Springer-Verlag, pp. 575–598.

Liberman MC, Brown MC (1986). Physiology and anatomy of single olivocochlear neurons in the cat. Hear Res 24:17–36.

Lichtman JW (1977) The reorganization of synaptic connexions in the rat submandibular ganglion during post-natal development. J Physiol (Lond) 273:155–177.

Lippe W (1994) Rhythmic spontaneous activity in the developing avian auditory system. J Neurosci 14:1486–1495.

Mariani J, Changeux J-P (1981) Ontogenesis of olivocerebellar relationships. I. Studies by intracellular recordings of the multiple innervation of Pukinje cells by climbing fibers in the developing rat cerebellum. J Neurosci 1:696–702.

McMullen NT, Goldberger B, Suter CM, Glaser EM (1988) Neonatal deafening alters nonpyramidal dendrite orientation in auditory cortex: a computer microscopic study in the rabbit. J Comp Neurol 267:92–106.

Mikaelian D, Ruben RJ (1965) Development of hearing in the normal CBA-J mouse. Acta Otolaryngol 59:452–461.

Mogdans J, Knudsen EI (1992) Adaptive adjustment of unit tuning to sound localization cues in response to monaural occlusion in developing owl optic tectum. J Neurosci 12:3473–3484.

Mogdans J, Knudsen EI (1994) Site of auditory plasticity in the brain stem (VLVp) of the owl revealed by early monaural occlusion. J Neurophysiol 72:2875–2891.

Moore DR, Irvine DRF (1981) Plasticity of binaural interaction in the cat inferior colliculus. Brain Res 208:198–202.

Moore DR, Kitzes LM (1985) Projections from the cochlear nucleus to the inferior colliculus in normal and neonatally cochlea-ablated gerbils. J Comp Neurol 240:180–195.

Moore DR, Kowalchuk NE (1988) Auditory brainstem of the ferret: effects of unilateral cochlear lesions on cochlear nucleus volume and projections to the inferior colliculus. J Comp Neurol 272:503–515.

Mourek J, Himwich WA, Mysliveĕk J, Callison DA (1967) The role of nutrition in the development of evoked cortical responses in rat. Brain Res 6:241–251.

Nordeen KW, Killackey HP, Kitzes LM (1983) Ascending projections to the inferior colliculus following unilateral cochlear ablation in the neonatal gerbil, *Meriones unguiculatus*. J Comp Neurol 214:144–153.

Palmer AR, King AJ (1982) The representation of auditory space in the mammalian superior colliculus. Nature 299:248–249.

Parks TN (1979) Afferent influences on the development of the brain stem auditory nuclei of the chicken: otocyst ablation. J Comp Neurol 183:665–678.

Parks TN (1981) Changes in the length and organization of nucleus laminaris dendrites after unilateral otocyst ablation in chick embryos. J Comp Neurol 202:47–57.

Poon PW, Chen X (1992) Postnatal exposire to tones alters the tuning characteristics of inferior collicular neurons in the rat. Brain Res 585:391–394.

Poon PW, Chen XY, Hwang JC (1990) Altered sensitivities of auditory neurons in the rat midbrain following early postnatal exposure to patterned sounds. Brain Res 524:327–330.

Potashner SJ (1983) Uptake and release of D-aspartate in the guinea pig cochlear nucleus. J Neurochem 41:1094–1101.

Preyer W (1908) *Die Seele des Kindes*. Leipzig, Germany: Grieben Verlag.

Pujol R (1969) Developpement des responses a la stimulation sonore dans le colliculus inferieur chez le chat. J Physiol (Paris) 61:411–421.

Pujol R (1972) Development of tone burst responses along the auditory pathway in the cat. Acta Otolaryngol 74:383–391.

Pujol R, Carlier E. Devigne C (1978) Different patterns of cochlear innervation during the development of the kitten. J Comp Neurol 177:529–536.

Rakic P (1976) Prenatal genesis of connections subserving ocular dominance in the rhesus monkey. Nature 261:467–471.

Rakic P, Bourgeois J-P, Eckenhoof MF, Zecevic N, Goldman-Rakic PS (1986) Concurrent overproduction of synapses in diverse regions of the primate cerebral cortex. Science 232:232–235.

Ramón y Cajal S (1908) Terminación periférica del nervio acústico de las aves. Trab Lab Invest Biol Univ Madrid 6:161–176.

Ramón y Cajal S (1919) La desorientación inicial de las neuronas retinianas de axon corto. (Algunos hechos favorables a la concepción neurotrópica.) Trab Lab Invest Biol Univ Madrid 17:65–86.

Redfern PA (1970) Neuromuscular transmission in new-born rats. J Physiol (Lond) 209:701–709.

Retzius G (1884) Das Gehörorgan der Wirbeltiere. II. Das Gehörorgan der Reptilien, der Vögel und Säugetiere. Stockholm: Samson and Wallin.

Ricard GC, Hafter ER (1973) Detection of interaural time differences in short duration, low frequency tones. J Acoust Soc Am 53:335.

Riggs GH, Walsh EJ, Schweitzer L (1995) The development of glycine-like immunoreactivity in the dorsal cochlear nucleus. Hearing Res 89:172–180.

Riquimaroux H, Gaioni SJ, Suga N (1992) Inactivation of the DSCF area of the auditory cortex with muscimol disrupts frequency discrimination in the mustached bat. J Neurophysiol 68:1613–1623.

Robertson D, Gummer M (1985) Physiological and morphological characterization of efferent neurones in the guinea pig cochlea. Hear Res 20:63–77.

Romand R (1971) Maturation des potentiels cochleaires dans la periode perinatale chez le chat et chez le cobaye. J Physiol Paris 63:763–782.

Romand R (1984) Functional properties of auditory-nerve fibers during postnatal development in the kitten. Exp Brain Res 56:395–402.

Romand R, Marty R (1975) Postnatal maturation of the cochlear nuclei in the cat: a neurophysiological study. Brain Res 83:225–233.

Romand R, Granier M-R, Marty R (1973) Dévelopment postnatal de l'activité provoquée dans l'olive supérieure latérale chez le chat par la stimulation sonore. J Physiol Paris 66:303–315.

Rübsamen R (1992) Postnatal development of central frequency maps. J Comp Physiol A Sens Neural Behav Physiol 170:129–143.

Ryan AF, Woolf NK, Sharp FR (1982) Functional ontogeny in the central auditory pathway of the mongolian gerbil. Exp Brain Res 47:428–436.

Sanes DH (1984) A developmental analysis of neural activity in the mouse auditory system: its onset, the ontogeny of stimulus following, and its role during the maturation of frequency tuning curves. PhD thesis, Princeton University, Princeton, NJ.

Sanes DH (1990) An in vitro analysis of sound localization mechanisms in the gerbil lateral superior olive. J Neurosci 10:3494–3506.

Sanes DH (1993) The development of synaptic function and integration in the central auditory system. J Neurosci 13:2627–2637.

Sanes DH, Constantine-Paton M (1985a) The development of stimulus following in the cochlear nerve and inferior colliculus of the mouse. Dev Brain Res 22:255–268.

Sanes DH, Constantine-Paton M (1985b) The sharpening of frequency tuning curves requires patterned activity during development in the mouse, Mus musculus. J Neurosci 5:1152–1166.

Sanes DH, Chokshi P (1992) Glycinergic transmission influences the development of dendritic shape. NeuroReport 3:323–326.

Sanes DH, Rubel EW (1988a) The ontogeny of inhibition and excitation in the gerbil lateral superior olive. J Neurosci 8:682–700.

Sanes DH, Rubel EW (1988b) The development of stimulus coding in the auditory system. In: Jahn E, Santos-Sacchi J (eds) Physiology of the Ear. New York: Raven Press, pp. 431–455.

Sanes DH, Siverls V (1991) The development and specificity of inhibitory axonal arborizations in the lateral superior olive. J Neurobiol 22:837–854.

Sanes DH, Takacs C (1993) Activity-dependent refinement of inhibitory connections. Eur J Neurosci 5:570–574.

Sanes DH, Wooten GF (1987) Development of glycine receptor distribution in the lateral superior olive of the gerbil. J Neurosci 7:3803–3811.

Sanes DH, Markowitz S, Bernstein J, Wardlow J (1992) The influence of inhibitory afferents on the development of postsynaptic dendritic arbors. J Comp Neurol 321:637–644.

Saunders JC, Coles RB, Gates GR (1973) The development of auditory evoked responses in the cochlea and cochlear nuclei of the chick. Brain Res 63:59–74.

Saunders JC, Dolgin KG, Lowry LD (1980) The maturation of frequency selectivity in C57BL/6J mice studied with auditory evoked response tuning curves. Brain Res 187:69–79.

Schnupp JWH, King AJ, Smith AL, Thompson ID (1995) NMDA-receptor antagonists disrupt the formation of the auditory space map in the mammalian superior colliculus. J Neurosci 15:1516–1531.

Schweitzer L, Cant NB (1985b) Differentiation of the giant and fusiform cells in the dorsal cochlear nucleus of the hamster. Brain Res. 352:69–82.

Schweitzer L, Cecil T (1992) Morphology of HRP-labelled cochlear nerve axons in the dorsal cochlear nucleus of the developing hamster. Hear Res 60:34–44.

Schweitzer L, Cecil T, Walsh EJ (1993) Development of GAD-immunoreactivity in the dorsal cochlear nucleus of the hamster: light and electron microscopic observations. Hear. Res 65:240–251.

Seeburg PH (1993) The molecular biology of mammalian glutamate receptor channels. Trends Neurosci 16:359–365.

Shatz CJ (1990) Impulse activity and the patterning of connections during CNS development. Neuron 5:745–756.

Sher, AE (1971) The embryonic and postnatal development of the inner ear of the mouse. Acta Otolaryngol Suppl 285:5–77.

Shnerson A, Willott JF (1979) Development of inferior colliculus response properties in C57Bl/6J mouse pups. Exp Brain Res 37:373–385.

Silverman MS, Clopton BM (1977) Plasticity of binaural interaction. I. Effect of early auditory deprivation. J Neurophysiol 40:1266–1274.

Simmons DD, Manson-Gieseke L, Hendrix TW, McCarter S (1990) Reconstructions of efferent fibers in the postnatal hamster cochlea. Hear Res 49:127–139.

Smith ZDJ, Gray L, Rubel EW (1983) Afferent influences on brainstem auditory nuclei of the chicken: n. laminaris dendritic length following monaural conductive hearing loss. J Comp Neurol 220:199–205.

Snead CR, Altschuler RA, Wenthold RJ (1988) A comparison of GABA- and glycine-like immunolocalization in the developing rat lower auditory brainstem. Assoc Res Otolaryngol Abstr 11:51.

Snyder RL, Rebscher SJ, Cao K, Leake PA, Kelly K (1990) Chronic intracochlear electrical stimulation in the neonatally deafened cat. I: Expansion of central representation. Hear Res 50:7–34.

Sommer B, Keinanen K, Verdoorn TA, Wisden W, Burnashev N, Herb A, Kohler M, Takagi T, Sakmann B, Seeburg PH (1990) Flip and flop: a cell-specific functional switch in glutamate-operated channels of the CNS. Science 249:1580–1585.

Spitzer NC (1981) Development of membrane properties in vertebrates. Trends Neurosci 4:169–172.

Spitzer NC (1991) A developmental handshake: neuronal control of ionic currents and their control of neuronal differentiation. J Neurobiol 22:659–673.

Sretavan DW, Shatz CJ (1986) Prenatal development of retinal ganglion cell axons: segregation into eye-specific layers within the cat's lateral geniculate nucleus. J Neurosci 6:234–251.

Tello JF (1917) Génesis de las terminciones nerviosas motrices y sensitivas. I. En al sistema locomotor de los vertebrados superiores. Histogénesis muscular. Trab Lab Invest Biol Univ Madrid 15:101–199.

Thorpe LA, Schneider BA (1987) Temporal integration in infant audition. Soc Res Child Dev Abstr 273.

Tobias JV, Zerlin S (1959) Lateralization threshold as a function of stimulus duration. J Acoust Soc Am 31:1591-1594.

Trune DR (1982) Influence of neonatal cochlear removal on development of mouse cochlear nucleus. II. Dendritic morphometry of its neurons. J Comp Neurol 209:425-434.

Wada T (1923) Anatomical and physiological studies on the growth of the inner ear of the albino rat. In: Memoirs of the Wistar Institute of Anatomy and Biology, Vol. 10. Philadelphia, PA: Wistar Institute, pp. 1-174.

Walsh EJ, McGee J (1986) The development of function in the auditory periphery. In: Altschuler RA, Hoffman DW, Bobbin RP (eds) Neurobiology of Hearing: The Cochlea. New York: Raven Press, pp. 247-269.

Walsh EJ, McGee J (1987) Postnatal development of auditory nerve and cochlear nucleus neuronal responses in kittens. Hear Res 28:97-116.

Walsh EJ, McGee J (1988) Rhythmic discharge properties of caudal cochlear nucleus neurons during postnatal development in cats. Hear Res 36:233-248.

Walsh EJ, McGee J, Wagahoff D, Scott V (1985) Myelination of auditory nerve, trapezoid, and brachium of the inferior colliculus axons in the cat. Assoc Res Otolaryngol 8:33.

Walsh EJ, McGee J, Javel E (1986) Development of auditory-evoked potentials in the cat. II. Wave latencies. J Acoust Soc Am 79:725-744.

Walsh EJ, McGee J, Fitzakerley JL (1990) GABA actions within the caudal cochlear nucleus of developing kittens. J Neurophysiol 64:961-977.

Walsh EJ, McGee J, Fitzakerley JL (1993) Development of glutamate and NMDA sensitivity of neurons within the cochlear nuclear complex of kittens. J Neurophysiol 69:201-218.

Walsh EJ, McGee J, Fitzakerley JL (1995) Activity-dependent responses of developing cochlear nuclear neurons to microionophoretically-applied amino acids. Hear Res 84:194-204.

Warren EH III, Liberman MC (1989) Effects of contralateral sound on auditory-nerve responses. II. Dependence on stimulus variables. Hear Res 37:105-122.

Watson CS, Gengel RW (1969) Signal duration and signal frequency in relation to auditory sensitivity. J Acoust Soc Am 46:989-997.

Webster DB (1983) Late onset auditory deprivation does not affect brain stem auditory neuron soma size. Hear Res 12:145-147.

Webster DB, Webster M (1979) Effects of neonatal conductive hearing loss on brain stem auditory nuclei. Ann Otol Rhinol Laryngol 88:684-688.

Wenstrup JJ, Ross LS, Pollak GD (1985) A functional organization of binaural responses in the inferior colliculus. Hear Res 17:191-195.

Wenthold RJ (1978) Glutamic acid and aspartic acid in subdivisions of the cochlear nucleus after auditory nerve lesion. Brain Res 143:544-548.

Wenthold RJ (1979) Release of endogenous glutamic acid, aspartic acid and GABA from cochlear nucleus slices. Brain Res 162:338-343.

Wenthold RJ (1980) Glutaminase and aspartate aminotransferase decrease in the cochlear nucleus after lesion of the auditory nerve. Brain Res 190:293-297.

Wilmington D, Gray L, Jarsdoerfer R (1994) Binaural processing after corrected congenital unilateral conductive hearing loss. Hear Res 74:99-114.

Wisden W, Seeburg PH (1993) Mammalian ionotropic receptors. Curr Opin Neurobiol 3:291-298.

Wise LZ, Irvine DRF (1985) Topographic organization of interaural intensity difference sensitivity in deep layers of cat superior colliculus: implications for auditory spatial representation. J Neurophysiol 54:185–211.

Withington-Wray DJ, Binns KE, Keating MJ (1990a) The developmental emergence of a map of auditory space in the superior colliculus of the guinea pig. Dev Brain Res 51:225–236.

Withington-Wray DJ, Binns KE, Keating MJ (1990c) The maturation of the superior collicular map of auditory space in the guinea pig is disrupted by developmental visual deprivation. Eur J Neurosci 2:682–692.

Withington-Wray DJ, Binns KE, Dhanjal SG, Brickley SG, Keating MJ (1990b) The maturation of the superior collicular map of auditory space in the guinea pig is disrupted by developmental auditory deprivation. Eur J Neurosci 2:693–703.

Woolf NK, Ryan AF (1984) The development of auditory function in the cochlea of the Mongolian gerbil. Hear Res. 13:277–283.

Woolf NK, Ryan AF (1985) Ontogeny of neural discharge patterns in the ventral cochlear nucleus of the mongolian gerbil. Dev Brain Res 17:131–147.

Woolf NK, Ryan AF (1988) Contributions of the middle ear to the development of function in the cochlea. Hear Res 35:131–142.

Wu SH, Oertel D (1987) Maturation of synapses and electrical properties of cells in the cochlear nuclei. Hear Res 30:99–110.

Xie Z-P, Poo M-M (1986) Initial events in the formation of neuromuscular synapse: rapid induction of acetylcholine release from embryonic neuron. Proc Natl Acad Sci USA 83:7069–7073.

Yost WA (1991) Auditory image perception and analysis: the basis for hearing. Hear Res 56:8–18.

Young S, Poo M-M (1983) Spontaneous release of transmitter from growth cone of embryonic neuron. Nature 305:634–637.

Zhou N, Parks TN (1992) Developmental changes in the effects of drugs acting at NMDA or non-NMDA receptors on synaptic transmission in the chick cochlear nucleus (nuc. magnocellularis). Dev Brain Res 67:145–152.

Zhou N, Parks TN (1993) Maintenance of pharmacologically-immature glutamate receptors by aberrant synapses in the chick cochlear nucleus. Brain Res 628:149–156.

7

Structural Development of the Mammalian Auditory Pathways

Nell Beatty Cant

1. Introduction

Highly specific patterns of structural organization underlie all aspects of auditory function. An unusually large number of cell groups make up the auditory brain stem, and each forms precisely organized connections with other brain stem nuclei or with the auditory structures of the forebrain. A detailed knowledge of the development of the relevant structures and circuitry is required before the mechanisms leading to that development can be understood (Morest 1968; Rubel 1978). As Morest (1968) has emphasized, theories intended to explain ontogenesis must account for the *sequence* of events that give rise to the adult structures because it is likely that early events constrain the later ones. With the advent of new insights and new techniques for manipulating and labeling specific structures in the embryonic nervous system, it has become possible to experimentally analyze all stages of development from conception to the attainment of adult morphology and function. The goal of this chapter is to describe the sequence of events in the morphological development of the mammalian auditory system as currently understood. Development of the cochlea is reviewed by Fritsch, Barald, and Lomax (Chapter 3), Pujol, Lavigne-Rebillard, and Lenoir (Chapter 4), and Lippe and Norton (Chapter 5) and is referred to here only when relevant for issues of central development. The development of physiological response properties is summarized by Sanes and Walsh (Chapter 6). Where appropriate, relationships between anatomical and physiological development are indicated, but the physiological literature is not discussed in detail in this chapter. Related reviews covering various aspects of auditory development are listed in Table 7.1.

To relate data from the different species used in developmental auditory research, it is necessary to define common points of reference. The onset of hearing as measured in various physiological and behavioral studies represents one obvious common reference point and allows division of the developmental sequence into two phases. The first phase, termed the "silent period" by Morey and Carlile (1990), extends from the beginning of cell

315

TABLE 7.1. Reviews of aspects of development of the auditory system.

Author(s) and Year	Subject reviewed
Aitkin 1986	Inferior colliculus
Brugge 1983, 1988, 1992	Auditory brainstem
Clopton and Snead 1990	Effects of experience
Coleman 1990	Structure of the auditory pathways
Ehret 1988, 1990	Behavior
Gottlieb 1971	Development of auditory system relative to other sensory systems
Henderson-Smart et al. 1990	Development of BAERs in humans
Kelly 1992	Orientation to sound
King and Moore 1991	Plasticity of auditory maps
Kitzes 1990	Physiological properties of central auditory system
Moore 1985, 1988, 1990c, 1991, 1992b	Central auditory system; consequences of auditory deprivation
Parks, Jackson and Conlee 1987	Avian auditory system
Payne 1992	Auditory cortex
Pujol and Uziel 1988	Peripheral structures
Rubel 1978	Comprehensive review of literature to 1976: All species
Rubel 1985	Development of peripheral and central nervous system; extrinsic influences
Rubel and Parks 1988	Avian auditory system
Rübsamen 1992	Postnatal development of frequency maps
Ryan and Woolf 1992	Auditory brainstem in gerbil
Sanes 1992	Structural correlates of functional development
Willard 1995	Structure of auditory pathways

genesis to the onset of hearing. The second phase, which is more open ended, begins at hearing onset. It can be subdivided into a period of rapid changes just at the time hearing begins and a subsequent longer, slow period lasting until the adult state is reached. The literature reviewed in this chapter is summarized in the form of time lines for each of the major auditory nuclei in seven mammalian species that have been studied in some detail (Tables 7.2–7.8).[1] When considered together, the data from the different species allow us to begin to construct a coherent picture of the relative timing and sequence of developmental events leading to the emergence of a functional adult system. The species that will be considered in most detail include four species of rodents (rat, *Rattus norvegicus,* Table 7.2; mouse, *Mus musculus,* Table 7.3; gerbil, *Meriones unguiculatus,* Table 7.4; and hamster, *Mesocricetus auratus;* Table 7.5), two species of carnivore (cat, *Felis catus,* Table 7.6 and ferret, *Mustela putorius,* Table 7.7), and a variety of marsupials (Table 7.8). (Limited data are also available for the pig, sheep, dog, guinea pig, bat, monkey, and human and will be discussed where appropriate.)

---

[1]The survey of the literature was completed in August 1994. Work published since that time may not be cited.

To provide a background for what follows, a brief review of the neuroanatomy of the relevant adult auditory pathways is presented in Section 2. Section 3 is devoted to a consideration of normal ontogeny up to the time of onset of hearing. The events occurring after the onset of hearing are considered in Section 4. The material covered in the chapter is summarized briefly in Section 5.

2. The Neuroanatomy of the Auditory System in the Adult Mammal

Only those aspects of the circuitry that are relevant to this review of the developmental literature are described here. For recent detailed descriptions of the neuroanatomy of the adult auditory pathways, see Webster, Popper, and Fay (1992). Two aspects of the organization of mammalian auditory circuitry bear emphasis. The first is its topographic arrangement: the receptor surface of the cochlea is represented systematically in almost all of the nuclei of the auditory pathways. The other is that each structure in the system comprises several populations of distinct neuronal types, each with specific connections and functions. Wherever data are available, development of the topography and the developmental history of specific cell types will be addressed in this review.

An outline of the basic circuitry of the auditory system is presented in Figure 1. This diagram is highly simplified; all of the structures represented comprise numerous subdivisions, each with its own set of specific connections. The many distinct pathways that arise in the different subdivisions are collapsed here to emphasize the overall pattern of connectivity. The central processes of cochlear spiral ganglion cells terminate in the cochlear nuclear complex. (The peripheral processes innervate the hair cells.) The cochlear nuclei give rise to two types of pathways to the midbrain. Direct pathways from the dorsal and ventral cochlear nuclei to the inferior colliculus (filled arrows) are predominately contralateral with small ipsilateral components in most species. Indirect pathways to the inferior colliculus from the ventral cochlear nucleus relay via the superior olivary complex (as well as via the nuclei of the lateral lemniscus, which are not illustrated). Cells in the cochlear nucleus project to both the ipsilateral and contralateral superior olivary complex. Nuclei of the superior olive, in turn, project to both the ipsilateral and contralateral inferior colliculi. The inferior colliculus projects mainly to the medial geniculate nucleus, the cell group in the dorsal thalamus that provides input to the auditory cortex. Clearly, at all levels of the auditory pathways beyond the cochlear nuclei, there are ample opportunities for binaural interactions.

The auditory nuclei are organized topographically such that the receptor surface of the cochlea from its basal to apical ends is mapped systematically

TABLE 7.2. Rat (*Rattus norvegicus*).

Age and Selected Embryological and Functional Time Points	Structural Development of the Auditory Pathways
E5–11. E5–7, implantation (1). E11 neurulation begins (1, 2). E10–12, peak neurogenesis of brain stem motor neurons (3).	*CN.* E11–13, generation of DCN pyramidal cells and PVCN octopus cells (4). *SOC and NLL.* E11, generation of MSO neurons (4, 5). E11–12, generation of LSO cells that project to contralateral IC (5). E11–13, generation of DNLL cells (6). E11–15, generation of SOC neurons (3–5). *IC.* E11, neuroepithelium only; no axons (7).
E12–14. E12, peak neurogenesis of spiral ganglion cells (from E11 to E13) (2). E12–16, cerebral hemispheres advance from neuroepithelium to cortical plate stage (8). E14, otocyst and vestibulocochlear ganglion present (9).	*CN.* E13–14, generation of most AVCN cells (4, 10). E13–15, generation of DCN small cells (inner layer) (4). E14, central processes from spiral ganglion enter myelencephalon; CN not yet recognizable (9). E14–15, two-thirds of CN neurons generated (10). *SOC and NLL.* E12, neurons of SPN generated (4, 5). E13–15, VNLL cells generated (6). LSO cells that project to ipsilateral IC generated (5). E14, 90% of MNTB cells generated (4). *IC.* E12, IZ appears (7). E13–14, IZ contains large bundles of axons and young neurons (7). E13–P0, IC neurons generated in a gradient (11). E14, axons from brain stem nuclei enter IC anlage (Fig. 7D of Ref. 7). *MG and AC.* E12–20, generation of neurons that will form AC (12). E13–14, generation of MG cells (peak E13); posterolateral to anteromedial gradient (13–15). E14, appearance of early thalamocortical systems (16); auditory and visual fibers lag somesthetic (17). E14–15, MG identifiable (14, 17, but see 16). E14, peak generation of AC layer 6 cells (12).
E15–17. E15, neuroepithelial layer accounts for 2/3 of total thickness of cerebral wall (18). E15–16, cortical plate identifiable (8, 18). E16, cochlear and vestibular nerve roots distinguishable (9).	*CN.* E15–17, axon outgrowth from CN to SOC, LL, and IC; no collateralization (19). E16–18, DCN small cells (outer layer) generated (4). E16, AVCN, PVCN, DCN identifiable; primary fibers enter VCN but not DCN; few collaterals (9). *SOC and NLL.* E15–17, axons from CN traverse SOC and LL (19). *IC.* E16–17, external, pericentral, and central nuclei distinguishable (7). Axons from contralateral CN invade IC but do not form collaterals (19). E17–19, temporal gradient of CN innervation from rostral to caudal (19). *MG and AC.* E15, no synapses in cortex (18). Peak generation of layer 5 cells (12). E16–19, first cortical synapses located in marginal layers above and below cortical plate (18). E16, peak generation of layer 4 cells (12). E16/17, peak generation of layer 3 cells (12). E17, peak generation of layer 2 cells (12).

E18-21.
E18, synaptic potentials (to electrical stimulation) evoked in SOC (20).

E22/P0.
Birth/P0, electrical stimulation of MG yields cortical-evoked response (23). Brain weight 11-16% of adult value; grows rapidly until P14 (24).

P1-5.
By P2, electrical stimulation of ventral acoustic stria leads to PSPs in SOC (20).

CN. E20, CN shape adultlike; complex terminal structures on primary fibers in DCN and VCN but no end bulbs (9). E21, DCN fusiform cells have adultlike morphology but lack dendritic spines on apical dendrites; growth cones present on distal dendrites (10).
SOC and NLL. E19, number of collateral branches on axons from CN increases markedly (19). E20, projection pattern from CN to both SOCs resembles adult (19). E21, collateralization of CN fibers in VNLL (19).
IC. E18, collateralization of CN fibers in contralateral IC; CN fibers enter ipsilateral IC (19). IC has adult shape (8).
MG and AC. E19, terminals from IC arborize in MG; some extend past MG (21, 22). E20, nontelencephalic fibers enter diencephalon (16).

CN. P0, first fibers project from one CN to the other (19). P0-6, auditory terminals occasionally form synapses with large dendrites in neuropil; no synapses on cell bodies (25, 26).
SOC and NLL. P0, SOC cells well developed cytologically (27); OC cells project into cochlea (28); efferent synapses from OC cells present beneath inner hair cells (29); collateralization of CN fibers in DNLL (19). P0-4, some OC cells die (28); MSO cells have relatively small somas (30). P0-7, MNTB neurons do not stain for CaBP; LSO neurons are CaBP+ (31). P0-7, LSO cells coupled by gap junctions (32). P0-14, MSO cells increase rapidly in size and number of dendrites (30).
IC. P0, CN and SOC project to central nucleus (19, 27, 33, 34); topography adultlike (27). P0-2, transient retinal fibers to IC; much sparser at P2 than at P0; disappear by P6 (35). P0-7, extracellular space about 2 times that of adult (36).
CN. P1-5, development to adultlike pattern of commissural projections (19). By P4, expression of mRNA for metabotrophic glutamate receptor present in DCN (Fig. 10c in Ref. 37).
SOC. P3, precursors of calyces in MNTB (19).
IC. P1, synapses present; cells close together (36). P3, axons from CN restricted to ventral half of IC (19); projections from AC first seen (38). P3-21, transient expression of tyrosine hydroxylase (39). P4, axons from CN distributed throughout (19). P5, adultlike pattern of projections (19, 34). P5-24, number of small blood vessel profiles (<10 μm) increases six-fold/unit area (40).
MG and AC. P3-7, sharp increase in cell growth in MG; maximum postnatal expansion of AC (34, 41). P5-11, number of spines on apical dendrites of layer 5 pyramidal cells increases gradually (42).

(Continued)

Table 7.2. (*Continued*)

Age and Selected Embryological and Functional Time Points	Structural Development of the Auditory Pathways
P6-9. P7/8, ABR elicited to high-intensity stimulation via bone conduction but not via air (43). P8, electrical stimulation of the auditory nerve gives rise to cortical evoked potential (44). P8/9, first CM recorded (45).	*CN.* P6-10, primary afferent fibers contact long appendages arising from spherical cells (25, 26). P7, somatostatin-positive cells at peak in VCN; decrease after this; few positive cells in adult (46). *SOC and NLL.* P6-12, efferent synapses from OC cells form on outer hair cells (29). P7, somatostatin-positive cells at peak in NLL; dense network of somatostatin-positive fibers in LSO, MSO, and MNTB; decreases after this (46). P8, soma sizes reach peak value in MSO (30). P8-18, MNTB somas increase in staining for CaBP; CaBP staining lost in LSO; CaBP$^+$ terminals appear in LSO, MSO, and SPN (31). *IC.* P8-16, only proximal and intermediate dendritic segments lengthen (47). P8-20, dendritic trees in central nucleus oriented in sagittal plane; undergo reorientation in the frontal plane (47). *MG.* P7, all subdivisions of MG recognizable (48).
P10-15. P11, ABR evoked by airborne acoustic stimuli (43). P11-12, first nerve AP (45, 49); behavioral responses to sound (50). P12, orientation to HF tone (51). P12-13, auditory meatus opens (52). P13-20, pups approach social call (53). P14, ABR (immature) evoked by LF tones; no response to HF tone until P16 (54); orientation to LF tone (51); brain growth rate slows (55). P15, maturation of BAER (56). P14-23, frequency shift demonstrated behaviorally (57); CM adult-like (45); conduction apparatus of ear mature (43).	*CN.* P10, end bulb forms synaptic contacts with cell body; three types of nonprimary terminals contact soma (25, 26). P10-14, expression of mRNA for FGF increases rapidly (58). P10-16, volume of VCN increases 42%; DCN 34% (59, 60). P12, end bulb covers most of the dendritic pole of the cell; somatic appendages disappearing (25). P14, Fos immunoreactivity indicates adultlike frequency representation in all nuclei (61). P14-16, end bulb resembles adult (25). *SOC.* P10-14, expression of mRNA for FGF increases rapidly (58). P12-14, dendrites of MSO cells branch extensively (30). P14, calyces in MNTB adultlike (19); size of MSO cells decreases and dendrites mature (decreased branching) (30); Fos immunoreactivity indicates adultlike frequency representation in all nuclei (61). P14-17, expression of mRNA for FGF like adult (58). *IC.* P14, Fos immunoreactivity indicates adultlike frequency representation (61); cells separated by areas of neuropil 10-30 μm wide (36). P15-26, increase in concentration of cerebroside (myelin protein) (62). *MG and AC.* P11, number of secondary and tertiary branches on basal dendrites of layer 5 cells in AC appears to stabilize (34); subdivisions of MG show adultlike relative proportions; cells are smaller and more densely packed than in adult (48). P11-15, synapses at all levels of auditory cortex (63). P11-16, second growth spurt in MG cells (34, 41); marked increase in spine density on layer 5 pyramidal cells in AC (34, 42).

P16–adult.

P16, nerve AP has mature shape (49, 64). P17, resting 2-DG uptake lower and more uniform than in adult (65). P18, frequency shift not seen behaviorally (57). P20, OHC adultlike (66). P21, pups weaned.

CN. P16 until at least P36, volume of CN continues to increase (59, 60).

SOC. P16–18, MSO neurons acquire adult characteristics; loss of second through fifth-order dendritic branches; decrease in cell size and total dendritic branching (30). P18, no indication of gap junctions in LSO (32). P28, CaBP staining adultlike (31).

IC. P16–20, distal dendritic segments increase in length (47). P24, adult level of vascularization (40).

MG and AC. P21, "exuberant" connections from IC to thalamus eliminated by end of third week (22). P30, terminal arborization of projections from IC to MG less branched than at younger ages (22). P35, maximum spine density on layer 5 pyramidal cells in AC (34, 42).

E, embryonic day; CN, cochlear nucleus; DCN, dorsal cochlear nucleus; PVCN, posteroventral cochlear nucleus; SOC, superior olivary complex; NLL, nuclei of the lateral lemniscus; MSO, medial superior olivary nucleus; LSO, lateral superior olivary nucleus; IC, inferior colliculus; DNLL, dorsal nucleus of the lateral lemniscus; AVCN, anteroventral cochlear nucleus; VNLL, ventral nucleus of the lateral lemniscus; MNTB, medial nucleus of the trapezoid body; IZ, intermediate zone; MG, medial geniculate nucleus; AC, auditory cortex; LL, lateral lemniscus; P, postnatal day; OC, olivocochlear; CaBP, calcium binding protein; ABR, auditory brain stem response; CM, cochlear microphonic; SPN, superior paraolivary nucleus; AP, action potential; HF, high frequency; LF, low frequency; BAER, brain stem auditory-evoked response; mRNA, messenger RNA; FGF, fibroblast growth factor; 2-DG, 2-deoxyglucose; OHC, outer hair cell.

Most authors consider the day of conception to be E0. Others, however, consider that day to be E1. In the latter cases, dates have been shifted back one day in Tables 2–8 (i.e., their day E12 would be given here as E11). A few authors do not clarify their usage of dates; for those authors, the dates are presented here as estimates. Because of inherent difficulties and uncertainty in identifying the precise time of conception, a variation in reported times is to be expected. This variation can result in times differing by as much as a day from study to study (discussed, e.g., by Gardette, Courtois, and Bisconte 1982). Conclusions reached in this chapter are not much affected by variability in reported times because developmental events in the auditory system have been studied at relatively few time points. Similar considerations apply to the postnatal ages (i.e., the day of birth is considered here to be P0).

1, Hebel and Stromberg 1986; 2, Altman and Bayer 1982; 3, Nornes and Morita 1979; 4, Altman and Bayer 1980a; 5, Kudo et al. 1992; 6, Altman and Bayer 1980b; 7, Repetto-Antoine and Meininger 1982; 8, Marin-Padilla 1988; 9, Angulo, Merchán, and Merchán 1990; 10, Friauf and Kandler 1993; 11, Altman and Bayer 1981; 12, Bayer and Altman 1991; 13, McAllister and Das 1977; 14, Altman and Bayer 1979; 15, Altman and Bayer 1989; 16, Coggeshall 1964; 17, Ströer 1956; 18, König, Roch, and Marty 1975; 19, Kandler and Friauf 1993; 20, Kandler and Friauf 1995; 21, Asanuma et al. 1986; 22, Asanuma et al. 1988; 23, Hassmannová and Mysliveček 1967; 24, Wilkinson 1986; 25, Neises, Mattox, and Gulley 1982; 26, Mattox, Neises, and Gulley 1982; 27, Friauf and Kandler 1990; 28, Harvey, Robertson, and Cole 1990; 29, Lenoir, Shnerson, and Pujol 1980; 30, Rogowski and Feng 1981; 31, Friauf 1993; 32, Rietzel and Friauf 1994; 33, Maxwell et al. 1988; 34, Coleman 1990; 35, Kato 1983; 36, Pysh 1969; 37, Shigemoto, Nakanishi, and Mizuno 1992; 38, Maxwell and Coleman 1989; 39, Jaeger and Joh 1983; 40, Andrew and Paterson 1989; 41, Coleman et al.1989; 42, Coleman et al. 1987; 43, Geal-Dor et al. 1993; 44, Tokimoto, Osako, and Matsuura 1977; 45, Uziel, Romand, and Marot 1981; 46, Kungel and Friauf 1995; 47, Dardennes, Jarreau, and Meininger 1984; 48, Clerici, Maxwell, and Coleman 1988; 49, Carlier, Lenoir, and Pujol 1979; 50, Wada 1923; 51, Kelly, Judge, and Fraser 1987; 52, Crowley and Hepp-Raymond 1966; 53, Kelly 1992; 54, Blatchley, Cooper, and Coleman 1987; 55, Dobbing and Sands 1971; 56, Church, Williams, and Holloway 1984; 57, Hyson and Rudy 1987; 58, Luo et al. 1995; 59, Coleman 1981; 60, Coleman, Blatchley, and Williams 1982; 61, Friauf 1992; 62, Shah, Bhargava, and McKean 1978; 63, König and Marty 1974; 64, Puel and Uziel 1987; 65, Clerici and Coleman 1986; 66, Pujol, Carlier, and Lenoir 1980.

TABLE 7.3. Mouse (*Mus musculus*).

Age and Selected Embryological and Functional Time Points	Structural Development of the Auditory Pathways
E8–10. E8, neurulation begins (1). E9, neural tube closed except for posterior neuropore (1, 2); first motor neurons generated (3). E10–11.5, central fibers project out of eighth nerve ganglion (4, 5).	**CN.** E10–11, large neurons of DCN, large multipolar neurons of PVCN generated (6, 7). **SOC and NLL.** E9–12, neurons of MSO and periolivary nuclei generated (3). E9–13, neurons of VNLL generated (peak at E10) (3); neurons of DNLL generated (peak at E11) (3). E9–14, neurons of LSO generated (3). E10.5, a few efferent fibers enter the eighth nerve root (8). **MG and AC.** E10–18, all neurons of cerebral isocortex generated (9, 10). E10, cells in extreme ventromedial part of MG generated (11).
E11–13. E11, first spiral ganglion cells and hair cells generated (12). E12, appearance of pontine flexure (6). E13, peak genesis of spiral ganglion cells, hair cells (range for both E11–15) and supporting cells in cochlea (12). E11–13, cerebral hemispheres advance from neuroepithelium to cortical plate stage (1).	**CN.** E11, auditory nerve fibers appear to have entered brain stem (6); neurons of acoustic nerve nucleus generated (6, 7). E12–13, medium-sized neurons of DCN and spherical, globular, and multipolar cells of VCN generated (6, 7). E12/13, cochlear nerve enters cluster of proliferating cells in myelencephalon (13). E12–18, granule cells generated (6). **SOC and NLL.** E11–12, neurons of MNTB generated (3). E12, a few cells in brain stem project to both otocysts; fibers present in sensory epithelium (8). **IC.** E11–13, neurons of central nucleus generated (peak at E12) (3, 10). E11–14, neurons of external nucleus generated (peak at E12/13) (3, 10). E11–17, neurons of dorsal cortex generated (3, 10). E12–13, IC distinguishable from superior colliculus (14). E13.5, peak production of neurons of rostral IC (10). **MG and AC.** E11, cells throughout MG generated (11). E13, cells in lateral half of MG generated (11).
E14–18. E16–17, CC appears; earliest fibers cross the midline (15).	**CN.** E14/15, anlage of CN situated lateral to restiform body (13). E14.5, small cells generated (6, 7). E16–17, CN anlage divisible into DCN and VCN (13). **IC.** E14/15, peak production of neurons of dorsal cortex (3). E16.5, peak production of neurons of caudal IC (10).

E19/P0.
Birth, brain weight 15–20% of adult value; rapid growth until P11–15 (16).

P1–5.
P4, bushy and stellate cells in VCN and cells in DCN respond to electrical stimulation with EPSPs and IPSPs (17).

CN. P1–3, little change in volume of DCN and VCN or in size of globular cells (18); no neuronal growth (19). P3–24, multipolar, globular, and spherical cells grow rapidly to reach peak size at P12, then drop to adult values (19).
SOC. P1–3, no changes in neuron size (19). P3–24, cells in LSO and MSO grow to adult size (19).
IC. P1–3, no changes in neuron size (19). P3–45, cells in central nucleus grow to peak size at P24, then drop to adult values (19).

P6–8.
P7, bushy and stellate cells in VCN and cells in DCN can be distinguished electrically (17).

CN. P6–12, DCN and VCN more than double in volume; globular cells double in somatic area (18), number of VCN cells containing Nissl substance doubles (18). P7, granule cells express mRNA for $\alpha6$ subunit of $GABA_A$ receptor (20).
SOC. P7, all periolivary cell groups recognizable cytoarchitectonically (21). P7–14, single globular cell axons often form two calyces in MNTB (rare in older animals) (22). P7–21, neuronal packing density in LSO decreases by more than 25% (21).

P9–13.
P9, no response to tones; spontaneous activity present (23). P10–11, some strains responsive to pure tones at medium-high intensity (24). P11–15, brain growth rate slows dramatically; continued slow growth to adult values (16, 25, 26). P12, ear open (24, 27). Preyer reflex elicited (range P9–14) (24, 28). IC units respond only to LF (29).

CN. P9, peak cell number in VCN (30). P11–13, peak cell number in DCN (30). P12, adult level of cells in VCN; 20–25% loss from peak at P9 (30). Globular, spherical, multipolar, and octopus cells in VCN have reached peak size, some decrease slightly in size after this age (19); rate of growth in volume slows but continues through at least P90 (18).
SOC. P10, first efferent synapses identified in cochlea (31). P12, cells of LSO and MNTB have attained adult size (19).
IC. P9–20, volume of IC does not change over this time (32).

(Continued)

TABLE 7.3. (Continued)

Age and Selected Embryological and Functional Time Points	Structural Development of the Auditory Pathways
P14–16.	*CN.* P16, adult level of cells in DCN; 30–35% loss from peak at P11/13 (30).
P14, behavioral responses to wide range of frequency and intensity (24). P15, brain weight 90% of adult value (26). P15–17, tuning curves in IC like adult (29, 33). P16, mature behavioral audiogram (24). Cochlea structurally mature (34).	

CC, corpus callosum; GABA, γ-aminobutyric acid; EPSP, excitatory postsynaptic potential; IPSP, inhibitory postsynaptic potential.
There are strain differences in the timing of developmental events in the auditory system of the mouse (24). For example, the cochlea of the CBA/J mouse is near maturity at P10, but parts of the cochlea of C57BL/6J mice continue to develop through P14. At P10, some mice of the NMRI strain respond to pure tones of moderately high intensity, but CBA/J mice do not respond until P11 and C57BL/6J are not responsive until P12–13. A number of different strains were used in the studies included here. As the events occurring at each time point are understood in more detail, comparisons *within* a strain will become increasingly important.

1, Theiler 1989; 2, Kaufman 1992; 3, Taber Pierce 1973; 4, Carney and Silver 1983; 5, Poston et al. 1988; 6, Taber Pierce 1967; 7, Martin and Rickets 1981; 8, Fritzsch and Nichols 1993; 9, Angevine and Sidman 1961; 10, Gardette, Courtois, and Bisconte 1982; 11, Angevine 1970; 12, Ruben 1967; 13, Hugosson 1957; 14, Otis and Brent 1954; 15, Silver et al. 1982; 16, Wilkinson 1986; 17, Wu and Oertel 1987; 18, Webster 1988a; 19, Webster and Webster 1980; 20, Varecka et al. 1994; 21, Ollo and Schwartz 1979; 22, Kuwabara, DiCaprio, and Zook 1991; 23, Romand and Ehret 1990; 24, Shnerson and Pujol 1983; 25, Hahn et al. 1983; 26, Kobayashi 1963; 27, Mikaelian and Ruben 1965; 28, Alford and Ruben 1963; 29, Shnerson and Willott 1979; 30, Mlonyeni 1967; 31, Kikuchi and Hilding 1965; 32, Romand and Ehret 1990; 33, Willott and Shnerson 1978; 34, Kraus and Aulbach-Kraus 1981.

in each nucleus (e.g., Merzenich et al. 1977). The representations of the cochlea in each structure take the form of bands or sheets of neurons that receive inputs directly or indirectly from a restricted extent of the cochlear receptor surface. A given auditory structure may contain several representations of the cochlea. For example, it is usually stated that the cochlea is represented in the cochlear nuclear complex three times, once each in the dorsal, posteroventral, and anteroventral nuclei. (In fact, the number of representations is probably higher because each subdivision of the anteroventral and posteroventral cochlear nuclei might be considered to receive its own representation.) Topographic organization in the auditory system can be related to tonotopic organization in the adult. The underlying basis for the topographic and tonotopic organization is the orderly and restricted geometry of axonal arborizations and dendritic ramifications at all levels of the system (Morest 1973). How the topography and the underlying neuronal structure arise is an important question addressed in developmental studies.

A diagram like that in Figure 7.1, even though it indicates the general direction of flow of information in the auditory system, is misleading because it ignores the fact that different components of the pathways indicated may arise from distinctly different neuronal populations and may terminate in different parts of the target structure. Almost every subdivision of each auditory structure contains several neuronal types, each of which may have not only quite distinct projection patterns but also unique sources and pattern of inputs. Ideally, a diagram like that in Figure 7.1 would illustrate the circuitry in terms of the projections of each of these different populations of cells. Unfortunately, present knowledge does not allow a complete depiction of the auditory pathways based on the connectivity of individual cell types, although current research by many investigators is leading to considerable progress along these lines.

Distinct neuronal types in the cochlear nucleus and superior olivary complex and some of the known details of their connectivity are illustrated in Figure 7.2. All of the illustrated neuronal types have been studied developmentally and will be discussed in the following sections. The inferior colliculus and forebrain structures are also made up of numerous neuronal types but less is known of the specific circuitry established by each of them.

3. Development from Conception to the Onset of Hearing. The Establishment of the Basic Circuitry.

During the first major phase of development, the neurons of the auditory system are generated, migrate to their adult locations, send out axons, undergo dendritic differention, and begin to establish synaptic connections. Also, during this time, some neurons undergo programmed cell death.

TABLE 7.4. Gerbil (*Meriones unguiculatus*).

Age and Selected Embryological and Functional Time Points	Structural Development of the Auditory Pathways
P0. Birth, brain weight is 16% of adult weight (1); EPSPs and IPSPs can be elicited in LSO (2).	*SOC.* By P0, axons from VCN project to ipsilateral LSO, contralateral MNTB, and both MSOs; axons in MNTB do not exhibit calyces (3, 4). P0, synapses have begun to form between axons from VCN and cells in MSO and MNTB (3). Parvalbumin staining in MNTB (5).
P1–10. P3, stimulation of MNTB elicits hyperpolarizing potentials in LSO (6). P10, no neurons in VCN are responsive to acoustic stimulation (7).	*CN.* P4, GABA-IR fibers and boutons found only in molecular layer of DCN (8). P8, GABA-IR fibers and terminals in granule cell layer of DCN (8). P10, synaptic terminals sparse in DCN up to this time (9). *SOC.* P1–3, cells in LSO highly branched (10). P2/3, MNTB axons in LSO have restricted arbors and give rise to both boutons and growth cones (11). P4–10, increase in glycine-receptor concentration in all areas of LSO (12). P5, calyces have begun to form in MNTB (3, 4). P6–9, total length of MNTB axons in LSO increases; many arbors give rise to diffuse or discontinuous terminal areas (11). P7, synaptic terminals from VCN to MSO have increased in size and contain more vesicles than those at P0; axosomatic contacts are infrequent; synapses are formed on small dendritic profiles (3). P7–14, single globular cell axons often form two calyces in MNTB (13). P10, pattern of projections from AVCN like that of adult (4). P10–13, total arbor length of MNTB axons in LSO continues to increase; discontinuous pattern still present (11). P10–21, glycine-receptor concentration fairly constant in dorsal LSO; decreases in ventral LSO (12); number of branches on LSO dendrites increases, then decreases to almost adult value; total length of HF dendrites increases, then decreases and continues to fall through P90; length of LF axons remains contant (10).

P11–15.

P12, about 15% of VCN cells respond to acoustic stimulation (7). 2-DG labeling in AVCN to high-intensity wide-band noise (14, 15). P12/13, CM recorded in midbasal turn of cochlea (16, 17). P13–16, some LSO neurons respond to two distinct frequency ranges (18). P14, first nerve AP (7, 19); first replicable ABR (20); majority of neurons in VCN respond to acoustic stimulation (7); reflex responses to sound (19). P15/16, mature unit responses in LSO (18).

P16–25.

P16, organ of Corti well developed (19); 2-DG labeling in IC (14, 15). P16–19, approach responses to social call first seen (23). P17/18, brain weight 80% of adult value; rate of growth slows abruptly; continues at a slower rate into adulthood (1). P18, single-unit responses in VCN adultlike (7).

CN. P12–14, large increase in number of synapses in DCN, immature in appearance (9). P13, GABA-IR fibers and terminals appear in deep DCN (8). P14–16, synapses in DCN continue to increase in number; morphology matures (8, 9). P15, DCN and VCN have reached 60% of their adult size (21).

SOC and NLL. P11–15, parvalbumin staining in VNTB (5). P12–25, refinement of terminal field arbors of MNTB axons in LSO (11). P13, total dendritic length and number of branch points in LSO at a peak (10). P13–14, tuning characteristics of excitatory and inhibitory inputs in LSO match closely (18); volume of LSO is about one-half of its adult value (22). P14–16, calyces in MNTB reach mature form (3); nuclei have reached 60–70% of their adult value (21). P15–19, parvalbumin staining in DNLL (5). P15/16, MNTB fibers in LSO more refined; end of major growth phase; no fibers with discontinuous arbors (11).

IC. IC has reached 60–70% of its adult size (21). P15–19, parvalbumin staining in IC (5).

MG and AC. Medial geniculate nucleus has reached 75% of its adult size (21).

CN. GABA-IR fibers and terminals in VCN are adultlike (8). P22, morphology and synaptic organization of DCN essentially adultlike (8, 9).

SOC. P18–25, MNTB fibers in LSO mature; fewer boutons than earlier (11). P19, mature pattern of parvalbumin staining (5). P21, glycine-receptor distribution in LSO adultlike (12); dendritic arbors of HF neurons become spatially constrained along the frequency axis (10).

GABA-IR, GABA immunoreactivity.

1, Wilkinson 1986; 2, Sanes 1993; 3, Kil et al. 1995; 4, Russell and Moore 1995; 5, Seto-Oshima et al. 1990; 6, Sanes 1992; 7, Woolf and Ryan 1985; 8, Ryan and Woolf 1992; 9, Schwartz and Ryan 1985; 10, Sanes, Song, and Tyson 1992; 11, Sanes and Siverls 1991; 12, Sanes and Wooten 1987; 13, Kuwabara, DiCaprio, and Zook 1991; 14, Ryan, Woolf, and Sharp 1982; 15, Ryan et al. 1985; 16, Harris and Dallos 1984; 17, Woolf and Ryan 1984; 18, Sanes and Rubel 1988; 19, Finck, Schneck, and Hartman 1972; 20, Smith and Kraus 1987; 21, Rübsamen et al. 1994; 22, Sanes, Merickel, and Rubel 1989; 23, Kelly and Potash 1986.

TABLE 7.5. Hamster (*Mesocricetus auratus*).

Age and Selected Embryological and Functional Time Points	Structural Development of the Auditory Pathways
E16/P0 (birth).	*CN.* P0, no GAD+ fibers or boutons in DCN (2); cells in DCN not in layers; cells have radiate dendritic fields (3); auditory nerve fibers have entered VCN but not DCN; branching of axons in PVCN increases over first weeks (4).
P1-9.	*CN.* P1-5, no significant increase in volume of DCN (5). P3, cochlear axons invade deepest DCN; no branches; a few terminals with round vesicles are present in the neuropil (4). P3-4, GAD+ axons in deepest DCN only (2). By P3, projections from ipsilateral IC present; axons restricted to deepest border of DCN (6). P5, no layers in DCN but most superficial cells in deep cell mass have taken on a bipolar appearance (3); cochlear axons ramify in deep DCN; pleomorphic synapses present throughout DCN (as in adult) (4); IC projections to DCN have grown in bilaterally but are not yet mature (6); terminals with small round vesicles present in superficial layers of DCN; terminals with large round vesicles found only in deepest DCN (4). P5-8, GAD+ fibers and terminals evenly distributed throughout DCN (2). P5-25, sharp increase in volume of DCN to about adult size (5). P6, GAD+ terminals apposed to dendrites and somas in DCN (2); descending axons ramify in same layers as in adult and exhibit puncta (6).
	SOC. P1, 20% of total CGRP+ cells in SOC are in LSO (3% of LSO cells) (7). P1-23, cell density in LSO decreases; cell size increases; cell number does not change (7).
	IC. By P2, projections from contralateral DCN present (6). P3, projections from ipsilateral DCN present (6). P5, projections from DCN appear fully mature (6).

P10–15.

P14, no response to acoustic stimulation, even at high sound levels (8).

P16 to adult.

P16, first BAERs (8). P22, threshold for evoked response like adult (8).

CN. P10, fusiform cells segregated from large cells; layers can be identified for first time (3); adult pattern of cochlear input in DCN; axons highly branched; terminals with large, round vesicles widely distributed (4). P10–15, fastest elongation of apical dendrites of fusiform cells in DCN; dendrites grow ~ 17 μm/day (9). P12, dense band of GAD$^+$ puncta in fusiform cell layer (2). P15, laminae in DCN fully formed; fusiform and giant cells adultlike in form but apical dendrites of fusiform cells not yet studded with spines (10) and not yet oriented (9, 11). P15–60, continued elongation of apical dendrites of fusiform cells in DCN; average length increases by ~2.5 times; number of branch points increases by greater than 2 times; the length of individual dendritic segments does not increase (9).

CN. P16–21, GAD$^+$ density increases throughout DCN but relatively more in molecular layer (2). P21, adultlike GAD$^+$ labeling in DCN (2). P25, tendency toward adult orientation of apical dendrites of fusiform cells (9, 11); volume of DCN reaches adult value (5). P25–60, orientation of apical dendritic tree of fusiform cells becomes fully established (9).

SOC. P23, more than 70% of CGRP$^+$ cells in SOC are in LSO (18% of all LSO cells) (7).

GAD$^+$, positive for glutamic acid decarboxylase; CGRP, calcitonin gene-related peptide.

In the Chinese hamster, which has a gestation length of 21 days, neurogenesis of neurons of MG occurs during E13–18; at E18, MG can be identified (as can the lateral geniculate nucleus; the rest of the dorsal thalamus has a homogeneous appearance) (1).

1, Keyser 1972; 2, Schweitzer, Cecil, and Walsh 1993; 3, Schweitzer 1990; 4, Schweitzer and Cant 1984; 5, Schweitzer and Cecil 1992; 6, Schweitzer and Savells 1991; personal communication; 7, Simmons, Raji-Kubba, and Strauss 1992; 8, Schweitzer 1987; 9, Schweitzer 1991; 10, Schweitzer and Cant 1985b; 11, Schweitzer and Cant 1985a.

TABLE 7.6. Cat (*Felis catus*).

Age and Selected Embryological and Functional Time Points	Structural Development of the Auditory Pathways
Before birth. E19, cerebral hemispheres first appear as distinct swellings at anterior end of neural tube (e.g., Ref. 3). E19–25, cortical plate first appears (4). E22, internal capsule is recognizable (3).	*CN.* E22, cochlear nerve fibers enter brain (1). E57, cochlear nerve shows first signs of myelination (2). *MG and AC.* E30–57, neurogenesis of cells that will form *visual* cortex (5). E35, first ascending thalamic fibers enter internal capsule; reach subplate at E46 (3). E43, CC links the hemispheres (3, 6). E54, presumptive MG undifferentiated; adult subdivisions not distinct (3, 7). E54–56, AC projects to MG (7). By E55, MG projects to AC (7). E55 to P4, all major projections from cortex to subcortical structures are present and topographically organized (8). E56, cells that project into CC widespread, located in superficial layers (6).
P0 (birth). P0, entire cochlea is immature (9). AC responds to electrical stimulation of MG (10). P0–7 glutamate receptors on CN cells can be activated by NMDA (11).	*CN.* P0, major subdivisions are recognizable; granule cells still form an external layer on DCN (12); GAD$^+$ puncta present in deep DCN (13); fusiform cells in DCN small with short dendritic processes; soma and dendrites covered with appendages (14). P0–3, few synapses in OCA (15). *SOC.* P0, distribution of cells giving rise to olivocochlear projection is adultlike (16). *IC.* P0, characteristic lamination present in central nucleus; different neuronal types recognizable on basis of dendritic branching, although the patterns are not mature (17); neurons have extensive dendritic trees (18). *MG and AC.* P0, pyramidal cells have very sparse dendrites (18); topography of connections between cortex and subcortical structures is adultlike, but patterns of corticocortical connections are immature (6, 19); axons and dendrites of subplate cells in posterior ectosylvian gyrus reach into the cortical plate; they decrease in numbers during the first two postnatal weeks (20). P0–38, transitory projections project from AI and AII to visual areas 17 and 18 (21).

P1–5.

P1, CM elicited to high-intensity stimuli (22, 23 [2 days before birth]). P3, base of cochlea begins to mature (9). P4, BAERs first recorded to clicks (24). P5, earliest definite behavioral responses to sound (25); earliest that cortical potentials can be evoked to intense clicks (26); no sound-evoked 2-DG labeling in IC (27).

P6–10.

P6, nerve AP recorded; high threshold; no tuning (32). P7–14, AERs attain adult sensitivity (33–35). P8–12, IID and IPD sensitivity in IC adultlike (36). P10, external meatus opens (30); frequency-specific 2-DG labeling in IC to all but very HF (27). P10–12, base of cochlea reaches maturity (9).

P11–17.

P11–16, orientation to sounds (38). P12, external auditory canal reaches maximum depth (25); first BAER elicited to clicks (39). P12–15, mature CM, nerve AP (23, 32). P15, apex of cochlea mature (9); OHC adultlike (41). P16, 75% of animals able to locate sounds (25, 41).

CN. P2, only octopus and globular cells in VCN can be recognized based on Nissl criteria (28). P2–10, end bulb in AVCN is spoon shaped with filopodia (28). P4, diffuse GAD+ label in DCN (13). P4–7, synaptic endings from cochlea present in OCA; dendrites begin to mature (15). P5, adult subdivisions of AVCN recognizable (28).

SOC. P2, cells in LSO fusiform with oriented dendrites similar to adult (29).

IC. P0–7, orientation of dendritic trees of disk-shaped neurons established (17). P2, main subdivisions distinguishable; cell bodies small (30). P3–4, projections from layer 5 of AC present (8, 31).

MG and AC. P1, callosal connections present but very immature (3, 19); cortex is 750–900 μm thick; during first 3 months, increases to adult thickness of 1,200–1,800 μm (6). P2, gross appearance of brain like adult but smaller (3); callosal axons extend throughout cortex (no alternating columns, no laminar specificity) (19). P3, MG projects to cortical areas AI and EP (31).

CN. P7–12, GAD+ distribution in DCN attains adultlike pattern (13). P10, ~50% of end bulbs have progressed from immature stage to become fenestrated (28).

IC. P7, spiny, disk-shaped neurons profusely covered with fine hairy appendages, which begin to disappear during the second week (17). P7–14, dendrites of large stellate neurons reach adult length; number of dendritic appendages increases during second and third week and then begins to decline (17).

MG and AC. P8–18, segregation of callosal afferents into vertical bands (19). P9, chandelier cells first recognized in AC (37).

CN. P12, fusiform cells in DCN essentially adultlike, although dendrites continue to mature for some time (14). P14, by 2–3 weeks, all CN cell types have adult features (although not adult size) (12).

IC. P12, orientation of dendritic trees not like that of adult (42).

MG and AC. P15–23, chandelier cells in AC develop vertically elongated axonal complexes (37).

(Continued)

TABLE 7.6. (Continued)

Age and Selected Embryological and Functional Time Points	Structural Development of the Auditory Pathways
P18 to adult. P18–24, consistent approach response to sounds (43). P21, essentially all responses in AVCN adultlike except for phase locking (44, 45). P31, pinna fully developed (41).	*CN.* P20, immature end bulbs left in AVCN, many of intermediate and adult type (28); CN has reached 75% of adult length (28). P20–35, type 3 synapses first appear in OCA (15). P21, OCA reaches cytological maturity; all three synaptic types appear on somas and basal dendrites (15). P45, 87% of end bulbs adultlike (28); CN has reached 85% of adult length (28). P45–95, endings from IC identified in DCN; glomeruli and synaptic nests present (15). P120, full maturation of fusiform cell layer as seen in EM (14). *SOC.* P20, in MSO, synapses well established on proximal dendrites; more distal dendrites not mature; dendritic and somatic spines are adultlike by the end of the third week (46, 47). *IC.* P30, density of dendritic spines at a maximum; decreases after this to the adult value (17). *MG and AC.* P18, laminar pattern of callosal terminals not mature until sometime after this (3). P42, adult morphology attained in some chandelier cells but not all (37). P98, callosally projecting cells do not yet form aggregates (3, 6, 19).

NMDA, N-methyl-D-aspartate; OCA, octopus cell area; AI, primary auditory cortex; AII, secondary auditory cortex; EP, posterior ectosylvian auditory area; AER, auditory evoked response; IID, interaural intensity difference; IPD, interaural phase difference.

1, Windle 1933; 2, Romand et al. 1976; 3, Payne 1992; 4, Marin-Padilla 1971; 5, Luskin and Shatz 1985; 6, Payne, Pearson, and Cornwell 1988a; 7, Payne, Pearson, and Cornwell 1988b; 8, Cornwell, Ravizza, and Payne 1984; 9, Pujol and Marty 1970; 10, Hassmannová and Mysliveček 1967; 11, Walsh, McGee, and Fitzakerley 1993; 12, Larsen 1984; 13, Schweitzer, Cecil, and Walsh 1993; 14, Kane and Habib 1978; 15, Schwartz and Kane 1977; 16, Warr, White, and Nyffeler 1982; 17, González-Hernández, Meyer, and Ferres-Torres 1989; 18, Pujol 1969; 19, Feng and Brugge 1983; 20, Valverde and Facal-Valverde 1988; 21, Innocenti and Clarke 1984; 22, Mair, Elverland, and Laukli 1978; 23, Pujol and Hilding 1973; 24, Shipley et al. 1980; 25, Olmstead and Villablanca 1980; 26, Rose, Adrián, and Santibáñez 1957; 27, Webster and Martin 1991; 28, Ryugo and Fekete 1982; 29, Romand, Granier, and Marty 1973; 30, Aitkin and Moore 1975; 31, Ravizza, Garlitz, and Cornwell 1978; 32, Carlier, Lenoir, and Pujol 1979; 33–35, Walsh, McGee, and Javel 1986a,b,c, respectively); 36, Blatchley and Brugge 1990; 37, deCarlos, Lopez-Mascaraque, and Valverde 1985; 38, Norton 1974; 39, Jewett and Romano 1972; 40, Pujol, Carlier, and Lenoir 1980; 41, Villablanca and Olmstead 1979; 42, Meininger and Baudrimont 1981; 43, Clements and Kelly 1978; 44, Brugge, Javel, and Kitzes 1978; 45, Kettner, Feng, and Brugge 1985; 46, Schwartz 1972; 47, Schwartz 1977.

TABLE 7.7. Ferret (*Mustela putorius*).

Age and Selected Embryological and Functional Time Points	Structural Development of the Auditory Pathways
Before birth. E 42 to P0 (birth).	SOC. E28, fibers from cochlear nucleus reach SOC (1). CN. P0, major subdivisions of CN are indistinct, neurons poorly differentiated; auditory nerve fibers arborize throughout VCN and DCN (2). SOC. P0-8, many MSO cells lack clear bipolar morphology; few synapses in MSO; the ones present are immature and found in neuropil (3); bipolar orientation of LSO cells apparent at birth; transient spines and appendages abundant on LSO cells (4). IC. P0, connections from LSO like adult except that a greater proportion of inputs may come from contralateral sources (5); spatial distribution of cells that project from both LSOs resemble adult pattern (5).
P1-26. P2/3, neural discharges recorded in midbrain (7).	MG and AC. P0-13, cells in layers 1-3 of *visual* cortex generated (6). CN. P18, ratio of ipsi to contra projections from CN to IC is 1:33 (8). P26, ratio of ipsi to contra projections from CN to IC is 1:29 (8). SOC. P7, glycine-positive cell bodies first appear, distributed as in adult except that none occupy LSO (9). P8-10, transient appendages on soma and dendrites of MSO cells, first proximal, then distal, persist through P30 (10). P8-30, MSO cells increase in size (10). P10-15, rapid increase in synaptogenesis in MSO (1, 3). P12-28, synaptogenesis in MSO proceeds to adult pattern (3). P16-32, cross-sectional area of cells in MSO increases by 60% (3). By P17, most cells in MSO are bipolar, with oriented dendritic fields (10).
P27-35. P27-35, BAER appears and attains adult thresholds (P27-35, Ref. 11; P32, Ref. 7, 12). P32, ear canals open; startle response appears (7).	SOC. P28, first glycine-positive cells in LSO (9). P28-30, tufts of distal dendritic branches appear on MSO cells (10). P28-56, tufts of fine tendrillike processes at end of LSO cell dendrites (4). P35, glycine-positive puncta observed in LSO neuropil; earliest distribution greatest in HF regions; density gradient decreases over next 2 weeks (9). IC. P35, clear tonotopic organization (7).
P36 to adult. P39-42, adultlike sensitivity in IC (7). P51, acoustics of auditory periphery adultlike (13).	CN. Ratio of ipsi to contra projections from CN to IC is 1:45 (as in adult) (8). SOC. Cells of MSO achieve mature size (3).

1, Brunso-Bechtold, Henkel, and Linville 1987; 2, Moore 1991; 3, Brunso-Bechtold, Henkel, and Linville 1992; 4, Henkel and Brunso-Bechtold 1991; 5, Henkel and Brunso-Bechtold 1993; 6, Jackson, Peduzzi, and Hickey 1989; 7, Moore 1982; 8, Moore and Kowalchuk 1988; 9, Henkel and Brunso-Bechtold 1995; 10, Henkel and Brunso-Bechtold 1990; 11, Morey and Carlile 1990; 12, Moore and Hine 1992; 13, Carlile 1991.

TABLE 7.8. Marsupials.

Species	Structural Development of the Auditory Pathways
Brush-tailed opossum (*Trichosurus vulpecula*) (1)	*E17 to P0* (birth). Four to five months spent in pouch; cells of CN and SOC generated some time before P5. P5, generation of large neurons in CN, ventrolateral central nucleus of IC and layer 6 of cortex. P5–9, most intense neurogenesis in MG, completed after P12. P9, CN not yet recognizable. P9–18, neurogenesis in dorsolateral and dorsomedial central nucleus and dorsal cortex of IC. P9–21, neurogenesis of granule cells; intense on P9–21, less on P28–46. P19, CN recognizable. P25, occasional genesis of granule cells; continues even after P46. P46, layer 2 of cortex generated. P100, eyes open.
Brazilian opossum (*Monodelphis domestica*) (10–12)	*E13 to P0* (birth; approximately equivalent to E13/14 mouse). Before P5, first hair cells recognizable. P6 (approximately equivalent to E16 mouse), migrating cells have just begun to enter DCN; fasicles of primary afferent fibers present throughout presumptive VCN; presumptive DCN contains lightly labeled fibers. P8 (approximately equivalent to E18 mouse), adult number of turns in cochlea; primary fibers in VCN organized topographically; giant and fusiform cells migrating into DCN; fine-caliber auditory nerve fibers extend into the forming cell mass. P15 (approximately equivalent to P3/4 in mouse), hair cells not mature; topographic order in CN well defined. P34, external auditory meatus opens; first responses to sound.
Virginia opossum (*Didelphis marsupialis*) (2–11)	E9–11, neurulation; posterior neuropore closes on E11. E12–13 to P0 (birth), DCN and VCN not identifiable. P0–20, neurons migrate into CN from medial cytogenetic zone. P3, VCN well defined. P5, neurons migrating to CN send axons to midbrain; neurons in SOC project to IC. P10, some migrating neurons have reached DCN; neurons projecting from VNLL to IC first seen; migratory neuroblasts have reached presumptive MNTB and have just begun to differentiate; calyceal axons have reached MNTB. P12, little indication of DCN before this. P15 (approximately equivalent to P1 mouse), vertically oriented neurons present in DCN; pattern of projections from brain stem to IC is like adult. P15–40, laminae develop in DCN. P15–75, size of DCN increases 1.4 times every 20 days. P20, DCN neurons short with thick primary dendrites and many secondary and tertiary branches; dendrites covered with filopodia and numerous beadlike varicosities, growth cones on their tips; neurons migrating

from medial cytogenetic zone have reached VCN. P20-33, projections from CN to IC topographic; topography well defined by P33. P32, all layers of DCN present. P48, parallel fibers first seen in DCN; cells in granule cell layer closely resemble principal cells of adult but soma and dendrites covered with thin, delicate filopodia. P48-77, frequency range that elicits behavioral responses increases from ~3 to 7 octaves; greatest increase is during P51-54. P50, first reflex responses to sounds in midfrequency range. P55, interneurons are present in DCN, but their dendrites are very immature; short, hairlike filopodia, numerous varicosities; principal cells adultlike; filopodia seen occasionally in distal dendritic tree. P70, all cells have adultlike shape, but small cells still have thin filopodia on dendritic shafts and sometimes on soma; growth cones on tips of dendrites of all types of neurons. P77, frequency range still increasing.

Northern quoll
(*Dasyrus hallucatus*) (13-16)

Stage 13, closure of neural tube. E21 to P0 (birth; stage 15 in Carnegie series; approximately equivalent to 35-38 days postconception in humans), 50-60 days spent entirely in pouch. Before P3, neurons of VCN generated; neurons of DCN generated over a "prolonged period." P3-9, neurons of MG generated. P5-7, neurons of SOC generated, peak at P7. P7-22, neurons of IC generated. P9, (stage 20), CN and SOC begin to appear. P9-42, neurogenesis of AC begins; complete on P42. P13 (stage 21), DCN, VCN, and components of SOC recognizable. P17, migrating cells from germinal zone have not yet reached IC. P23 (stage 23, end of embryogenesis), IC first recognizable; almost all migration of cells from ventricular germinal zone has been completed; central nucleus recognizable as a distinct nucleus. P36, peak cell number in IC. P45, young leave pouch for brief periods. P45-50, decrease in packing density of cells in IC. P50, organ of Corti formed; cell number in IC falls to a minimum and then rises to adult values (due to proliferation of glia) by P75-80. P55, external ear duct patent. P75, eyes open. P75-80, young remain out of pouch most of the time. P81, adult cytoarchitecture of IC is clear.

McCrady (1938) suggested that during the first 10.5 days of development, the rabbit and North American opossum develop at the same rate, but that after that time, the phases of development become relatively prolonged in the opossum. Development in the marsupial appears to slow considerably after birth. The *sequence* of events in the developing auditory system, however, appears to be the same as in eutherian mammals.

1, Sanderson and Aitkin 1990; 2, McCrady 1938; 3, McCrady, Wever, and Bray 1937; 4, McCrady, Wever, and Bray 1940; 5, Morest 1969a; 6, Morest 1969b; 7, Willard and Martin 1986; 8, Willard and Martin 1987; 9, Willard and Martin (unpublished results); 10, Willard 1993; 11, Willard 1995; 12, Willard and Munger (unpublished results); 13, Aitkin et al. 1991; 14, Nelson 1992; 15, Aitkin et al. 1994; 16, Gemmell and Nelson 1992.

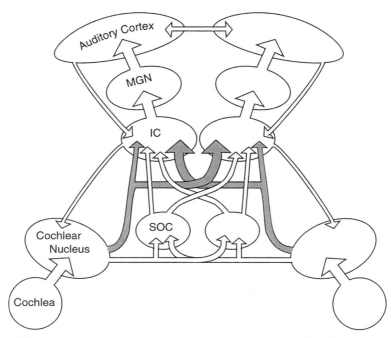

FIGURE 7.1. Overview of the mammalian auditory pathways. Spiral ganglion cells, located in the cochlea, project into the cochlear nucleus via the eighth nerve. The cochlear nucleus projects to the inferior colliculi (IC; the projection is represented in shading) and to the superior olivary nuclei (in the superior olivary complex [SOC]) on both sides of the brain. (The major projection to the IC is contralateral; the presence and size of the ipsilateral projection appear to vary among species.) Superior olivary nuclei also project to the IC on both sides. (The nuclei of the lateral lemniscus, which are not illustrated, have a pattern of connections similar to those of the superior olivary nuclei.) The output of the IC is conveyed to the medial geniculate nucleus (MGN) via the brachium of the IC. The MGN, in turn, projects to the auditory cortex via the auditory radiations. The auditory cortices on the two sides of the brain communicate through the corpus callosum. Descending axons arise in the auditory cortex and project to the ipsilateral inferior colliculus. Projections from the IC descend to the cochlear nuclei on both sides (only the ipsilateral component of the projection is illustrated). Other descending projections, mentioned in this chapter but not illustrated here, arise in the SOC and travel out to the cochlea to innervate the hair cells in the organ of Corti. Adding to the complexity of the pathways as illustrated is the fact that each structure in the auditory pathways is made up of multiple cell types; different components of the pathways arise from different populations of neurons (see Fig. 7.2).

These developmental events lead to the very early establishment of the basic pattern of circuitry that will be characteristic of the adult. In eutherian mammals (Tables 7.2–7.7), most of these events occur and are well advanced before birth; in metatherian species (Table 7.8), much of the

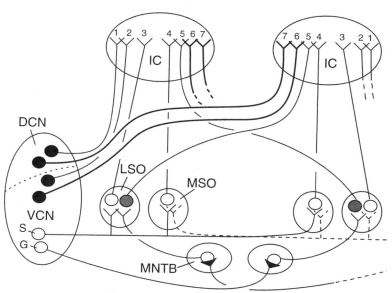

FIGURE 7.2. Projection patterns of some of the neuronal types found in the cochlear nucleus and SOC. This illustration is highly simplified in that a number of the known cell types and connections are not shown; the neurons represented here are those the development of which is discussed in this chapter. Neurons in the dorsal cochlear nucleus (DCN) include the fusiform and giant cells. (Each filled symbol in the DCN represents both types; for the sake of clarity they are not represented separately.) Both of these neuronal types project to the contralateral IC (termination 7); in many species, both types also give rise to a small ipsilateral component (termination 1). Multipolar neurons in the ventral cochlear nucleus (VCN) have a similar pattern of contralateral (termination 6) and ipsilateral (termination 2) projections to the IC. (Each filled symbol represents several types of multipolar cells as defined morphologically and physiologically.) Spherical cells (S) of the VCN project to nuclei in the SOC, including the ipsilateral lateral and medial superior olivary nuclei (LSO and MSO, respectively) and the contralateral MSO. The globular cells (G) of the VCN project to the contralateral medial nucleus of the trapezoid body (MNTB), where they form the very large synapses known as the calyces of Held. A number of neuronal types in the SOC project to the IC. These include neurons in the MSO (termination 4) and two types of neurons in the LSO, one group of which projects to the ipsilateral IC (open circle, termination 3) and the other of which projects to the contralateral IC (shaded circle, termination 5). In the SOC itself, the cells of the MNTB that receive input from the globular cells of the contralateral VCN project to the ipsilateral LSO. All of the projections illustrated appear to be topographically organized. Dashed lines represent axons that arise from the cochlear nucleus that is not illustrated. Each eighth nerve fiber makes synaptic contact with each of the cell types in the cochlear nucleus. The synapse formed with the spherical cells is very large and is known as the end bulb of Held.

development is postnatal. All of these developmental events occur concurrently; some neurons undergo dendritic differentiation and synaptogenesis before others have even been generated. It is somewhat arbitrary to separate the discussion of neurogenesis and migration from that of cell differentiation and axon outgrowth because the timing of these two aspects of development may overlap considerably. For example, cells may give rise to axons before they leave the neuroepithelial zone, and their axons may reach the region of their targets before the parent neurons have migrated to their final positions. Nevertheless, the different aspects of development have generally been addressed in separate studies and will be discussed separately here. Although the basic circuitry of the auditory system appears to be established by the time of onset of hearing, changes in dendritic and axonal morphology continue over a protracted time period, well beyond the onset of hearing and possibly into adulthood (e.g., Moore 1988). These later aspects of normal development are considered in Section 4.

3.1 Neurogenesis and Migration

When the neural tube closes, its wall consists of a thin epithelium of dividing cells, many of which will become neurons. The location of this "neuroepithelium" is referred to as the ventricular zone (Angevine et al. 1970) because the lumen of the neural tube will give rise to the ventricular system of the adult. Migration away from the ventricular zone by neurons that have undergone their last division gives rise to a subventricular zone where cells sometimes appear to "wait" before migrating further and to a mantle or intermediate zone, which is the precursor of the mature brain and spinal cord. In areas where a cortex forms, a cortical plate arises external to the mantle zone. Finally, an external fiber layer or marginal zone, made up of axons growing away from their cell bodies, forms the most superficial aspect of the neural tube. Ultimately, all neural cells migrate away from the ventricular zone so that only a thin ependymal cell layer is left between the ventricle and the neural tissue. As neurons undergo their terminal division, sometimes even before they begin to migrate, they give rise to axons. Axon outgrowth and cell migration leads ultimately to the establishment of the adult pattern of auditory structures. (Axon outgrowth, although it occurs concurrently with cell migration, is discussed in Section 3.2.)

Neurons destined to give rise to auditory structures arise from distinct parts of the neuroepithelium all along the length of the neural tube. Although different neuronal types in a particular part of the auditory system appear to be generated at different times, neurogenesis proceeds in auditory structures at all levels of the neuraxis during essentially the same time period (summarized in Table 7.9; see also Tables 7.2, 7.3, and 7.8). The first auditory neurons generated form the cochlear nucleus, superior olivary complex, and medial geniculate nucleus. Genesis of the neurons that form the inferior colliculus and auditory cortex begins slightly later; the

TABLE 7.9. Time of genesis of neurons in the auditory system of five mammalian species.

	Spiral Ganglion	Cochlear Nucleus	SOC and NLL	Inferior Colliculus	MG	Auditory Cortex
Brush-tailed opossum (1)		<P5	<P5	P5-18	P5-9	P5-46
Northern native cat (Northern quoll) (2)		<P3	P5-7	P7-22	P3-9	P9-42
Rat (3)	E12*	E11-18	E11-15	E13 to P0	E13-14	E12-20
Mouse (4)	E13*	E10-15	E9-13	E11-17	E10-14	E10-16
Rabbit (5)		E12-17.5	E12-15	E15-18.5		

1, Sanderson and Aitkin 1990 (see for a similar table); 2, Aitkin et al. 1991; 3, McAllister and Das 1977; Altman and Bayer 1979, 1980a,b, 1981, 1982, 1989; Nornes and Morita 1979; Bayer and Altman 1991; Kudo et al. 1992; Friauf and Kandler 1993; 4, Angevine and Sidman 1961; Ruben 1967; Taber Pierce 1967, 1973; Angevine 1970; Martin and Rickets 1981; Gardette, Courtois, and Bisconte 1982; 5, Oblinger and Das 1981.
*The date of peak neurogenesis of spiral ganglion cells.

time course of neurogenesis for these structures may be prolonged compared to the other areas. The appearance of spiral ganglion cells and hair cells in the cochlea appears to lag behind the appearance of neurons of the central auditory nuclei by a slight margin (Tables 7.2 and 7.3). In some marsupials, neurogenesis of much of the auditory system occurs after birth, although it occurs at a *stage* of development roughly equivalent to the stage of development at which neurogenesis occurs in the mouse (cf. Tables 7.3 and 7.8). In the ferret, neurogenesis of the superficial layers of the *visual* cortex occurs during the first two weeks after birth (Table 7.7). The genesis of the auditory cortex is probably slightly earlier (see Section 3.1.4) but could also occur postnatally. In the rodents studied, essentially all neurogenesis has ceased before birth, although there may be continued production of granule cells in the cochlear nucleus. A peak of cell division occurring during the perinatal period has been attributed to the production of glial cells (Martin and Rickets 1981).

3.1.1 The Cochlear Nuclei

Neurogenesis of neurons of the cochlear nuclei begins shortly after the neural tube begins to close and the motor columns of the brainstem begin to form (Tables 7.2 and 7.3). At least two neuroepithelial sites in the rhombencephalon give rise to neurons that will eventually mix to form the cochlear nuclei (Harkmark 1954a,b; Altman and Bayer 1980a; Willard and Martin 1986). One of these, the rhombic lip, lies at the point of attachment of the alar plate and the roof plate of the neural tube (His 1888; Blake 1900). The other lies medial to the solitary tract at the level of the sulcus limitans (Willard 1995). Specific parts of the rhombic lip give rise to neurons destined to form specific brain stem nuclei (including the inferior olive and the pontine and raphe nuclei) in a definite sequence (Harkmark 1954a). Neurons generated in the rhombic lip follow longitudinal and/or circumferential migratory paths that may take them to a final resting place far from their site of origin (e.g., Essick 1912; Taber Pierce 1966). Most authors (e.g., Taber Pierce 1967; Nornes and Morita 1979; Martin and Rickets 1981; Altman and Bayer 1987) agree that the rhombic lip per se is a source of auditory neurons, although Harkmark (1954a) concluded that the acoustic nuclei are derived from the neuroepithelium medially adjacent to the rhombic lip. The neurogenetic zone is at the level of the lateral recess; that for the dorsal cochlear nucleus is caudal to that for the ventral cochlear nucleus (Taber Pierce 1967). The most lateral neuroepithelium of the lateral recess of the fourth ventricle is also the major source of granule cells in the cochlear nucleus (Altman and Das 1966; Taber Pierce 1967).

In general, the neurons that migrate away from the rhombic lip are in a "very immature state" (Harkmark 1954b). As the cells migrate, they become oriented in the direction of the migration. They are still immature on arrival at their destination. Nissl bodies are not seen in the neurons until some time after migration has concluded. Taber Pierce (1967) followed the medial migration of the neurons generated on embryonic day 12 (E12) in the mouse (neurons of the acoustic nerve nucleus and spherical, globular, and multipolar cells of the ventral cochlear nucleus; Taber Pierce 1967; Martin and Rickets 1981). The E12-generated neurons appear first in the acoustic area of the rhombic lip at the level of the lateral recess and, over the course of 96 hours, migrate ventrolaterally, moving lateral to the fibers of the restiform body and the spinal trigeminal tract. All cells generated on E12 reveal a similar pattern except for the granule cells, although the migration distances vary. By E14 and 7 hours, a cell mass lateral to the restiform body can be identified as the presumptive cochlear nucleus in the mouse, and by E16.5, the anlage of the dorsal and ventral cochlear nuclei can be distinguished (Hugosson 1957).

The other cytogenetic zone that gives rise to cochlear nuclear cells is located in the neuroepithelium near the sulcus limitans at the pontomedullary junction (Willard 1995). In the North American opossum, the fusiform and giant cells of the dorsal cochlear nucleus and at least some multipolar cells of the ventral cochlear nucleus (all of which project, in the adult, to the inferior colliculus) are generated in this zone (Willard and Martin 1983, 1986). In the newborn opossum (a stage approximately equivalent to an E13/14 mouse; Willard 1993) a band of horizontally oriented neurons extends from the cytogenetic zone to, but not into, the presumptive cochlear nucleus (Figure 7.3). At this age, separate dorsal and ventral cochlear nuclei are not distinguishable and only a few scattered cells lie dorsolateral to the vestibular nerve root in the future location of the cochlear nucleus. From birth until about postnatal day 20 (P20), the band shifts laterally as neurons migrate from the cytogenetic zone in the most medial part of the alar plate over the dorsal boundary of the vestibular nerve and into the cochlear nucleus. While the neurons are migrating, their axons reach the inferior colliculus. Migration probably occurs via perikaryal translocation through long horizontally disposed leading and trailing processes characteristic of the primitive migratory cells (Morest 1970; Book and Morest 1990; Willard and Martin, personal communication). Three groups of neurons separate in the cochlear nuclei, forming the principal and giant neurons of the dorsal cochlear nucleus and at least some of the large multipolar neurons of the ventral cochlear nucleus. The neurons entering the presumptive ventral cochlear nucleus arrive after the more laterally generated spherical and globular cells have already arrived. Subsequent to the entry of the large cells, small cells generated in the neuroepithelium of the rhombic lip enter the dorsal cochlear nucleus (Willard and Martin, personal communication).

The mixing of neurons from different neuroepithelial sources occurs in other parts of the brain stem as well (inferior olive: Ellenberger, Hanaway, and Netsky 1969; nucleus reticularis tegmenti pontis: Taber Pierce 1966; raphe nuclei: Harkmark 1954a) and may reflect a general rule of brain stem development. It is of great interest to determine whether neurons derived

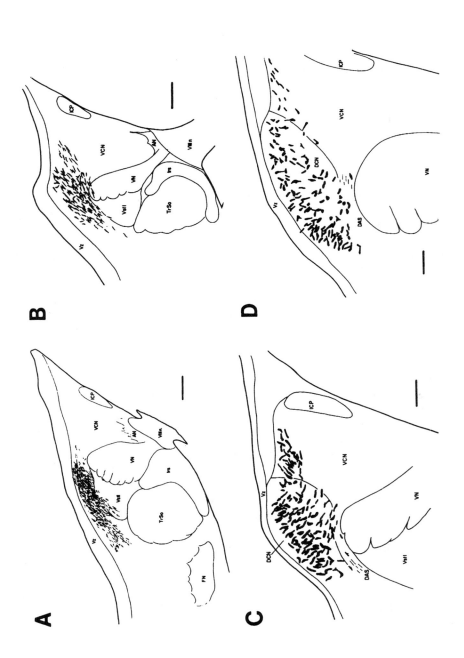

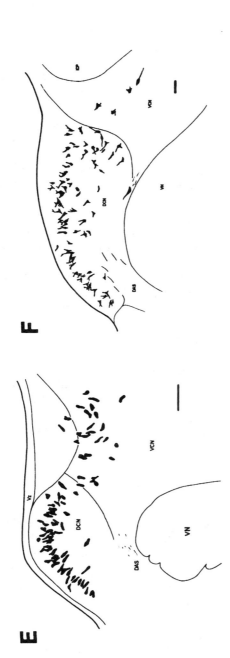

FIGURE 7.3. Drawings illustrating the positions of labeled neurons after injections of horseradish peroxidase into the contralateral IC of North American opossum pouch young. A: estimated postnatal day (P) 5; B: P10; C: P15; D: P20; E: P35; F: adult. Scale bars in each panel represent 100 μm. Neurons that are generated in a medial cytogenetic zone and labeled through their axons in the midbrain migrate laterally to form a part of the presumptive cochlear nucleus. After the cells arrive in the presumptive cochlear nucleus, they separate into three groups—the fusiform and giant cells of the DCN and multipolar cells of the VCN. Vz, ventricular zone; Vstl, lateral vestibular nucleus; TrSO, spinal trigeminal nucleus, pars oralis; FN, facial nucleus; ICP, inferior cerebellar peduncle; VN, vestibular nerve; trs, tract of descending spinal trigeminal nucleus; AN, auditory nerve; DAS, dorsal acoustic stria. (From Willard and Martin [1986], reprinted by permission of the authors and John Wiley and Sons, Inc.)

from the different sources have different patterns of connectivity. The results cited above would indicate that such is the case because at least some of the cell types generated in the medial cytogenetic zone have different projection patterns from some of those generated in the rhombic lip.

In general, in any particular part of the central nervous system, large neurons are generated earlier than small ones (Angevine 1970; Jacobson 1991). This pattern is evident in the generation of neurons of both the dorsal and ventral cochlear nuclei in rats (Nornes and Morita 1979; Altman and Bayer 1980a), mice (Taber Pierce 1967; Martin and Rickets 1981), and the Dutch rabbit (Oblinger and Das 1981). In both the dorsal and ventral cochlear nuclei, there is sequential production of large, intermediate, and, finally, small neurons. Neurons generated at different times show no tendency to segregate according to time of origin. Thus the order of generation is a function of cell size and not of location in the adult nuclei (Taber Pierce 1967). The production of the smallest neurons, the granule cells, continues well after production of other neuronal types is complete (Altman and Das 1966; Taber Pierce 1967). Many are generated postnatally in rat and mouse. The granule cells appear to migrate over the surface of the dorsal cochlear nucleus as an external granular layer and then to migrate radially into the deeper layers (Ochs and Brunso-Bechtold 1982; Larsen 1984; Schweitzer, Bell, and Slotkin 1987). The migratory pattern of the cochlear nuclear granule cells is similar to that of the granule cells destined for the cerebellar cortex.

Generation of neurons of the dorsal cochlear nucleus extends over a relatively long period of time compared to that of neurons of the ventral cochlear nucleus. In rats, nongranule neurons of the dorsal cochlear nucleus are produced from E11 through E18 and granule cells continue to be produced after that (Altman and Bayer, 1980a; Friauf and Kandler 1993; also mouse, Taber Pierce 1973). A relatively prolonged production of neurons is also seen in the inferior colliculus (Section 3.1.3) to which many of the cells in the dorsal cochlear nucleus project.

3.1.2 The Superior Olivary Complex and the Nuclei of the Lateral Lemniscus

Neurons of the superior olivary complex and nuclei of the lateral lemniscus are generated as early as or earlier than neurons of the cochlear nucleus. The neuroepithelial source or sources of these nuclear groups are not entirely clear. Altman and Bayer (1980a) suggest that the neurons are produced in the dorsal aspect of the medial neuroepithelium and then migrate ventrally. Nornes and Morita (1979), however, conclude that some nuclei of the superior olivary complex derive from the rhombic lip. In the human, Streeter (1912) identified the superior olivary complex by the eighth week. He speculated that the cells migrate from the rhombic lip in the same way that cells of the inferior olive do. In the chick, neurons of the superior olivary complex are generated in the rhombic lip in the same part of the neuroepithelium that gives rise to the cochlear nuclei; the neurons of the superior olivary complex then migrate ventrally (Hugosson 1957). It may be that the neurons of the superior olive, like those of other nuclei of the rhombencephalon, are derived from more than one neuroepithelial source (Section 3.1.1). Neurons of the superior olivary nuclei and those of the nuclei of the lateral lemniscus may be produced in contiguous sites in the neuroepithelium but then migrate in different directions. Streeter (1912) noted that the nuclei of the lateral lemniscus appear to develop as a forward extension of the superior olive in humans. The neurons destined for these two areas are generated during the same time period (Tables 7.2, 7.3, and 7.9).

Neurons of the different nuclei of the superior olivary complex are generated in a specific order. In studies in rats (Altman and Bayer 1980a; Kudo et al. 1992), mice (Taber Pierce 1973), and rabbits (Oblinger and Das 1981), the same general sequence is reported. The medial superior olivary nucleus is among the earliest nuclei to be generated in any part of the auditory system.[2] Generation of neurons of the lateral superior olivary nucleus begins about the same time but lasts for a longer period. Generation of neurons of the medial nucleus of the trapezoid body begins several days later than that of neurons of the other nuclei. Gradients of cell production were seen in most nuclei of the superior olivary complex with a medial-to-lateral gradient dorsally and a lateral-to-medial gradient ventrally (Altman and Bayer 1980a). The neurogenetic gradients of neurons of several cell groups are in register with the topographic gradients of those nuclei (Altman and Bayer 1980a). In the lateral superior olivary nucleus, the high-frequency neurons are generated earliest, but in the medial nucleus of the trapezoid body (which projects to the LSO), low-frequency neurons appear to be the earliest generated. Altman and Bayer (1980a) suggest that the order of production of the neurons can be related to whether the predominant projections to the nucleus are ipsilateral or contralateral.

Kudo et al. (1992) demonstrated that within the lateral superior olivary nucleus neurons with different patterns of projections are generated at different times. They used neuroanatomical tracing techniques in combination with birth dating to demonstrate that the neurons that project to the contralateral inferior colliculus in the adult are generated on E11–E12, whereas those that project to the ipsilateral inferior colliculus are generated

---

[2]Altman and Bayer (1980a) misidentified some of the nuclei of the superior olivary complex. The nucleus that they called the medial superior olivary nucleus is actually the superior paraolivary nucleus; the nucleus that they called the lateral nucleus of the trapezoid body is actually the medial superior olivary nucleus. The correct nomenclature is used here and in Table 7.2. Thus, although they give the date of generation of the medial superior olivary nucleus as E12, the actual date, based on my interpretation of their figures, is E11. (These dates have also been changed to conform to the convention that the day of conception be considered to be E0 rather then E1.)

on E13–E15. There is almost no overlap in the times of generation of the two groups.

The neurons of the nuclei of the lateral lemniscus are generated during the same time period as those of the superior olivary complex (Taber Pierce 1973; Altman and Bayer 1980a,b; Oblinger and Das 1981). Results differ with respect to the relative times of generation of the neurons of the ventral and dorsal nuclei. Taber Pierce (1973) and Oblinger and Das (1981) report that generation of the two nuclei is essentially concurrent but that the peak of generation of the neurons of the ventral nucleus of the lateral lemniscus is a little earlier than that of the dorsal nucleus. On the other hand, Altman and Bayer (1980b) reported in rats that the cells of the dorsal nucleus were generated from E11–E13, whereas those of the ventral nucleus were generated from E13–E15. As for nuclei in the superior olivary complex, there appears to be a relationship between frequency representation in the dorsal nucleus of the lateral lemniscus and the order of generation of its neurons (Altman and Bayer 1980b).

3.1.3 The Inferior Colliculus

Neurons of the inferior colliculus are derived from the neuroepithelium of the posterior recess of the cerebral aqueduct (Altman and Bayer 1981; Cooper and Rakic 1981). Generation of neurons of the inferior colliculus occurs relatively late compared to that of other auditory areas (including the forebrain; Section 3.1.4) in all species studied (Table 7.9; also, rhesus monkey: Cooper and Rakic 1981; Rakic, personal communication).

In the inferior colliculus, neurogenesis appears to follow the general rule that large cells are generated earlier than small ones (Altman and Bayer 1981; Oblinger and Das 1981). The cells of the inferior colliculus appear to migrate in an outside-in pattern such that the cells generated earliest lie most laterally, rostrally, and ventrally in the adult inferior colliculus, whereas those generated last lie medially, caudally, and dorsally (Altman and Bayer 1981; Cooper and Rakic 1981; Oblinger and Das 1981; Gardette, Courtois, and Bisconte 1982; Aitkin et al. 1991). Altman and Bayer (1981) concluded that there was a correspondence between this cytogenetic gradient and the tonotopic gradient seen in the adult. However, as Coleman (1990) points out, this can be considered to be true only in the grossest sense (cf. Clerici and Coleman 1986; also Aitkin et al. 1991, especially their Fig. 4; Faye-Lund and Osen 1985). Faye-Lund and Osen suggested that the neurogenetic gradient is more correctly thought of as perpendicular to the frequency axis of the nucleus (see Section 3.1.6).

Cooper and Rakic (1981) labeled newly generated neurons by injecting tritiated thymidine into embryos of E40–E41 rhesus monkeys, a time near the peak of production of inferior colliculus neurons (E30–E56 of a 165-day gestation). At this time, there is a prominent pontine flexure in the rhombencephalon but little evidence of midbrain development. If the animals were sacrificed 1 hour after injection, the labeled cells were, as expected, located in the ventricular and subventricular zones. By 3 days, many cells had reached the outer edge of the developing central gray matter but had not moved into the presumptive colliculus. By 7 days, the cells had completed their migration. Based on a study of Golgi-impregnated rat inferior colliculus, Repetto-Antoine and Meininger (1982) identified two modes of migration related to the final destination of the neurons. When they are first generated, all neurons have a trailing process directed toward the ventricular surface and a leading process that is attached to the pial surface. Around E14–E15, the trailing processes of the cells destined for the central nucleus of the inferior colliculus detach from the ventricular surface. Then, the nucleus of each neuron moves through the leading process, which is still attached to the pial surface. As the neuron reaches its destination in the intermediate or mantle zone (the presumptive central nucleus of the inferior colliculus), an axon arises from the leading process or from the cell body and grows toward the marginal fiber layer. Migration into the dorsal cortex is similar until the cell nucleus reaches the boundary of the intermediate layer. At this point, the cell body continues to translocate through the leading process past the point of emergence of the axon and enters the superficial part of the developing midbrain to form a cortical plate. These modes of migration are like those described in the cerebral hemispheres of the opossum (Morest 1970).

3.1.4 The Medial Geniculate Nucleus and the Auditory Cortex

Neurons of the medial geniculate nucleus arise from the dorsal neuroepithelium of the third ventricle in the posteroventral thalamus (Altman and Bayer 1979, 1989). After their final division, the cells migrate in a posterolateral direction (Altman and Bayer 1989). Generation of the neurons of the medial geniculate starts about the same time as generation of the neurons of the cochlear nucleus and the superior olivary complex and begins and ends before the completion of generation of the neurons of the inferior colliculus (Tables 7.2, 7.3, 7.8, and 7.9). The neurons of the medial geniculate are among the earliest neurons to appear in the dorsal thalamus (Keyser 1972). In the rat, the cells migrate in an outside-in pattern such that the cells generated earliest lie in the lateral and caudal aspect of the nucleus in the adult, whereas cells generated later lie medially and rostrally (McAllister and Das 1977; Altman and Bayer 1989). The cells of the magnocellular subdivision of the medial geniculate appear to be the last cells generated (Altman and Bayer 1979). The gradient in the mouse appears to be different; the cells generated earliest reside in the ventromedial part of the medial geniculate and cells generated later land in the lateral part of the nucleus (Angevine 1970).

General principles of early ontogenesis of the cerebral cortex have been reviewed in comprehensive articles by Marin-Padilla (1988) and Payne

(1992). The neocortex of the adult mammal develops from the dorsolateral walls of the cerebral vesicles, which evaginate from the walls of the prosencephalon shortly after neurulation is complete. Like the rest of the neural tube, the cerebral vesicle first consists of a thin, pseudostratified epithelium. Neurons migrate out of the epithelium toward the pial surface to form the cortical plate, the first morphological indication of the future cerebral cortex. The migration of the neurons from the ventricle to the incipient cortex follows an inside-out pattern such that the cells generated later lie in the more superficial layers of the cortex (e.g., Angevine and Sidman 1961). In the rat, the cells generated earliest reach the cortical plate in about two days. Cells originating later take longer as they head for "an ever-receding target" (Hicks and D'Amato 1968).

There appears to be a gradient of development that moves across the entire cortical surface (Gardette, Courtois, and Bisconte 1982; Smart 1983; Rakic et al. 1986; Sanderson and Weller 1990). Smart (1983) showed that the first neurons to begin to migrate in the mouse appear at E11 in a rostral location that corresponds to the oral representation in the adult somato-sensory cortex. A wave of commencement of migration then moves over the lateral telencephalic wall, spreading both caudally and dorsally. The same gradient can be demonstrated for establishment of the cortical plate, its thickness at any point in time, and the ingrowth of the first thalamic afferents. Thus development of the somatosensory cortex would precede that of all others; the auditory cortex would be next, followed by the visual cortex. Such a sequence is observed for the time of neurogenesis of the neurons of the auditory and visual cortices, with the neurons in any given layer generated about a day earlier in the auditory cortex compared to the visual cortex (Bayer and Altman 1991). Bayer and Altman also observed gradients of neurogenesis such that neurons of the primary auditory and visual areas were generated before those of the secondary areas.

3.1.5 Neurogenesis in Human Embryos

Although nothing is known about the timing of neurogenesis in the human, it is possible to make some guesses based on the animal literature. In rodents, the cells produced earliest in the auditory system appear about the time that neurulation is almost complete and the caudal neuropore is closing (Tables 7.2 and 7.3). In the human, neurulation begins around E22 and the neural tube closes about E26 (O'Rahilly and Gardner 1971; Larsen 1993; also Streeter 1912, who places the closure a little earlier). Based on a survey of the timing of a number of different identifiable events in the develop-ment of the nervous system, Otis and Brent (1954) arrived at the following equivalent ages for mouse and human

Mouse	Human
E10	E29
E11	E32

E12	E37
E15.5	E59
E16.5	E77

If the generation of the neurons of the auditory nuclei follows a similar time course relative to the events from which these equivalent stages were deduced, generation of the neurons of the cochlear nuclei, superior olivary complex, and medial geniculate nucleus begins at the end of the first month of gestation and ends near the end of the second month. The cochlear nucleus can be recognized in the 20- to 30-mm embryo brain (E40–48 [Otis and Brent 1954]; Streeter 1912). The medial geniculate nucleus is identifiable in the dorsal thalamus of the 29- to 39-mm human embryo (about E47–E54 [Otis and Brent 1954]; Gilbert 1935; Dekaban 1954.)

Generation of the neurons of the inferior colliculus and auditory cortex would begin at the same time or a little later and extend into the third month. The inferior colliculus is a recognizable structure in the human embryo by about 51 days (O'Rahilly and Gardner 1971). In the rat and mouse, the first cortical neurons to be generated appear about a day earlier than the first collicular neurons (Tables 7.2 and 7.3). In the rhesus monkey, the inferior colliculus is generated from E30 to E56 (Cooper and Rakic 1981), whereas the visual cortex (which probably lags generation of auditory cortex by some number of days) is not begun until E45 (generation of the deepest cells) and lasts through E102 of a 165-day gestation (Rakic 1974). The cortical plate first appears at about 54 days in the human (Bartelmez and Dekaban 1962; O'Rahilly and Gardner 1971; Molliver, Kostović, and van der Loos 1973), a stage a little later than the stage in the mouse at which the cortex is first recognizable (E13, Theiler 1989).

3.1.6 Neurogenesis and Migration: Discussion

A number of aspects of early development may hold clues for understanding the mechanisms leading to the emergence of the mature auditory system. Both the timing and location of neurogenesis appear to be important factors in the subsequent differentiation of cells into distinct neuronal types with characteristic connectivity. In addition, cell movements in the earliest stages of neural generation may be important. At the stage during which the neural tube closes in chick (i.e., at the beginning of neurogenesis) but not at later times, clones of neuroepithelial cells migrate widely along both the dorsoventral and rostrocaudal axes of the developing brain stem (Leber and Sanes 1995). Whether the members of a dispersed clone ultimately give rise to neurons related through connections is not known but stands as an intriguing possibility.

Different cell types in each of the auditory nuclei are generated at different times and even in distinctly different parts of the neuroepithelium. Sequential generation of neurons may determine the cytoarchitectural organization of the auditory nuclei, as the neurons formed earliest may have organizing influences on succeeding generations (Das and Nornes 1972). Furthermore, the sequential production of neurons may play a role in the establishment of orderly connections among neurons (Rubel 1978). In the auditory system of the chick, neurons of the nucleus magnocellularis (cells equivalent to the spherical cells of the mammalian cochlear nucleus) are generated early and appear to send processes into the neuroepithelial zone where their target cells (cells of nucleus laminaris, the avian equivalent of the medial superior olivary nucleus) are generated later (Young and Rubel 1986; Rubel and Parks 1988). It is possible that contact between these cells is made as cells of nucleus laminaris are generated and are maintained during their subsequent migration. Early connectivity is not always important, however. Although fibers of the eighth nerve enter the brain during the time that their target cells are being generated, presence of the nerve fibers does not appear to be critical in control of migration or early differentiation of the cells. In chicks, otocyst removal at stages before the eighth nerve enters the rhombencephalon has no effect on the number of neurons generated or on their pattern of migration (Levi-Montalcini 1949; Parks 1979).

Altman and Bayer (1981) reported that cytogenetic gradients tend to be roughly aligned with cochleotopic gradients in structures of the auditory system from the superior olivary complex to the auditory cortex. Such a relationship appears to hold in at least some of the individual superior olivary nuclei (Altman and Bayer 1980a), but Faye-Lund and Osen (1985) disputed this claim for the inferior colliculus. They suggested that in this structure temporal cytogenetic gradients result in the deposition of cells orthogonal to the cochleotopic organization rather than parallel to it. This raises the possibility that the internal organization of a given frequency region in the inferior colliculus could be determined, in part, by synchrony in timing of migration of specific neurons and ingrowth of axons from specific sources. This type of synchrony appears to obtain in the hippocampus, for example, such that the location of afferent terminals along the dendrites of principal cells is related to the time of origin of the neurons that give rise to them (Bayer 1980). A detailed understanding of the sites and timing of neurogenesis in terms of specific cell types may reveal something of the mechanisms leading to the establishment of appropriate connections.

Mechanisms guiding cell migration in the brain stem are still incompletely understood but are clearly important in determining the relationships among different groups of neurons as they move into presumptive auditory structures from different cytogenetic zones or at different times. Textbook discussions of neuronal migration are usually limited to the neural crest in the peripheral nervous system and cortical structures in the central nervous system (e.g., Purves and Lichtman 1985; Jacobson 1991). The migration patterns of the neurons that make up the cerebral and cerebellar cortices have been studied in detail (reviewed by Jacobson 1991). Early in development, cells in the cerebral hemispheres migrate radially away from the ventricular zone toward the pial surface. As development proceeds, however, the direction of migration of some cells becomes more complex as the routes begin to comprise both radial and tangential components (Hicks and D'Amato 1968; Austin and Cepko 1990; Bayer et al. 1991). Throughout cortical development, migratory neurons are aligned parallel to radial glial cells that span the distance between the ventricular and pial surfaces of the hemispheres; these glial cells appear to play an important role in guiding the translocation of the cells (e.g., Rakic 1972; Hatten and Mason 1990; Fishman and Hatten 1993).

Although some aspects of neuronal migration in the brain stem may be similar to those in cortex, the situation in the brain stem is more complex because many of the cell movements, even very early in development, are not in the radial direction (e.g., Harkmark 1954a,b; Domesick and Morest 1977a,b; Willard and Martin 1986; Book and Morest 1990; Leber and Sanes 1995). Some neurons that originate laterally in the rhombencephalon migrate circumferentially to settle medially, whereas other neurons that originate medially migrate laterally; some of these movements involve changes in rostrocaudal position as well. Cell movements also occur in the radial direction, orthogonal to the circumferential cell movements. Clearly, control of migration in the brain stem must depend on a number of factors, none of which is yet well understood. One characteristic of migratory cells that should provide clues about their mode of migration is the disposition of their leading processes (Morest 1969a, 1970; Book and Morest 1990). In many different parts of the developing central nervous system, young neurons elongate through growth of a leading process; migration occurs as the perikaryon moves through this process to settle finally in its adult location (reviewed by Book and Morest 1990). Studies of migration of the neurons destined to populate the pontine gray matter and the inferior olivary nuclei have revealed relationships between the leading processes of the migrating cells and other migratory neurons or bundles of axons (Bourrat and Sotelo 1988, 1990; Ono and Kawamura 1989, 1990). To date, little is known about the morphology of migrating cells in the mammalian auditory system. Perikaryal translocation through a leading process does appear to occur in neurons migrating into the dorsal cochlear nucleus in the opossum (Willard and Martin 1986, personal communication) and in neurons in the inferior colliculus of the rat (Repetto-Antoine and Meininger 1982). Because growth of the leading process determines the migratory route, the mechanisms controlling its growth are of fundamental importance in understanding migration. Factors influencing pathfinding by the leading process may be similar to those that guide pathfinding by axons (Book and Morest 1990) and are discussed further in Section 3.2.5.

Work in molecular biology may soon lead to insights about important developmental events that occur even before neurogenesis begins. A set of homeobox genes known as *HOX* genes appear to regulate cell identity along the anterior-to-posterior axis of the embryo. Deletion of the *HOX* 1.6 gene results in neurological defects in the part of the brain stem bounded by rhombomeres 4–7 (Chisaka, Musci, and Capecchi 1992). In homozygous animals, which die shortly after birth, there are defects in the formation of the external, middle, and inner ears and the cochlear nerve is absent. Animals assayed at E13.5–E19 exhibit absence or severe reduction in the superior olivary and facial nerve nuclei. Although the cochlear nerve is absent, the dorsal and ventral cochlear nuclei are present and said to be qualitatively normal. Deletion of another gene, the proto-oncogene *int*-1 results in absence of the inferior colliculus (along with the cerebellum and superior colliculus), but the forebrain and myelencephalon appear normal (McMahon and Bradley 1990; Thomas and Capecchi 1990). It is known that *int*-1 encodes for a protein that is secreted by cells and is associated with the extracellular matrix. At E8.5 in the mouse (the neural plate stage), the expression of *int*-1 is restricted to the midbrain and the edges of the anterior hindbrain (Wilkinson, Bailes, and McMahon 1987). Genetic control of the processes of regional determination is just beginning to be explored and may lead to insights about the earliest events in the establishment of neuronal systems.

3.2 Axonal Outgrowth and Development of Circuitry

A fundamental question concerning the development of the central nervous system is how the complex but highly specific circuitry characteristic of the adult is established. As neurons are generated and begin to migrate to their adult location, they also begin to send out the axons that will connect them to neurons in other parts of the brain. Many of these axons reach their targets while the brain is very small and simple in shape compared with the adult; as a consequence, the distances traveled are much shorter than those spanned in the adult (see, e.g., Larsen 1993). As an axon grows toward its target, its parent cell body may migrate "away" from the axon, further increasing the distance between the cell body and axon terminals. In many cases, much of the increase in axonal length that occurs as the brain grows occurs after the axon terminals have reached their targets.

The question of the mechanisms of axonal pathfinding in the auditory system, with its multiple brain stem nuclei and its complex patterns of inhibitory and excitatory convergence and crossed pathways, is particularly intriguing, but until recently very little was known even about the normal sequence of events. Now, with the advent of tracing techniques based on the movement of carbocyanine dyes (e.g., diI) within the lipid membranes of dead, fixed tissue (obviating the need for manipulation of living fetuses), it has become possible to explore the normal sequence of events in the development of circuitry in the auditory system from very early stages. Knowledge of the normal sequences will provide the understanding necessary to pose questions about the mechanisms. One obvious possibility, given the facts discussed in Sections 3.1.1 and 3.1.2 about the different sequences of generation and migration of specific cell types, is that the establishment of connections in the auditory system is related to the time at which axons reach their targets.

One structural feature that should have significant influence on the development of function is the degree of myelination of auditory pathways. This appears to occur long after connections have been established and, to a great extent, after hearing has begun; accordingly, it will be discussed in Section 4.

3.2.1 Ingrowth of the Auditory Nerve

Cochlear nerve fibers arise from spiral ganglion cells and enter the brain very early in development (Tables 7.2, 7.3, and 7.6), well before the hair cells or other structures of the cochlea are mature (Pujol and Hilding 1973; Lenoir, Shnerson, and Pujol 1980; Pujol, Carlier, and Lenoir 1980; Pujol 1985). In the mouse, the period during which the cochlear fibers enter the brain overlaps with the period of generation of neurons of the cochlear nucleus (Taber Pierce 1967). At this stage, the cochlear nucleus is not yet identifiable as such, and the cochlear fibers appear to enter the proliferative region of the rhombic lip. (Proliferative activity in the rhombic lip spreads into the thin ependymal floor of the lateral recess and converts it into a thick stratum that then fuses with the medulla; it is this stratum that the auditory nerve fibers penetrate [Shaner 1934].) In humans, cochlear fibers can be traced into the alar proliferative zone sometime around 33–35 days of gestation (Streeter 1906; Cooper 1948; age based on equivalency between embryo length and days of gestation given by Otis and Brent 1954). This appears to be about the same developmental stage during which cochlear fibers enter the brain in the mouse.

Poston et al. (1988; also Carney and Silver 1983) studied ingrowth of the auditory nerve in the mouse. Central fibers arising from ganglion cells project out of the spiral ganglion about E11. A "funnel" formed by neural crest cells reaches out from the ganglion and invades the brain, disrupting the basal lamina. Where the funnel cells abut the brain, processes of neuroepithelial cells appear to penetrate outward through the basal lamina and contact the funnel cells. The auditory nerve fibers enter the brain at this break in the basal lamina; the funnel disappears as the fibers grow in.

The pattern of growth of cochlear nerve fibers once they have entered the brain is known in some detail in opossums (Willard 1993, 1995), hamsters (Schweitzer and Cant 1984; Schweitzer and Cecil 1992), and rats (Angulo, Merchán, and Merchán 1990). In the opossum, some auditory nerve fibers reach the ventricular surface of the presumptive dorsal cochlear nucleus around postnatal day 6 (P6), before the large cells that they will presumably contact have migrated into place (Willard and Martin 1986; Willard 1993, 1995). The field of arborization of the axons is restricted to a narrow plane from the time of their earliest entry. Topographic order appears to be established in the cochlear nucleus as early as P8, some time before the cochlea has matured (Willard 1993; Willard and Munger, personal communication). In the hamster, cochlear nerve fibers have invaded the ventral cochlear nucleus at birth but have not yet entered the dorsal cochlear nucleus, although growth cones are present at its boundaries (Schweitzer and Cant 1984). At this time, many if not all of the large neurons that will become targets of the cochlear axons are already in place, although they are still quite immature and have not yet segregated into layers (Schweitzer 1990). Three to five days later, cochlear axons ramify densely in the deep dorsal cochlear nucleus, and by P10, the adult pattern has been established. Axonal ingrowth is complete in hamsters before the onset of hearing (P16), and after this time, the morphology of the axons remains unchanged (Schweitzer and Cecil 1992). In the rat by E18–E20, the primary afferent fibers are distinct and have an adultlike distribution in the cochlear nucleus, but they do not yet produce complex terminal branches (Angulo, Merchán, and Merchán 1990). These results in opossums, hamsters, and rats suggest that migrating neurons and ingrowing axons arrive in the cochlear nucleus during approximately the same time period; in some species, the fibers may enter first, and in others, the cells may arrive slightly earlier. Further study of the migratory histories of specific neuronal types and ingrowth patterns of axons from specific sources should provide a clearer picture of the relationships between the arrival times of these two elements of the system.

3.2.2 Circuitry of the Brain Stem

Just as auditory nerve fibers enter the brain stem while the neurons of the future cochlear nucleus are still being generated, so also do the axons of the brain stem nuclei begin to reach their target structures during genesis of the neurons of those structures. In the rat, axons from the cochlear nucleus enter the superior olivary complex and inferior colliculus during the time that many of the neurons that will populate them are being generated (Altman and Bayer 1980a, 1981; Repetto-Antoine and Meininger 1982; Kandler and Friauf 1993). In the human embryo, fibers can be traced from the area of the presumptive cochlear nucleus into the trapezoid body and lateral lemniscus at the 13-mm stage (~35 days [Otis and Brent 1954]; Streeter 1912; Cooper 1948), a time at which neurogenesis of auditory structures is likely to be ongoing (Section 3.1.5).

Axons undergo characteristic morphological changes as they grow into presumptive target structures. In a study of the development of a specific set of axons in the brain stem of opossum pouch young, Morest (1968) followed the sequence of changes in the calyciferous axons that project from the cochlear nucleus to the superior olivary complex. Axon growth proceeds in a sequence of characteristic morphological stages. First, there is the stage of the migratory growth cone as the axon makes its way to its presumptive target. When the axon approaches the recipient neuron, which itself has just completed its migration from the proliferative zone, the growth cone begins to give rise to many thin sinuous processes, which are, in turn, replaced by broad branches. This morphological transformation signals the beginning of the second stage, that of the protocalyx, during which a calycine ending forms on the framework supplied by these branches, giving rise ultimately to the stage of the young calyx. A similar sequence appears to obtain in the rat (Kandler and Friauf 1993) and gerbil (Kil et al. 1995). There is concurrent development of the neurons on which the calyces form synapses (Morest 1969a; Section 3.3). Thus, as axons grow into a target area, axonal form appears to change as the axons interact with the neurons and other elements there.

By birth, even in highly altricial species, the patterns of connections characteristic of the adult appear to have been established in the brain stem. In the newborn ferret, projections from the lateral superior olivary nucleus to the inferior colliculus are bilateral and ordered topographically along the presumptive frequency axis (Henkel and Brunso-Bechtold 1993). At P35, the time that hearing begins, there is a clear tonotopic organization (Moore 1982). However, the relative proportion of ipsilateral- and contralateral-projecting cells from both the cochlear nucleus and the superior olivary complex (or, possibly, the arborization patterns of some of the cells) does appear to change during postnatal development up to or shortly after the time of hearing onset (Moore and Kowalchuk 1988; Henkel and Brunso-Bechtold 1993). Ascending projections to the midbrain are present by P5 in the Virginia opossum (Willard and Martin 1987; Willard 1995). The projections from the superior olivary complex appear to be in place slightly before those of the ventral cochlear nucleus. By P25–P33 (the Virginia opossum hears at P50), narrow sheets of neurons in the cochlear nucleus project to restricted fibrodendritic laminae in the inferior colliculus. Topographic order in the projections to the inferior colliculus is also established before birth in the rat (Friauf and Kandler 1990; Kandler 1993; Kandler and Friauf 1993); the frequency representation appears to be adultlike at hearing onset (Friauf 1992).

The diI-tracing technique has been exploited to study the details of the development of brain stem circuitry in rats (Kandler and Friauf 1993; Kandler 1993), gerbils (Kil et al. 1995; Russell and Moore 1995), and hamsters (Schweitzer and Savells 1991). In embryonic and early postnatal rats, three stages of development lead to the establishment of connections between the neurons of the cochlear nucleus and their targets (Kandler 1993; Kandler and Friauf 1993; Fig. 7.4). The first stage encompasses E15–17 and accounts for the main period of axonal outgrowth. (E15 is only one day after the peak production of neurons in the cochlear nucleus and well before the end of neurogenesis [Table 7.2].) During this time, axons grow out of the cochlear nucleus, traverse the superior olivary complex, and enter the lateral lemniscus and inferior colliculus. The fibers end in growth cones but exhibit no collateralization in any of the nuclei that they enter. On

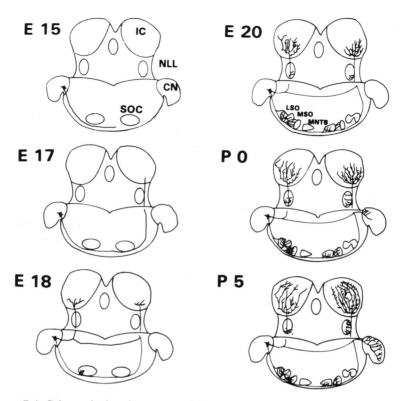

FIGURE 7.4. Schematic drawing summarizing the development of axonal projections arising in the cochlear nucleus of the rat. At each stage from embryonic day 15 (E15) to postnatal day 5 (P5), the cochlear nuclear axons and their branches are indicated to illustrate the state of development attained at that stage. NLL, nucleus of the lateral lemniscus; CN, cochlear nucleus. Details are given in the text. (Modified slightly from Kandler and Friauf [1993]; reprinted by permission of the authors and John Wiley and Sons, Inc.)

E18, a period of collateralization begins. From E18–P5, the number of projecting fibers increases, and there is pronounced collateral branching of the cochlear nuclear fibers in all targets, with the branching beginning in the inferior colliculus slightly earlier than in the superior olivary complex. During this time, there appear to be some differences in the rates of maturation of the axons in the different olivary cell groups. By E20, the basic projection pattern to both the ipsilateral and contralateral superior olives resembles that of the adult, although in the case of the medial superior olivary nucleus, the axons appear to accumulate along the margins rather than arborizing throughout the nucleus. Thus, an essentially mature pattern of connections in which axons terminate in restricted laminae is in place before birth in the rat, although at P5, growth cones are still present on the axonal tips. From P5–14 (hearing begins around P12), terminal structures differentiate and mature; growth cones are not seen on axons after P11. Kandler and Friauf (1993) observed gradients in the ingrowth of axons into all of the targets. Similar to other developmental gradients, those areas that will represent high frequencies in the adult were the first to mature.

In the gerbil, highly ordered projections from the ventral cochlear nucleus to the medial and lateral superior olivary nuclei bilaterally and to the medial nucleus of the trapezoid body contralaterally are present at birth (Kil et al. 1995). The axons from the cochlear nucleus make synaptic contacts containing large round synaptic vesicles (as in the adult) in all of the targets at birth (Section 3.4). Over the first two postnatal weeks, the terminal arbors continue to develop and the number of synapses increases. Although synapses are present in the medial nucleus of the trapezoid body, calyces are not identifiable until P5. The first mature calyces are present at P14–16, about the time hearing begins. In the hamster, projections from the dorsal cochlear nucleus to the inferior colliculus appear fully mature by P5 (Schweitzer and Savells 1991, Schweitzer, personal communication); hearing begins at approximately P16. At the level of resolution of the technique, exuberant growth that was later retracted was not observed in any of these studies (Kandler and Friauf 1993; Kil et al. 1995; Schweitzer and Savells 1991; see also Willard 1995, opossum).

The manner in which two inputs to a given neuron become topographically matched (i.e., become matched in frequency specificity) has not been studied systematically in the mammalian auditory system. Young and Rubel (1986) studied the ontogeny in the chick of the ipsilateral and contralateral inputs to the nucleus laminaris, the avian equivalent of the medial superior olivary nucleus of mammals. In adult animals, these two inputs (which arise from the nucleus magnocellularis, the equivalent of the anteroventral cochlear nucleus of mammals) are precisely matched in terms of their characteristic frequencies (Rubel and Parks 1975; Young and Rubel 1983). The ipsilateral axon exhibits the correct topography when it first arrives in the nucleus laminaris (before any sign of onset of function). The contralateral axons arrive a little sooner than the ipsilateral axons, but, although their arborizations are spatially restricted and oriented as in the adult, they do not exhibit the adult specificity. Over a period of ∼5 days, the contralateral terminal fields appear to shift position gradually to match the tonotopic specificity of the ipsilateral fields. Young and Rubel (1986) speculated that during this time (which overlaps with the onset of function), fine processes sprout and retract from the ventral arbors as they sample prospective targets in the nucleus laminaris. The mechanisms involved are not known, but it is possible that ultimate target selection depends on patterns of neuronal activity (Section 3.4.5).

Like the excitatory pathways described above, inhibitory circuitry is also established early in development, well before the onset of hearing. Inhibitory and excitatory postsynaptic potentials can be elicited shortly after birth by electrical stimulation in the cochlear nucleus of mice (Wu and Oertel 1987) and in the superior olivary complex of gerbils (DH Sanes 1993) and rats (Kandler and Friauf 1995). In brain stem slices taken from rats as early as E18, synaptic potentials can be evoked in the lateral superior olivary nucleus by electrical stimulation of both its ipsilateral and contralateral inputs (Kandler and Friauf 1995). The ipsilateral stimulation always leads to depolarizing postsynaptic potentials. Stimulation of the contralateral inputs yields depolarizing potentials from E18–P4; after P8, contralateral stimulation yields hyperpolarizing potentials, as in the adult. The contralaterally evoked potentials can be blocked by strychnine at all ages and so appear to be mediated by glycinergic inputs to the cells.

Sanes and Siverls (1991) traced the development of the inhibitory (glycinergic) axons that project from the medial nucleus of the trapezoid body to the lateral superior olivary nucleus in gerbils at ages P2–25 (hearing onset is about P14–16). At the earliest ages studied, the axons arborize in the appropriate topographic position. They are short and relatively restricted in extent and many terminate in growth cones. As development proceeds, some of them give rise to diffuse or topographically "inappropriate" projections. At P10–13, about one-third of the axons give rise to discontinuous terminal fields and the number of boutons per fiber is significantly higher than in more mature arbors. By P15–16, the arbors have been refined so that none exhibits discontinuous or diffuse projections; major growth in length of the axons appears to end about this time. The number of boutons per arbor continues to decrease until at least P25, although the boutons themselves increase in size (Section 4.3). Another set of presumably inhibitory projections, the descending inputs to the dorsal cochlear nucleus from the inferior colliculus, ramify throughout the same layers that they occupy in the adult and exhibit puncta by P6 in the hamster (Schweitzer and Savells 1991, Schweitzer, personal communication).

Brain stem projections to the cochlea also develop early. (Development of the cochlea itself is considered by Pujol, Lavigne-Rebillard, and Lenoir [Chapter 4] and Lippe and Norton [Chapter 5].) Based on the patterns and timing of efferent innervation of the inner and outer hair cells, inferences can be made about the development of olivary cells that project to the cochlea from the brain stem. At least some cells appear to send their axons to the cochlea shortly after they undergo their last division because efferent fibers are present in the nerve as early as E10.5 and are present in the sensory epithelium on E12 in the mouse (Fritzsch and Nichols 1993). However, in the rat, not all olivocochlear circuits are established at birth (Pujol, Carlier, and Devigne 1978). In the adult, efferent synapses present under the inner hair cells originate from neurons located in the lateral parts of the superior olivary complex, whereas those that contact outer hair cells arise from neurons in its medial part. At birth in the rat, efferent synapses are present under the inner hair cells, but the first efferent junctions are not seen on the outer hair cells until P6–12 (Lenoir, Shnerson, and Pujol 1980). Between P12 and P16, just as the animal is beginning to hear, the pattern becomes adultlike. Simmons, Raji-Kubba, and Strauss (1992) studied the organization of calcitonin gene-related peptide (CGRP)-positive cells in the developing superior olive of the hamster. (These cells are the ones that project to the inner hair cells as the lateral olivocochlear system.) Between P1 and P23, the number of CGRP-positive cells in the lateral superior olivary nucleus increases from 3% of the total to 18% of the total while the total cell number stays constant. The change in this marker could imply continuing maturation of the olivocochlear cells over the first three weeks after birth. Thus, although there is evidence for very early formation of connections from the brain to the cochlea, there is also evidence of continued change and rearrangement of those connections even after the onset of hearing.

3.2.3 Projections from the Inferior Colliculus to the Medial Geniculate Nucleus

In the rat, terminals from the inferior colliculus arborize in the medial geniculate nucleus as early as E19, but they are not restricted to the medial geniculate at this time (Asanuma et al. 1986). By the day of birth, the projecting axons arborize extensively throughout the medial geniculate (more so than in rats at P30), and a large number of axons extend rostrally beyond the medial geniculate into adjacent thalamic nuclei and even beyond the thalamus (Asanuma et al. 1988). These "exuberant" arbors disappear by the end of the third postnatal week. Widespread projections that ultimately become restricted to specific subdivisions of the thalamus are also characteristic of the somatosensory and cerebellar projections to the thalamus (Asanuma et al. 1986, 1988). In the human, acoustic fibers have reached the medial geniculate by the end of the second month and the brachium of the inferior colliculus is clearly recognizable by the end of the third month (Streeter 1912). Coggeshall (1964) reported that fibers did not enter the diencephalon from the brain stem until about E20 in the rat, some time after connections are established between the thalamus and cortex.

3.2.4 Projections from the Medial Geniculate Nucleus to the Cortex and Cortical Circuitry

Early in the development of the cerebral hemispheres, axons of thalamic neurons are identifiable in the internal capsule in the opossum (Morest 1970). In 30-mm pouch young (approximately P12 [Willard and Martin, personal communication]), these axons appear to enter into a relationship with the external processes of neuroepithelial cells that will become cortical neurons. In the rat, fibers extending between the thalamus and cortex can be seen on E13 or E14 (Coggeshall 1964). This is only a day or two after the generation of the first neurons of the cortex and medial geniculate nucleus and probably before many of them have completed their migration. As mentioned in Section 3.2.3, Coggeshall (1964) did not identify fibers entering the diencephalon from the brain stem until about E20, almost a week later (on the basis of Klüver, silver, and Weil stains). Altman and Bayer (1981) also infer, on the basis of neuronal birth dates, that the medial geniculate projects to the cortex before the inferior colliculus projects to the medial geniculate but provide no direct evidence. Recently available tracing techniques should allow a more definitive assessment of the relative timing of establishment of the pathways into and out of the medial geniculate nucleus. In the cat, most or all major projections between the cortex and its subcortical sources and targets are present by the eighth gestational week (Payne, Pearson, and Cornwell 1988b). Both the cells of origin and the axonal targets of these projections are arranged topographically before birth. The only connection that is present in the P4 kitten (just before the time of hearing onset) that is not present in the E55 animal is a projection from the auditory cortex to the inferior colliculus (Cornwell, Ravizza, and Payne 1984).

The corpus callosum is established relatively late in development. In the mouse, the first callosal fibers cross the midline on E17 (Silver et al. 1982). (The role of specialized glial cells in the development of the corpus callosum has been described in detail by Poston et al. [1988].) In the auditory cortex of the cat, cortico-cortical connections are present at birth, but the axons form a dense, continuous band that extends throughout the cortex; there are no alternating vertical columns and no laminar bands as in the adult (Feng and Brugge 1983; Payne 1992). Development of the adult pattern does not begin until after the onset of hearing (P8 [Feng and Brugge 1983]; discussed further in Section 4.4). The widespread innervation of the cortex by callosal fibers during early development appears to be a general phenomenon and is seen in other parts of cortex as well (Innocenti 1988). Thus, whereas the connections of the cortex with subcortical structures appear adultlike in newborn cats, the patterns of connections linking cortical areas are immature, and the period of refinement of these projections is very protracted (Payne, Pearson, and Cornwell 1988a; discussed further in Section 4.3).

In newborn kittens, bilateral transitory axons project from the auditory cortex to visual cortical areas (Innocenti and Clarke 1984; Dehay, Kennedy, and Bullier 1988). Reciprocal projections to the auditory cortex are not seen (Clarke and Innocenti 1986). Distinct neuronal populations in the auditory cortex project to different parts of the visual cortex with a rough topographic organization (Innocenti and Clarke 1984). The transitory axons appear to be confined to white matter and deep layers of the visual cortex (Innocenti, Berbel, and Clarke 1988; but see Dehay, Kennedy, and Bullier 1988).

3.2.5 Development of Circuitry: Discussion

There can be no doubt that the basic circuitry of the auditory system develops in the absence of acoustic stimulation and that its characteristic topographic (cochleotopic) organization is not dependent on acoustic input for its expression. The first connections between auditory areas are made during the time that neurons are still being generated; indeed, the timing of neurogenesis may influence the establishment of connections (Section 3.1.6). By the time an animal begins to hear, the pattern of brain stem auditory circuitry is like that of the adult, although some aspects of cortical connectivity still appear grossly immature. That the circuitry is functional even at birth, some time before hearing begins in many mammals, is indicated by the responsiveness of neurons to electrical stimulation during the early postnatal period (reviewed by Mysliviček 1983; also Wu and Oertel 1987; Sanes and Rubel 1988; DH Sanes 1993; Kandler and Friauf 1995).

Two essential elements in the establishment of circuitry, pathfinding and appropriate target selection by axons, are among the most studied phenomena in developmental neurobiology. (An introduction to current thinking about the mechanisms involved is provided by the following reviews: Poston et al. 1988; Udin and Fawcett 1988; Constantine-Paton, Cline, and Debski 1990; Jacobson 1991; Frank and Wenner 1993; Goodman and Shatz 1993; JR Sanes 1993). In general, three phases of axonal growth and development can be identified. These are exemplified in the auditory system by the results of Kandler and Friauf (1993; Section 3.2.2). The initial phase involves directed growth of axons over relatively long distances. In the course of this growth, the axons approach or pass by structures that they will eventually innervate, but during this time, there is essentially no branching. In many cases, this phase begins very early, even before the parent neurons have completed their migration. Environmental cues such as differential expression of cell adhesion molecules and their receptors on the surfaces of the axonal growth cones and cellular elements in their path could play a role in guidance of the axons (e.g., Goodman and Shatz 1993). Temporal regulation of such expression could lead to precise control of the pathways taken by specific axons. Transient, specialized neuroepithelial structures in both the central and peripheral nervous systems also appear to play an important role in axonal guidance by forming either channels that promote growth or barriers that prevent growth (Singer, Nordlander, and Egar 1979; Whitehead and Morest 1985; Poston et al. 1988). (As suggested in Section 3.1.6, many of the mechanisms guiding axonal pathfinding might also be important in the directed growth of the leading processes through which migrating neurons move.)

A second phase of growth commences when axons begin to form branches and collateralize within their target(s). This may not occur until after a "waiting period" that may be as long as 3–5 days (Kandler and Friauf 1993). Axon outgrowth occurs so early that some axons reach their target structures before all of the neurons destined for that structure have been generated; waiting periods may reflect the time during which potential target neurons are still in the act of migration. Evidence is accumulating that diffusible signals produced by target structures induce branching and collateralization in specific axons that pass near the structures (e.g.,

Tessier-Lavigne et al. 1988; Heffner, Lumsden, and O'Leary 1990; Goodman and Shatz 1993). Kandler and Friauf (1993) noted that collateralization of cochlear nuclear projection neurons occurs in different nuclei at different times. For example, on E19, some nuclei in the superior olivary complex receive a large number of axonal branches from cochlear nuclear fibers, whereas others do not receive collaterals from the axons passing by them for several more days. Differences in the production of attractive signals as well as differences in the time of origin and rates of maturation of the target cells are only two factors that could underlie the differences in the timing of collateralization. Axons invade their targets in a systematic way that, in the auditory system, is related to the cochleotopic organization seen in the adult (e.g., Schweitzer and Cant 1984; Sanes and Siverls 1991; Kandler and Friauf 1993; Willard 1993). The initial topographic organization appears to be established in central nervous structures based on positional cues, perhaps related to gradients of molecular markers (e.g., Frank and Wenner 1993; JR Sanes 1993). In both the mammalian and avian auditory brain stem, initial connections appear to be established with great precision. The final phase in the establishment of circuitry encompasses a period of maturation of synaptic terminals; this phase is discussed further in Section 3.4.5.

Alterations in the pattern of normal afferent input to a nucleus can lead to changes in the pattern of connectivity that is established (Kitzes et al. 1995; Russell and Moore 1995). In the normal gerbil pup, the projections of the cochlear nucleus to the medial superior olivary nucleus are present at birth and exhibit the adult pattern, in which the lateral aspect of the nucleus is innervated by the ipsilateral cochlear nucleus and the medial aspect is innervated by the contralateral cochlear nucleus. After a lesion of the cochlea at P2, axons from the contralateral cochlear nucleus approach the side of the medial superior olive that they do not normally innervate. This sprouting is seen within 24 hours of the lesion and continues to develop over at least 11 days. The projections are topographically appropriate and persist in the adult. In addition to the abnormal projection to the medial superior olivary nucleus, calyces are formed in the ipsilateral medial nucleus of the trapezoid body by the large axons that normally innervate only the contralateral side. A lesion at P10 or later has little or no effect, implying that the mechanisms involved operate early and in the absence of acoustic input. These results lead to the conclusion that the innervation patterns established by cochlear nuclear axons are normally constrained by the presence of axons from the other cochlear nucleus (and, presumably, from other sources as well). The mechanisms underlying the presumed interactions are not known, but there is increasing evidence for a role of neuronal activity in the establishment of connections. Before hearing begins, spontaneous activity is present in auditory structures and so could play a role in their development. Because establishment of functional circuitry involves dendritic development and synaptogenesis as well as axon growth, further discussion of mechanisms leading to its mature expression is deferred to Section 3.4.5.

3.3 Differentiation of Cells and Dendrites

During the time that auditory circuitry is first established and for some time after that, neurons undergo the process of differentiation that takes them from the stage of immature postmigratory neuroblasts to the adult phenotype. Differentiation, including both changes in cell size and the growth and maturation of dendrites, begins after migration is completed and after axonal outgrowth has occurred. In fact, in many cases, differentiation of the mature dendritic processes may not even begin until the axon of the cell has grown out to its destination (Ramón y Cajal 1909; Morest 1970; Repetto-Antoine and Meininger 1982; Jacobson 1991). In an analysis of dendritic differentiation in many different parts of the brain, Morest (1969b) noted a number of consistent features. Growing dendrites are characterized by growth cones and filopodia at their tips. Dendrites of specific cell types differentiate in a fixed sequence during a specific time period and across specific spatial gradients. The dendrites of large cells and projection neurons differentiate before those of small cells and local circuit neurons. Dendritic and other postsynaptic surfaces appear to differentiate in conjunction with the afferent axons that will form synapses with them (see also Jacobson 1991). In many neuronal types, developing dendrites and cell bodies are covered with immature spines and appendages that are later lost. (Some of these may play a functional role in synaptogenesis [Section 3.4]). In all of the work to be described below, patterns of growth of neurons have been assessed in animals only after birth. Newly available methods, including labeling of developing neurons in fixed tissue with carbocyanide dyes, should allow study of the processes of differentiation that occur prenatally.

3.3.1 The Cochlear Nuclei

Larsen (1984) in her study of the development of the kitten cochlear nuclei identified two growth phases of neurons after birth. First, there is a period of about one month during which the average cross-sectional areas increase rapidly. This is followed by a slower growth phase that lasts up to 12 weeks. (Hearing onset occurs early in the first month in kittens.) Different cell types mature at different rates. The octopus cells are the first to reach adult size and morphology. The spherical and globular cells take the longest to reach adult size (about postnatal week 12). In the mouse, there is little change in the volume of the dorsal and ventral cochlear nuclei or in the size of neurons over the first three days after birth (Webster and Webster 1980; Webster 1988a). From P6 to P12 (P12 is about the time of hearing onset; Table 7.3), the dorsal and ventral cochlear nuclei more than double in volume (Webster 1988a). During this time, the number of identifiable neurons also increases rapidly to a peak and then decreases to adult levels (Mlonyeni 1967). All of the identifiable cell types in the cochlear nucleus have reached their adult sizes (or are even slightly larger) at this time (Webster and Webster 1980; Figure 7.5). At P15 in the gerbil (about the time of hearing onset), the auditory nuclei of the brain stem and the medial geniculate have attained 60–70% of their adult size; full adult size is not reached until after three months or more (Rübsamen et al. 1994).

The first neurons in the auditory system to begin dendritic differentiation are those in the cochlear nucleus (Morest 1969b; 1970). Willard and Martin (1986, personal communication) followed the development of specific cell types in the dorsal cochlear nucleus of the North American opossum from the time the neurons first enter the nucleus. As large cells migrate into the dorsal cochlear nucleus (Section 3.1.1), they are characterized by long leading and trailing processes that extend medially along the migratory route (the presumptive dorsal acoustic stria). These processes are covered with filopodia, and growth cones are found on their tips. As the large neurons settle into the dorsal cochlear nucleus, they lose the elongated processes and enter a radiate stage during which their primary dendrites give rise to many secondary and tertiary branches, also covered with filopodia and numerous varicosities and terminating in growth cones. Finally, the cells that will become fusiform cells begin to assume a vertical orientation and layers begin to form as these cells appear to migrate radially toward the pial surface of the dorsal cochlear nucleus. Other cells remain radiate in shape and become the giant cells. By the time of hearing onset (about P50), the large cells are essentially adultlike, although filopodia are still occasionally present on the distal dendrites. The more proximal dendrites of the fusiform cells become adorned with spines about this time. The fusiform cells in the dorsal cochlear nucleus of cat undergo a similar sequence, developing from small, spicule-laden somata with short dendrites at birth to an essentially adultlike morphology by P12, although the dendrites continue to mature for some time after this (Kane and Habib 1978).

Schweitzer (1990, 1991) and Schweitzer and Cant (1985a,b) have emphasized the differences in the development of the two large neuronal types,

---→

FIGURE 7.5. Mean (±SD) cross-sectional areas of neurons at seven different ages in normal mice (filled circles connected by lines). *, 45-Day acoustically deprived mice; □, 45-day conductive-loss mice; o, 90-day deprived-reversal mice (mice that were raised for 45 days in a sound-attenuated environment followed by 45 days in a normal environment). MP, multipolar cells of the VCN; OCT, octopus cells; GLOB, globular cells; SSC, small spherical cells; LSC, large spherical cells; CRDCN, central region of the DCN; CNIC, central nucleus of the inferior colliculus. (Redrawn from Webster and Webster [1980]; reprinted by permission of the authors and Annals Publishing Co., St. Louis, MO.)

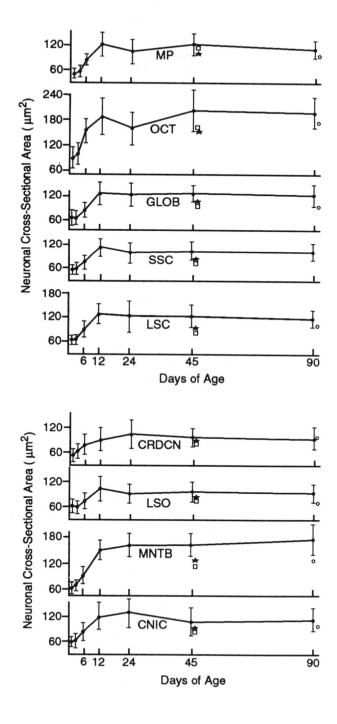

fusiform and giant cells, in the dorsal cochlear nucleus of the hamster (hearing onset at approximately P16 [Schweitzer 1987]). At birth, different neuronal types cannot be distinguished in the dorsal cochlear nucleus and layers have not been established. Thin, wispy neurites extend in all directions from the cell bodies of the immature cells; some exhibit large expansions at their tips. By P5, the neurites have grown in length and diameter and have begun to branch to a limited extent. Some neurons begin to take on a bipolar appearance. By P10, there is a marked increase in the branching of all dendrites, and fusiform cells can be distinguished from giant cells by their location in the most superficial aspect of the cell mass and their increasingly bipolar appearance. The length of individual secondary and higher order branches at P10 is not different from that in 60-day-old animals ("adults"), but the number of branch points and total dendritic length continue to increase after P10. By P15 (about the time of hearing onset), the large neurons in the dorsal cochlear nucleus appear grossly mature except that the apical dendrites of the fusiform cells are not yet covered with spines as they will be in the adult. Development of the apical dendrites of the fusiform cells occurs over a longer time course than that of the basal dendrites, and the mature phenotype of the two is different. Schweitzer (1990) has speculated that the different developmental patterns and time course seen for the two sets of dendrites on the same cell is related to the differing types and times of arrival of the inputs to those dendrites.

In the human, the progression from the first identifiable neurons in the brain stem auditory nuclei to neurons with characteristic adult morphology (in cell body stains) occurs from ~19–28 weeks of gestation (estimated to be equivalent to P2–10 in rat [Perazzo and Moore 1991]).

3.3.2 The Superior Olivary Complex

Differentiation of several specific cell types in the superior olivary complex has been described. Using Golgi-impregnated material, Morest (1969a) followed the development of the principal cells of the medial nucleus of the trapezoid body from the point at which they first enter the presumptive nucleus (Fig. 7.6). Initially, the cells are recognizable as elongated neuroblasts at the outer surface of the medulla. A primitive process, covered with filopodia, grows dorsally into the presumptive medial nucleus and gives rise to an axon. At the point of emergence of the axon, afferent fibers from the cochlear nucleus have already begun to differentiate (Section 3.2.2). Thus the first sign of differentiation of the principal neuron is the outgrowth of its axon, which occurs as the internal process moves into the area occupied by the growth cones of the ingrowing calyceal axons (Morest 1969a). The soma of the cell then moves through the internal primitive process to the point of the emergence of the axon. After this perikaryal translocation, dendrites begin to develop, passing through the typical stages with transitory filopodia and growth cones to reach the mature form.

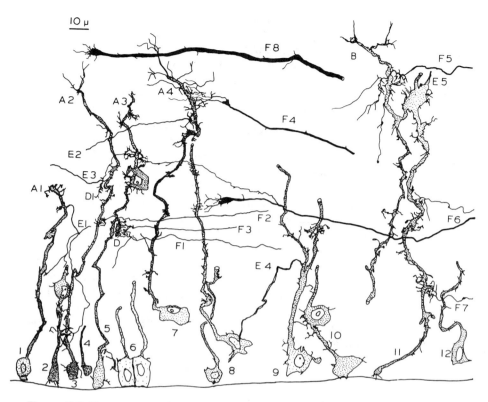

10 µ

FIGURE 7.6. Transverse section through the primordial medial nucleus of the trapezoid body in a North American opossum pup (27-mm crown-rump length = P10 as estimated by Willard and Martin, personal communication). The outer limiting pial layer is at the bottom of the drawing. 1–12, Neuroblasts at various stages of development; the more mature neuroblasts are located more medially in the drawing (i.e., cells 1, 3, 5, and 11 represent progressively more mature neurons); A1–A4, terminal growth cones of primitive internal processes; B, terminal dendritic growth cone; D and D1, preterminal dendritic growth cones; E1–E5, efferent axons; F1–F8, afferent axons. The larger axons will give rise to calyces (Morest 1968). For further details, consult Morest [1969a]. (From Morest (1969a); reprinted by permission of the author and Springer-Verlag New York, Inc.).

In the medial superior olivary nucleus of the rat, there is a rapid increase in cell size and dendritic proliferation from P0 to P14 (Rogowski and Feng 1981). At the time of ear opening (about P14), the dendrites reach their peak exuberance with many appendages, filopodia, and growth cones. After this time, the dendrites appear to enter a "stabilization" phase during which there is a decrease in cell size and dendritic branching and the dendrites mature. The loss of dendrites is in the second- to fifth-order branching. Neurons in the equivalent nucleus in the chick (nucleus laminaris) undergo a very similar developmental sequence with a period of prolific dendritic growth followed by a period of elimination of branches (Smith 1981). Bilateral otocyst removal has no effect on this sequence of events, although there is some decrease in total dendritic length (Parks, Gill, and Jackson 1987; see also Parks and Jackson 1984).

In ferrets, Golgi impregnation reveals growth of cells in the medial superior olivary nucleus from about P8 or P10 to one month (Henkel and Brunso-Bechtold 1990; time of hearing onset is about P32 [Moore 1982]). The dendrites of these cells are not oriented during the first postnatal week but develop their characteristic bipolar appearance by P17. During this time, transient appendages, first proximally and then more distally, appear on the soma and dendrites (Henkel and Brunso-Bechtold 1990). At P28–30, tufts of tertiary dendritic branches first appear on the ends of the dendrites. In the lateral superior olivary nucleus of the ferret, cells exhibit a bipolar appearance at birth (Henkel and Brunso-Bechtold 1991) and transient spines and other appendages are abundant on somas and dendrites. These are gradually lost during the first postnatal month. At P28 and extending past the time of hearing onset, tufts of fine processes appear at the tips of the dendrites. In the lateral superior olivary nucleus of the gerbil at P1–3, the neurons exhibit a profusion of processes with many branches (Sanes, Song, and Tyson 1992). At P13–14 (about the time of hearing onset; Table 4), the total dendritic length and number of branches of presumptive high-frequency neurons in the lateral superior olivary nucleus reaches a peak value, after which it drops to the adult value (see Section 4.3). There is no change in the total length of the dendrites of presumptive low-frequency neurons, although the total number of branches decreases after hearing onset. The number of primary dendrites and the size of the soma of these cells is adultlike before hearing onset (Sanes, Song, and Tyson 1992).

3.3.3 The Inferior Colliculus

In the mouse, the inferior colliculus does not change much in size during the early postnatal period (Romand and Ehret 1990; also bat, Rübsamen 1987), although during this time the dorsal and ventral cochlear nuclei increase greatly in size (Webster, Sobin, and Anniko 1986). The mean cross-sectional area of the neurons of the central nucleus do increase in size, however (Webster and Webster 1980). From P3 to P12, cell size increases rapidly and then continues to increase slowly to a peak at P24 (Fig. 7.5). Unlike that of the mouse, the inferior colliculus of the gerbil does grow during the immediate postnatal period, increasing steadily from 20 to 60% of its adult volume by P15 (Rübsamen et al. 1994).

In the inferior colliculus, as in other structures, the large neurons differentiate before the smaller ones (Morest 1969b). The dendrites of neurons of the central nucleus begin to grow and differentiate before those of the dorsal cortex (Morest 1969b). In the cortex, there appears to be a gradient of development such that the neurons of the deeper layers differentiate before those of the more superficial layers. Differentiation in the central nucleus also appears to occur in gradients (Repetto-Antoine and Meininger 1982). Different neuronal types are recognizable in the central nucleus of the cat at birth because of differences in their dendritic branching patterns (González-Hernández, Meyer, and Ferres-Torres 1989). The spiny, disk-shaped neurons in the central nucleus of the cat mature during the first postnatal month. At birth, the processes of the neurons are almost smooth, but at the end of the first week, they are profusely covered with "fine hairy and spiny appendages." These appendages disappear during the second week of life, and the density of spines increases (hearing onset occurs at about P5–7, Table 6). The dendrites appear to become oriented before hearing begins. At birth, the dendritic tree is symmetric but becomes disk shaped during the first postnatal week. Dardennes, Jarreau, and Meininger (1984) also found changes in the orientation of dendritic fields from P8 to P20 in the rat (which spans the period of hearing onset). They concluded that the reorientation of dendrites in the central nucleus of the inferior colliculus is accomplished by changes in the terminal or peripheral segments of the dendrites. Meininger and Baudrimont (1981) compared the dendritic trees in kittens at P12 and in adults and reached a similar conclusion. The dendrites of the large stellate neurons in the central nucleus are covered with long, hairy appendages at birth (González-Hernández, Meyer, and Ferres-Torres 1989). They reach their adult length during the second postnatal week, at which time distal dendritic "branchlets" appear.

3.3.4 The Medial Geniculate Nucleus and the Auditory Cortex

In the medial geniculate nucleus of both the cat and opossum, dendrites of the principal cells differentiate in the ventral nucleus before the dorsal nucleus (Morest 1969b). As in other structures, the dendrites of the small neurons differentiate later than those of the principal neurons in their respective subdivisions. In the newborn rat, the ventral and dorsal divisions of the medial geniculate nucleus are already distinguishable (Coleman 1990). Cells in all parts of the medial geniculate undergo a rapid growth spurt from P3 to P7. Cell processes and somas are covered with appendages, and the dendrites exhibit growth cones at P3. A second growth spurt occurs from P11 to P15 as hearing commences (Coleman et al. 1989; Coleman 1990).

The cells of the auditory cortex appear to be the last cells of the auditory system to undergo differentiation (Morest 1969b). In the rat, the auditory cortex increases greatly in thickness from P0 to P7, with the greatest growth from P5 to P7 (Coleman 1990), but the full thickness is not reached until about P30. From P11 to P14, layers 3 and 4 enlarge disproportionately relative to the other layers. In the human at 11–12 weeks of gestation, the cortical plate has been established and postmigratory neurons with developing dendrites are present. Migratory neurons can be identified deep to the cortical plate, and by 12–13.5 weeks, cortical layers can be identified (Krmpotić-Nemanić et al. 1979). Morphological development of neurons in the motor cortex of humans was described in detail by Marin-Padilla (1970a,b, 1988; also cat, 1971, 1972). Comparable descriptions of the development of the auditory cortex of humans are not available, although Marin-Padilla (1970a) suggests that his findings are representative of the development of neocortex in general.

What is known about the differentiation of auditory cortical neurons indicates that their stages of development are similar to those of the neurons in other auditory nuclei. The large pyramidal cells mature first; the apical dendrites emerge and differentiate before the basal dendrites (Fox 1968; Coleman et al. 1987). At birth in the dog, the layer 5 cells are more developed than those in layers 2–4 (Fox 1968). From P0 to P10, there is little change in the morphology of the pyramidal cells, although increasing numbers of neurons develop apical dendrites during this period. Many neurons have no basal dendrites at birth. By P10, short, unbranched basal dendrites begin to appear, and by P28, most cells have well-developed basal dendrites. (The first auditory evoked potentials were elicited at P12–14 in these dogs, the time at which the external auditory meatus opened [Fox 1968].) McMullen, Goldberger, and Glaser (1988) followed the development of nonpyramidal cells in layers 3 and 4 of the rabbit (in which hearing onset occurs around P6; Fig. 7.7). At birth, layers 2–4 have not formed. Presumptive layer 4 cells have short, vertically oriented dendrites tipped with growth cones. They have soma areas equal to 34% of the adult value and dendritic lengths equal to only 10% of their adult value. By P6, a six-layered cortex is first apparent and dendritic length has increased four times. At the time that hearing begins, the dendritic fields are more highly branched than in the adult, although the total dendritic length is considerably less than that in the adult. After P6, the increase in total length is entirely due to growth of the terminal dendrites. The basal dendrites of the layer 5 pyramidal cells are relatively well differentiated at P5 in rabbit (König and Marty 1974). The growth of the dendrites is highly directional throughout development (McMullen, Goldberger, and Glaser 1988). As discussed in Section 4, dendritic development in the cortex continues for some time after hearing begins.

3.3.5 Differentiation of Cells and Dendrites: Discussion

Growth in the auditory system takes place on a background of brain growth in general. The entire brain grows rapidly from birth until about the end of the second postnatal week in rats (Dobbing and Sands 1971), mice (Kobayashi 1963; Hahn et al. 1983), and gerbils (Wilkinson 1986). At the end of this "growth spurt," the rate of growth falls off dramatically in all three species, although growth continues at a slow, steady rate for some

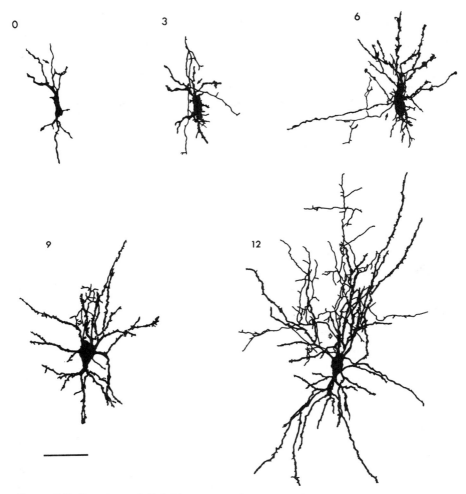

FIGURE 7.7. Drawings of Golgi-impregnated nonpyramidal cells from the presumptive auditory cortex of the rabbit at P0, 3, 6, 9, and 12. (The rabbit begins to hear about P6.) There are increases in both terminal and preterminal growth cones on the pially directed dendrites at P6. Axons are indicated by arrows. Scale bar = 50 μm. (Modified slightly from McMullen, Goldberger, and Glaser [1988]; reprinted by permission of the authors and John Wiley and Sons, Inc.)

time thereafter. Human brain weight increases by about four times from birth through three years of age (Dekaban 1978). It then continues to increase at a slower rate until it peaks at about 15 years. After 45–50 years, human brain weight declines steadily at a (mercifully) slow rate. Growth involume of the auditory nuclei that occurs after birth (rat: Coleman, Blatchley, and Williams 1982; mouse: Webster 1988a; gerbil: Rübsamen et al. 1994; human: Konigsmark and Murphy 1972) could be due to a number of factors including neuronal growth, generation and growth of glial cells, myelination of fiber tracts (Section 4.2.4), and continuing angiogenesis (Section 4.2.5). Certainly, one important contribution to growth is the continued growth of dendrites and the axon arbors that contact them.

The morphological changes associated with maturation have been studied in detail in only a few cell types in the auditory system, but the sequence of dendritic differentiation is similar in all types studied. Unlike the establishment of the basic circuitry, which is accomplished during the period before hearing onset, dendritic development appears to continue long after hearing onset (Section 4.2.2). As in other parts of the central nervous system (Jacobson 1991), an initial phase of dendritic growth leads to the production of numerous branches; this is often followed by a phase during which branch number decreases. In the auditory system, dendrites appear to reach a maximum degree of branching just before hearing onset; at this stage, they are often covered with fine hairlike appendages, filopodia, and growth cones. Commonly, more proximal dendrites mature first, whereas terminal dendrites continue to grow and may remain in a immature state for a long period of time (e.g., Schwartz 1972; Scheibel and Scheibel 1974; deCarlos, Lopez-Mascaraque, and Valverde 1985; Henkel and Brunso-Bechtold 1990, 1991).

Growth and differentiation of dendrites occur in concert with the ingrowth and maturation of the axon terminals that will form synapses with them (Morest 1969a,b, 1970; Jacobson 1991). Interactions between target neurons and their inputs may play an important role in the determination of their ultimate adult morphology. Signals mediating intercellular communication among developing neurons include growth factors, cytokines that regulate neuronal gene expression, steroid hormones, neurotransmitters, and cell surface markers (reviewed, e.g., by Parks, Jackson, and Conlee 1987; Lauder 1988, 1993; Mattson 1988; Purves, Snider, and Voyvodic 1988; Rubel, Hyson, and Durham 1990; Patterson and Nawa 1993). Factors regulating development may have the effect of suppressing or enhancing neurite outgrowth in specific sets of neurons. These effects may be mediated at the growing tips or growth cones of immature dendrites (e.g., Mattson 1988). As Patterson and Nawa (1993) point out, the transsynaptic exchange of signals may be highly specific because the circuitry that is formed among developing neurons allows for discrete, localized interactions. Among the traits that could be regulated are the overall shape and size of the dendritic tree, the types of neurotransmitters produced, the types of receptors expressed, and the types of synapses formed on the dendritic and somatic surfaces.

Circulating hormones and factors released by glial cells or cells of the immune system may also influence dendritic form (e.g., Patterson and Nawa 1993). Although presumably such effects would be less specific than those conveyed through neuronal transport, regional variations among glial cells do appear to exist. For example, Chamak et al. (1987) demonstrated that the ability of astrocytes to promote growth in specific sets of central neurons (in culture) depended on the region of the brain from which the astrocytes were taken. Indeed, the morphology of glial cells themselves appears to be regulated by cell-to-cell interactions (Canady and Rubel 1992). To date, factors affecting growth and differentiation have been studied largely in the peripheral nervous system or in culture systems, but the mechanisms involved are also potentially of great importance for understanding normal developmental interactions between neurons in the intact central nervous system. Studies in the chick auditory system indicate that factors influencing neuronal and dendritic development can be identified and related to metabolic activity in specific cells (e.g., Rubel, Hyson, and Durham 1990; Morest et al. 1993; Hyde and Durham 1994).

Cochlear input appears to play a role in determining cell number and cell size in the auditory system before the onset of auditory function. If one cochlea is removed from gerbil pups at P7, there is a 35% decrease in the number of cells in the anteroventral cochlear nucleus 2 days later and a 25% decrease in the size of the remaining neurons (Hashisaki and Rubel 1989; see also Trune 1982a,b; Nordeen, Killackey and Kitzes 1983; Kitzes 1986). Whether this influence is mediated via physical contact, synaptic activity, or trophic factors is not known. In older animals (after the onset of function), cessation of activity in the eighth nerve is sufficient to yield a decrease in cell size (Pasic and Rubel 1989); in these older animals, the effects of removal of activity are reversible (Pasic and Rubel 1991). The influence of eighth nerve activity on cell number appears to disappear before hearing onset. Cochlear removal at 4–8 weeks in gerbils has no effect on cell number (Pasic and Rubel 1991) In ferrets, cochlear ablation at P5 results in a loss of >50% of the large neurons in the ipsilateral cochlear nucleus, but ablation at P24 does not produce loss of neurons (Moore 1990a). This suggests that the influence on cell number wanes as hearing onset is approached. Cochlear ablation before hearing onset also leads to changes in the number or morphology of neurons in nuclei of the superior olivary complex in adult animals (Moore 1992a; Sanes et al. 1992), but it is not known whether the effect occurs before or after hearing begins.

One essential aspect of neuronal differentiation, the development of the mature complement of proteins specific for particular neuronal types is only now beginning to be explored. The developmental time course of expression of a number of proteins or the mRNAs that encode for them has been assessed in the brain stem auditory system of the rat and the gerbil (Tables 2 and 4). For example, an mRNA that encodes for fibroblast growth factor is expressed at very low levels during the first 10 postnatal days in rats but increases rapidly from P10 to P14 (Luo et al. 1995; Riedel et al. 1995). Adult levels are expressed at P14–17, a time that coincides with the onset of function. At present, such observations provide little insight into the mechanisms of development. However, as more is known about the morphological, physiological, and biochemical changes during the development of particular cell types, interrelationships among structure, function, and cell metabolism may become apparent.

3.4 Synaptogenesis

Synaptogenesis in the central nervous system begins very early, even before neurogenesis is complete (Jacobson 1991). Although ingrowth of axons into an area (Section 3.2), differentiation of the dendrites in that area (Section 3.3), and synaptogenesis are surely interdependent processes, few studies are available that address the relative timing of the different stages of each of these events. Some authors have attempted to relate the timing of axonal ingrowth and the first appearance of recognizable synaptic boutons, but, in most cases, conclusions have been based on a correspondence in time rather than definitive identification of the synapses from particular sources. It is possible, and even probable, that the timing of synaptogenesis vis á vis time of axon arrival is different and characteristic for different populations of axons. In all of the studies reviewed below, criteria for recognition of synapses have been essentially the same as those for recognition of synapses in the adult. Whether functional contact is made before (or after) synapses are identifiable morphologically is not known.

The process of synaptogenesis includes not only maturation of axonal terminals but also insertion of receptors into the appropriate parts of the postsynaptic membrane and their maturation. Some information about the development of postsynaptic receptors is available and is addressed below. One important goal of future research will be to establish the relationships between the development of specific types of presynaptic terminals and the postsynaptic receptors that respond to particular neurotransmitters.

3.4.1 The Cochlear Nuclei

Synaptogenesis in the mammalian auditory system has been studied in most detail in the cochlear nucleus. Neises, Mattox, and Gulley (1982) and Mattox, Neises, and Gulley (1982) provided a detailed description of the sequence of morphological changes in both the pre- and postsynaptic components of the synapses formed by end bulbs on the spherical cells in the rat. The initial interactions between ascending branches of auditory nerve fibers and the surfaces of spherical cells must take place before birth because synaptic contacts are present in the newborn animal. At birth, the primary afferent terminals are small, their terminal contours are irregular, and they insinuate among processes in the neuropil. They form asymmetric axodendritic contacts but do not contact the somas of spherical cells at this stage. From P7 through P10, the primary terminals, which at this age take the form of long, flat swellings, become enveloped by the long, fingerlike appendages that protrude from the somas of spherical cells. The somatic protrusions and the afferent terminals form numerous synapses. By P10, processes arising from the afferent terminal contact the soma itself, and by P12, the number of true somatic contacts increases as the somatic appendages become smaller and the end bulb becomes closely apposed to the soma. At P10, at least three types of nonprimary terminals appear in the neuropil; some of them contact the somas of spherical cells. The nonprimary contacts are not associated with the transient somatic appendages. Thus the basic synaptic arrangement between the spherical cell and the end bulb is established before hearing begins. Furthermore, assuming that most of the nonprimary contacts are inhibitory (e.g., Cant 1991), the spherical cell receives its inhibitory inputs only slightly later than it receives the excitatory ones. The development of inhibitory inputs (GABA-positive terminals) is also synchronized with the development of end bulbs on the cells of the nucleus magnocellularis of the chick (Code, Burd, and Rubel 1989).

Ryugo and Fekete (1982) followed the development of the same synapses at the light microscopic level in the cat (Fig. 7.8). The endbulbs pass through two developmental stages on their way to the adult configuration. In kittens younger than 10 postnatal days, the end bulbs consist of large, solid, spoon-shaped swellings with many thin threadlike processes emanating from them. By P10, ~50% of the end bulbs have progressed from this immature stage to become fenestrated, but they still appear immature. By P20, ~50% of the end bulbs are adultlike, and by P45, almost all of them are. Ryugo and Fekete suggested that the early nonfenestrated end bulb might play a functional role in shielding the spherical bushy cells from

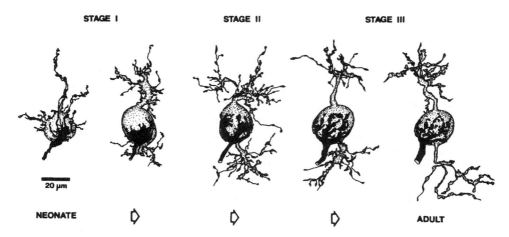

FIGURE 7.8. Drawings to represent a hypothetical sequence of development for a single endbulb in the cochlear nucleus and the postsynaptic spherical bushy cell with which it forms synaptic contact. The four drawings on the left span 45 days of postnatal development in the cat. (From Ryugo and Fekete [1982]; reprinted by permission of the authors and John Wiley and Sons, Inc.)

noncochlear endings. These would be allowed access to the cell body only after the end bulb becomes fenestrated. In this way, they suggest, the timing of maturation of the end bulb may influence the synaptic organization of the other inputs to the cell. Because electron-microscopic studies of end bulb development are not available for the cat and light-microscopic descriptions are not available for the rat, it is not possible to determine how the stages defined by Neises, Mattox, and Gulley (1982) and Ryugo and Fekete (1982) relate. It seems likely, given the relative maturity of the cat at birth, that the first stage of end bulb development described by Ryugo and Fekete corresponds to a stage after the time that the end bulb has contacted the somatic surface of the spherical cells (see also Jhaveri and Morest 1982 for the chick).

Electron microscopy and immunocytochemical methods have been used to study synaptogenesis in the dorsal cochlear nucleus of cats (Kane and Habib 1978; Schweitzer, Cecil, and Walsh 1993), hamsters (Schweitzer and Cant 1984; Schweitzer and Savells 1991; Schweitzer, Cecil, and Walsh 1993), and gerbils (Schwartz and Ryan 1985; Ryan and Woolf 1992, quoting Yu and Schwartz). In the hamster, the time of appearance of synaptic terminals of various types can be related to the time of ingrowth of axons from different sources. Growth of cochlear fibers into the dorsal cochlear nucleus occurs over the first 1.5 weeks after birth (Schweitzer and Cant 1984; Section 3.3.2). The first appearance of synaptic terminals with large, round vesicles, presumed to be the cochlear terminals, lags behind the fiber ingrowth by about two days. The first synapses formed appear to be on dendrites, which are themselves undergoing maturation during this time (Schweitzer 1990; Section 3.3.1). In the newborn cat, cochlear fibers have already invaded the dorsal cochlear nucleus, and terminals with large, round vesicles are present in the neuropil (Kane and Habib 1978). Later, these terminals also form synapses with cell somata.

Physiologists studying the development of the dorsal cochlear nucleus have concluded that excitatory and inhibitory inputs must mature together (e.g., Brugge and O'Connor 1984; Wu and Oertel 1987). Consistent with this conclusion, terminals with pleomorphic or flattened vesicles, which are presumed to be inhibitory, appear in the neuropil of the dorsal cochlear nucleus during the same time that the terminals with round vesicles appear (Kane and Habib 1978; Schweitzer and Cant 1984). Studies of the development of GABAergic synapses provide further structural evidence that inhibitory synapses mature during this time. Glutamic acid decarboxylase (GAD)-positive terminals are present at birth in the dorsal cochlear nucleus of cat and by P3 in that of hamster (Schweitzer, Cecil, and Walsh 1993). By hearing onset in both species, the pattern of GAD labeling is essentially mature. Electron microscopy confirms that the immunolabeled profiles are apposed to dendritic and somatic processes with well-developed pre- and postsynaptic membrane specializations. Schwartz and Ryan (1985) observed a similar, although somewhat delayed, pattern in the dorsal cochlear nucleus of the gerbil (hearing onset about P14). They found that synaptic terminals were sparse through P10 but that there was a major increase in synaptogenesis from P12 to P14. At first, these synapses appear immature but by P14–P16, there is maturation of the morphology. By P22, the morphology is essentially adultlike. GABA immunoreactivity also develops during this time (Yu and Schwartz quoted by Ryan and Woolf 1992). By P18, the pattern of GABA staining is adultlike. In the hamster, the time of arrival of presumably inhibitory descending inputs to the dorsal cochlear nucleus from the inferior colliculus is just before synaptogenesis involving terminals with pleomorphic vesicles (Schweitzer and Savells 1991, Schweitzer, personal communication). GABAergic terminals form during this same time, but the source of these terminals has not been established definitively.

Development of receptor efficacy in the cochlear nucleus has been addressed by Walsh and his colleagues physiologically in the kitten (Walsh and McGee 1987; Walsh, McGee, and Fitzakerley 1990, 1993). From the time of birth, ~80% of all units recorded respond to iontophoretically applied GABA (Walsh, McGee, and Fitzakerley 1990), a percentage that does not change with age. The application of GABA causes a decrease in both acoustically evoked and spontaneous activities, the latter even in neurons in which acoustically evoked activity is not present. This result implies that GABA receptors and associated structures are present and functional at least as early as the time of birth (see also Walsh and McGee 1987). Sensitivity to the excitatory amino acid transmitter glutamate and its agonist N-methyl-D-aspartate is also present in the caudal cochlear nucleus of newborn cat (Walsh, McGee, and Fitzakerley 1993). From P0 to P7, iontophoretically applied glutamate leads to an increase in the acoustically evoked discharge rate of ~80% of the neurons. As was the case for GABA, no developmental changes in the percentage of responsive neurons were seen. Acoustically unresponsive neurons (present until about P10) also responded to application of glutamate and the application made some but not all of them responsive to sound. Thus the excitatory and inhibitory receptors and their associated ionophores are present on the surfaces of neurons and are functional before the circuitry is completely operational. One important issue that may provide insight into the mechanisms of development is that of how receptor subtypes change over the course of development (see Section 3.4.5). An intriguing question is how the disposition of specific receptors in the postsynaptic membrane and the establishment of synapses of a particular type on that membrane are related.

3.4.2 The Superior Olivary Complex

Axons from the cochlear nucleus form synapses in all of the main nuclei of the superior olive at birth (Kil et al. 1995). Over the first 10 postnatal days, the synapses increase in size and in the number of synaptic vesicles that they contain. Synapses from the cochlear nucleus contain numerous round vesicles and form contacts with small dendritic profiles in the medial superior olivary nucleus. During this same time, the fibers that will give rise to the calyces in the medial nucleus of the trapezoid body form multiple synaptic contacts filled with round vesicles, although calyces do not form until P5 and are not mature until P14/16. By P10, each calyx forms multiple axosomatic contacts with a single principal cell. Glycine receptor concentration increases in the lateral superior olivary nucleus of the gerbil from P4 to P10, indicating development of inhibitory synapses during this time (Sanes and Wooten 1987). After P10, the concentration of receptors remains fairly constant in the dorsal part of the nucleus but decreases in the ventral region. By P21, the receptor distribution is similar to that in the adult.

In the ferret, input from the cochlear nucleus reaches the superior olivary complex by E28 (Brunso-Bechtold, Henkel, and Linville 1987), but the major period of synaptogenesis does not begin in the medial superior olivary nucleus until near the end of the second postnatal week (Brunso-Bechtold, Henkel, and Linville 1992). At birth, the few synapses that are present are immature and are located in the neuropil, which also contains many growth cones. Terminals with both round (potentially excitatory) and nonround (potentially inhibitory) vesicles are seen at this earliest stage of synaptogenesis. From P12 to P20, somatic synaptogenesis commences, and by P24, the soma is covered with contacts as in the adult. During this time, the somas give rise to elaborate appendages that snake out into the neuropil. After the third postnatal week, the synaptic organization of the medial superior olivary nucleus appears very similar to that in the adult. Thus the major period of synaptogenesis in this nucleus appears to precede by about two weeks the onset of hearing in the ferret.

3.4.3 The Inferior Colliculus

Synaptogenesis has not been studied in the inferior colliculus in any detail. The results reported by Pysh (1969) imply an increase in growth of the neuropil over the first two postnatal weeks in the rat. At P1, cell bodies are very close together (<7 μm). By P14, areas of neuropil 10–30 μm thick separate the cells. Synapses are present at birth but increase in number with age.

3.4.4 The Auditory Cortex

One of the earliest studies of synaptogenesis in any part of the cerebral cortex was devoted to a description of the development of temporal (auditory) cortex in the rat (König, Roch, and Marty 1975). On E15, there are no identifiable synapses in the nascent cortex, although desmosome-like junctions are present. At E16, the first cortical synapses appear, but they have relatively few vesicles and inconspicuous membrane thickenings.

These early synapses are located above and below the cortical plate, which is forming at this time. By the next day, synapses appear more mature, but they are still not found in the newly formed cortical plate. The earliest synapses in the subplate zone of the incipient cortex have been implicated in directing thalamocortical axons and descending cortical projections to their appropriate targets (e.g., Ghosh et al. 1990; McConnell, Ghosh, and Shatz 1994). Whether the early synapses described by König, Roch, and Marty (1975) play such a role remains to be determined. By P5 in the rabbit (hearing onset occurs about the end of the first week [Foss and Flottorp 1974]), synapses are found at every level of the auditory cortex (König and Marty 1974).

Few other studies have addressed synaptogenesis in auditory cortex specifically, although all available evidence indicates that its development is very similar to that of other cortical areas (reviewed by Jacobson 1991). Rakic et al. (1986) demonstrated that synaptogenesis occurs simultaneously in many areas and layers of cortex in the rhesus monkey. (The study included the visual, somatosensory, motor, and prefrontal cortices but not the auditory cortex.) During the final two months of a 165-day gestation, synaptic density increases at a rapid rate in all areas and proceeds concurrently in all cortical layers. An increase in synaptic density continues for 2–4 months postnatally.

Although the timing of synaptogenesis in the human auditory cortex per se has not been studied, Molliver, Kostović, and van der Loos (1973) studied the lateral cortex of human fetuses using the electron microscope. From 8.5 to 18 weeks of age, synaptic profiles containing synaptic vesicles are present above and below the cortical plate (as in the E16 rat) but never within it. The number of identifiable synaptic terminals increase with age. By 23 weeks, synaptic terminals with vesicles and asymmetric appositions are present, the time of onset being sometime between 19 and 23 weeks. These first synapses appear to be axodendritic. By 12–13.5 weeks, true cortical layers can be recognized and neurons with growing dendrites are present (Krmpotić-Nemanić et al. 1979). Krmpotić-Nemanić et al. (1980) demonstrated a developmental sequence of acetylcholinesterase staining in the subplate (E20 to 24 weeks) followed by staining in the cortical plate (24–26 weeks). By 28 weeks of gestation, a columnar pattern of acetylcholinesterase staining is seen throughout the auditory cortex. Whether this staining corresponds to ingrowth and formation of synapses by cholinergic fibers is not known.

3.4.5 Synaptogenesis: Discussion

As axons grow into their targets, the dendrites of neurons in those targets are growing and branching in patterns characteristic of each cell type. The formation of synapses between the appropriate axonal and dendritic partners during this time is the ultimate determinant of function in the mature nervous system. A large number of mechanisms that might control synaptogenesis have been proposed and discussed (see, e.g., reviews by Parks, Jackson, and Conlee 1987; Lauder 1988, 1993; Mattson 1988; Purves, Snider, and Voyvodic 1988; Vaughn 1989; Rubel, Hyson, and Durham 1990; Patterson and Nawa 1993). To date, most of the information available about synapse formation in the mammalian auditory system is qualitative; further systematic, quantitative studies of synaptogenesis between specific synaptic partners will be crucial for understanding the precise mechanisms.

Reliance on morphological studies to establish the time of onset of synaptogenesis is open to question because criteria for identification of the earliest functional synapses have not been developed. Synaptic transmission can occur before the morphological maturation of synapses is apparent (e.g., Landmesser and Pilar 1972), so synaptogenesis may go on for some time before it can be detected in the electron microscope. More precise knowledge of the sequence of ingrowth of the various sources of input to a given neuron would help to define the possible interactions and their time course. As discussed in Section 3.4.1, both excitatory and inhibitory receptors are present in the caudal cochlear nucleus of the cat at birth, about one week before activity can be evoked acoustically in the auditory system. Interactions among growing axons and membrane receptors in dendrites or growth cones could function to guide early development. Substances that act as neurotransmitters in the adult may inhibit or promote dendritic growth in developing neurons (e.g., Mattson 1988). In rat hippocampal neurons, for example, a balance between activity in glutamatergic and GABAergic systems regulates the level of mRNA coding for brain-derived neurotrophic factor and nerve growth factor (Zafra et al. 1991), two factors that could, in turn, regulate neuronal growth and differentiation.

One matter of much interest is the way in which topographic matching of synaptic inputs from different sources is accomplished. Based on physiological measures, the topography of both excitatory and inhibitory connections is established in the auditory system by the onset of hearing (reviewed by Sanes and Walsh [Chapter 6]; also Brugge 1988, 1992; Kitzes 1990). The characteristic frequencies of excitatory and inhibitory inputs on cells in the lateral superior olivary nucleus are very similar at the time of hearing onset in the gerbil (Sanes and Rubel 1988), implying that the appropriate topographical matching of inputs at the single-cell level has occurred before this time. There is continued improvement in the degree to which the excitatory and inhibitory characteristic frequencies correlate as the animals mature, and the maturation of excitatory and inhibitory inputs occurs over a nearly identical time course. Brugge (1988), commenting on the fact that the response areas of units in the cochlear nucleus of young animals are like those in adults, emphasized that the response areas reflect the interaction of excitatory and inhibitory inputs. Concluding that the circuits must develop as structural and functional units (see also Brugge and O'Connor 1984; Brugge, Reale, and Wilson 1988), he introduced the concept of "functional constancy" to emphasize that the initial establishment of most physiological responses to acoustic stimuli does not depend on auditory experience and arises before the elements of auditory circuits are completely mature structurally (i.e., before dendritic and axonal processes are fully developed).

A role for neuronal activity in the specification and refinement of synaptic connections is well established in developing neuronal sytems (reviewed, e.g., by Fawcett and O'Leary 1985; Rubel, Hyson, and Durham 1990; Shatz 1990a,b; Goodman and Shatz 1993; Katz 1993). Different temporal and spatial patterns of activity in input neurons can lead to their segregation to different parts of a target structure or to different types of target neurons. After the auditory system becomes functional, acoustically evoked activity appears to play an important role in the maintenance and refinement of the system (Section 4). It is likely that activity also plays an important role during the relatively long period of time after which synapses have been formed in the auditory system but before the system responds to acoustic stimuli. Synaptic potentials can be evoked in the cochlear nucleus and superior olivary complex electrically well before they can be evoked acoustically (Wu and Oertel 1987; DH Sanes 1993; Kandler and Friauf 1995), implying that synapses are functional shortly after axons have grown into their targets. Spontaneous activity has been recorded before hearing onset in the auditory nerve of newborn kittens (Carlier, Abonnenc, and Pujol 1975; Romand 1984); the cochlear nucleus of kittens (Romand and Marty 1975; Brugge and O'Connor 1984; Walsh and McGee 1988; Walsh, McGee, and Fitzakerley 1993) and gerbils (Woolf and Ryan 1985); the superior olivary complex of gerbils (DH Sanes 1993); and the inferior colliculus of ferrets (Moore 1982) and bats (Rübsamen and Schäfer 1990). The spontaneous activity in all parts of the auditory system takes the form of brief bursts of action potentials at intervals of 100–250 ms (ibid., also Pujol 1969, 1972; Lippe 1994). The bursting pattern of spontaneous activity is similar to that seen in developing animals in the optic nerve (Galli and Maffei 1988; Meister et al. 1991) and somatosensory fibers of the dorsal root (Fitzgerald 1987). In the visual system, coordinated temporal and spatial patterns of spontaneous activity in the retina, effected through a coupling of neighboring cells that is not seen in the adult, determine the organization of ganglion cell axon terminals in the lateral geniculate nucleus (Mastronarde 1983; Dubin, Stark, and Archer 1986; Sretavan, Shatz, and Stryker 1988; Meister et al. 1991; Wong, Meister, and Shatz 1993; Penn, Wong, and Shatz 1994). It has been suggested that spontaneous activity generated in the cochlea could play a similar role in the auditory system (e.g., Friauf and Kandler 1990; Friauf 1993; Lippe 1994).

It cannot be assumed that activity in the immature nervous system has the same effects as in the adult because the physiological properties of a developing synapse may be quite different from its properties in the adult. For example, postsynaptic potentials recorded from the lateral superior olivary nucleus in brain slices of newborn gerbils have a time course up to two orders of magnitude longer than those in mature animals (DH Sanes 1993). A brief train of impulses (as in the characteristic bursts of spontaneous activity described above) could therefore lead to sustained polarization of the postsynaptic membrane. Transmitters may also have different effects in the developing animal compared with the adult. In the superior olivary complex of rats from E18 to about P7, activation of strychnine-sensitive receptors via electrical stimulation of their inputs or application of glycine leads to depolarization; after P8, activation of these synapses leads to hyperpolarization as in the adult (Kandler and Friauf 1995). Such characteristics of immature synapses could be of functional importance in controlling developmental processes.

Experimental manipulation of the auditory system before hearing onset could provide information about the role of activity in regulating synaptogenesis. To date, however, most experimental protocols have been applied to animals about the time of or after the onset of hearing. Even in those cases where a manipulation was done before hearing onset, the effects have not generally been assessed until well after hearing would have begun.

4. Development After Hearing Onset. Refinement and Maturation of Axons and Dendrites.

Although the state of maturity of the auditory system at birth is highly variable across mammalian species (Tables 7.2–7.8), its development has reached essentially the same stage in all species at the time of hearing onset. Using development of the brain stem auditory evoked response as an index, Morey and Carlile (1990) demonstrated that a similar stage was reached at ~110% (range 104–115%) of the "silent" period (i.e., the period before hearing begins) in cats, ferrets, gerbils, hamsters, humans, rats, and sheep; as a percentage of gestation time, however, it varied from 79 to 225%. The most important determinant of the time at which hearing begins appears to be cochlear maturation (Pujol, Lavigne-Rebillard, and Lenoir [Chapter 4]; Lippe and Norton [Chapter 5]), although continued development of the neurons of the central auditory system probably accounts, at least to some extent, for the maturation of physiological properties such as sharpness of tuning and changes in dynamic range (e.g., Sanes and Rubel 1988; Sanes 1992). At the time of hearing onset, the basic structures of the auditory pathways are in place, and the basic patterns of circuitry have been established. Indeed, hearing onset appears to be delayed until the system is essentially intact and functional. It has been suggested that a *lack* of environmental influences may be just as important during early development as their presence is later (Marty and Thomas 1973).

4.1 Structural Changes After Hearing Onset

Structural maturation after hearing onset involves stabilization of cell size and continued growth of the auditory nuclei. Because neuronal number does not change appreciably after hearing onset, the increase in volume of auditory structures must be due to continued growth of axons and dendrites; synaptogenesis; glial growth, including myelination; and angiogenesis. The extent to which each of these factors contributes to growth to the adult size is not known. Studies of normal structural development in the auditory system after hearing begins are reviewed in this section.

4.1.1 Volume of Auditory Structures and Cell Size

For some time after hearing begins, the volume of the auditory nuclei continues to increase. (The entire brain continues to grow after this time, although the rate of growth is considerably slower than during the preceding weeks [Section 3.3].) The volume of the ventral cochlear nucleus increases from birth through middle age in humans (Konigsmark and Murphy 1972), through at least P90 in mice (Webster 1988a), and through at least P36 in rats (Coleman, Blatchley, and Williams 1982). In the gerbil, all auditory nuclei are ~60–70% of their adult size at hearing onset, with a slow increase to adult values thereafter (Rübsamen et al. 1994). The increase in volume of auditory nuclei is not due to increases in neuronal numbers. In the mouse, the number of cells in the ventral cochlear nucleus reaches its adult value at P12 (from a peak at P9) and in the dorsal cochlear nucleus at P16 (from a peak at P11–13; Mlonyeni 1967). In the human, there are no changes in cell numbers in the ventral cochlear nucleus after birth (Konigsmark and Murphy 1972). Indeed, the value at birth is maintained into old age.

Establishment of adult soma size has been studied in detail in the mouse (Webster and Webster 1980; Webster 1988a; Figure 7.5) and the cat (Larsen 1984). The results are different in the two species vis á vis the size of cells at hearing onset relative to the adult value. In mice, cells in the cochlear nucleus (multipolar, octopus, globular, and spherical cells), the superior olivary complex, and the inferior colliculus reach a peak size by P12 (about the time hearing begins). Some cells maintain that size into adulthood; others may decrease slightly in size after that. Neurons in the medial superior olivary nucleus of the rat reach a peak size at P9 and fall to adult levels at about the time hearing begins (Rogowski and Feng 1981). In the lateral superior olivary nucleus of gerbils, cell size does not change after P10/12 (Sanes, Song, and Tyson 1992). In cats, on the other hand, the several cell types in the cochlear nucleus are smaller at the end of the second week of life than in the adult (although they have attained their adult appearance in Nissl-stained preparations). Although the fastest growth occurs before hearing onset (Section 3.3.1), somatic growth continues for a number of weeks after that time. Small spherical cells, octopus cells, and some medium-sized cells in the posteroventral cochlear nucleus reach their adult size by the end of the first month, but large spherical cells and globular cells have attained only 70–75% of their adult size at three months of age. In ferrets, the cells in the medial superior olivary nucleus do not reach their mature size until P56, almost a month after hearing begins (Henkel and Brunso-Bechtold quoted by Brunso-Bechtold, Henkel, and Linville 1992). Given this limited information, it appears that rodents and carnivores may differ in the extent to which somatic size changes after hearing onset.

4.1.2 Growth and Retraction of Dendrites

The most obvious structural changes that take place in neurons after hearing onset involve their dendrites. These include changes in number and length of dendritic segments, changes in spatial orientation of the dendritic tree, and changes in the complement of dendritic appendages. Interneurons may undergo most of their differentiation and maturation after hearing begins, but they have been studied in detail in only a few parts of the auditory system (Willard and Martin, personal communication: small cells in the dorsal cochlear nucleus of opossum; deCarlos, Lopez-Mascaraque, and Valverde 1985: chandelier cells in the auditory cortex of cat). Few systematic, quantitative studies of dendritic growth in any specific cell type have been undertaken in mammalian species. One exception is a study of the development of the apical dendritic tree of fusiform cells in the dorsal cochlear nucleus of the hamster (Schweitzer 1991; Table 7.5). After P15, there is no change in the number of primary dendrites or in the length of individual dendritic segments on these cells, but until at least P60, there is continued elongation and addition of branches at the terminal ends. Terminal growth accounts for an increased total length of the dendritic tree; addition of branches and fanning out of the existing branches accounts for continued changes in its spatial orientation. Neurons whose dendrites continue to grow by increases in higher order branching and continued growth of terminal segments are also found in the inferior colliculus of the rat (Dardennes, Jarreau, and Meininger 1984) and the cat (Meininger and Baudrimont 1981; González-Hernández, Meyer, and Ferres-Torres 1989) and in the auditory cortex of the rabbit (McMullen, Goldberger, and Glaser 1988) and the dog (Fox 1968). Some neuronal types in the lateral and medial superior olivary nuclei of the ferret acquire terminal tufts of branches on their distal dendrites about the time hearing begins, and these continue to increase in length until the end of the second postnatal month (Henkel and Brunso-Bechtold 1990, 1991).

It is a common finding that neurons exhibit peak branching just at hearing onset. In the auditory cortex of the rabbit, there appears to be a significant proliferation of dendrites on nonpyramidal cells just at hearing onset (about P6 [McMullen, Goldberger and Glaser 1988]; Fig. 7.7). During the second postnatal week, some of these dendrites are lost. Dendritic growth involves increases in the length of the terminal segments of the remaining dendrites. Adult length is reached at P30. In the medial superior olivary nucleus of the rat, just before hearing begins, the dendrites branch extensively. At hearing onset, the total number of dendritic branches begins to fall, reaching the adult value by P18 (Rogowski and Feng 1981). The loss of dendrites is in the second- to fifth-order branching. Similarly, there is a decrease in number of branch points on neurons in the lateral superior olivary nucleus of the gerbil from a peak just at hearing onset to the adult value, which is reached during the third postnatal week (Sanes, Song, and Tyson 1992). This occurs most dramatically in the neurons of the high-frequency representation and has the effect of constraining their dendritic fields parallel to the frequency axis of the nucleus. In addition, the high-frequency neurons (but not the low-frequency ones) undergo a decrease in total dendritic length during this time. A similar, frequency-specific remodeling of dendritic structure occurs postnatally in the nucleus laminaris of the chick (Smith 1981).

As indicated by the study of Sanes, Song, and Tyson (1992), the differential loss or gain of dendritic branches and increases in dendritic length can lead to spatial orientation of the dendritic tree of specific neuronal types. Oriented dendritic fields develop after hearing onset in fusiform cells in the dorsal cochlear nucleus of the hamster (Schweitzer and Cant 1985a; Schweitzer 1991); in neurons in the central nucleus of the inferior colliculus of the rat (Dardennes, Jarreau, and Meininger 1984) and the cat (Meininger and Baudrimont 1981); and in nonpyramidal cells in the auditory cortex of the rabbit (McMullen, Goldberger, and Glaser 1988). Alterations in dendritic orientation that occur after the onset of responsiveness to acoustic stimulation could contribute to adult frequency selectivity (Sanes, Song, and Tyson 1992). Biased directionality in growth is also apparent in the case of entire auditory structures (lateral superior olivary nucleus: Sanes, Merickel, and Rubel 1989; cochlear nucleus: Rübsamen et al. 1994).

Dendrites change not only in length but also in width and in their complement of appendages. Immature spines and excrescenses are lost with maturation and appear to last the longest at the periphery of each dendrite (Scheibel and Scheibel 1974). In the inferior colliculus of the cat, dendrites continue to mature through the first postnatal month, at which time they become essentially adultlike, although the density of spines is at a peak and will later decline to the adult value (González-Hernandez, Meyer, and Ferres-Torres 1989). The dendrites of the large stellate neurons reach adult length during the second week; during the second and third weeks the number of appendages and spines increases and then begins to decline (González-Hernandez, Meyer, and Ferres-Torres 1989). Maximum spine density is achieved on pyramidal cell dendrites in the auditory cortex of rat at P35 (Coleman et al. 1987; Coleman 1990). In the rabbit, the number of spines on nonpyramidal cells peaks between P12 and P15 and then gradually declines until the cells are essentially spine free by P30 (McMullen, Goldberger, and Glaser 1988).

4.1.3 Axon Growth and Synaptogenesis

As terminal segments of dendrites change in length and mature, a concomitant change in axonal arborization patterns and continuing synaptogenesis would be expected to occur (cf. Morest 1969a,b). At present, however, little can be said about dendritic and axonal interactions after the onset of function. Many connections are described as adultlike at hearing onset but in only a few cases has the morphology of a specific set of axons been examined in any detail. After hearing begins in the hamster, the morphology of the cochlear axons that innervate the dorsal cochlear nucleus remains unchanged (Schweitzer and Cecil 1992). On the other hand, in the gerbil, the axonal arbors projecting from the medial nucleus of the trapezoid body to the lateral superior olivary nucleus give rise to arbors that are more restricted in their extent at P15/16 than at P13/14 (Sanes and Siverls 1991). This decrease in axonal arborizations appears to mark the end of a major growth phase of these axons and may serve to sharpen the tuning of the neurons that receive the inputs (Sanes and Rubel 1988). The number of boutons (presumably the sites of synaptic contacts) decreases on these axons from P18 to P25 (Sanes and Siverls 1991). There also appears to be some refinement of connections to the inferior colliculus. The proportion of ipsilateral- to contralateral-projecting cells to the inferior colliculus from both the cochlear nucleus (Moore and Kowalchuk 1988) and the lateral superior olivary nucleus (Henkel and Brunso-Bechtold 1993) continues to change until sometime after hearing begins. Whether this change is due to actual changes in the number of projecting cells or to a rearrangement of their axonal arborization patterns (making them more or less likely to take up tracers) is not clear, but, in either case, the result would suggest a change in connectivity that, because of its timing, is subject to environmental influences.

In the forebrain, axons do not appear to have reached a level of maturity at hearing onset comparable to that seen in the brain stem. The "exuberant" arbors of the colliculothalamic axons in the rat (Section 3.2.3) are not completely eliminated until the end of the third postnatal week (Asanuma et al. 1988). By that time, both the branching of the axons outside the medial geniculate nucleus and also some of the branching within the medial geniculate disappears. In the auditory cortex of the cat, substantial reorganization of corticocortical connections appear to take place for some time after hearing begins. During the second postnatal week, callosal afferents begin to become segregated into vertical bands, a process that ends at about P18 (Feng and Brugge 1983). The laminar pattern of callosal terminals characteristic of adult cats is not seen until sometime after P18 (Payne 1992), and the cells that form the callosal projections do not form clusters until after three months of age (Feng and Brugge 1983; Payne, Pearson, and Cornwell 1988a; Payne 1992). Perhaps associated with this clustering is a reported loss of about one million axons per day from the corpus callosum during the first postnatal month (Payne 1992); presumably, some of these axons represent projections from the auditory cortex. Transient projections from auditory cortical areas to visual areas 17 and 18 persist until the fifth postnatal week in the cat after which time they disappear, although the cells of origin do not (Innocenti and Clarke 1984; Innocenti, Berbel, and Clarke 1988; Clarke and Innocenti 1990).

At the level of synaptogenesis, an overview of the development of specific circuits comes from studies of the appearance and maturation of various transmitter-related molecules. Distribution of GAD attains its adult pattern from P7 to P12 in the dorsal cochlear nucleus of the cat and by P21 in that of the hamster (Schweitzer, Cecil, and Walsh 1993). GABA-like immuno-reactivity is adultlike in the dorsal cochlear nucleus of gerbils by P18 (Yu and Schwartz quoted by Ryan and Woolf 1992). In all three species, then, there is a short period after hearing onset when the distribution of this marker of inhibitory synapses is not mature. Likewise, the distribution of receptors for the inhibitory transmitter glycine changes after hearing onset in the lateral superior olivary nucleus of the gerbil (Sanes and Wooten 1987). From an initial state of uniform distribution, there is an apparent decrease in receptor concentration in the low-frequency parts of the nucleus.

At the electron-microscopic level, there have been no quantitative studies of rates or sites of synaptogenesis, so it is not known whether the processes of synaptogenesis that begin well before hearing onset (Section 3.4) continue unchanged or whether there are identifiable changes when the system becomes subject to acoustic stimulation. In the medial superior olivary nucleus of the cat, synaptogenesis is not complete until sometime after P20 (Schwartz 1972). At this time, the proximal dendrites are mature and receive an adult complement of synapses, but the more distal dendrites are still undergoing synaptogenesis. Seven synaptic types identified in the dorsal cochlear nucleus of adult cats appear sequentially in postnatal kittens, and some do not appear for over a month (Kane and Habib 1978). In the dorsal cochlear nucleus, the first endings to appear are from the cochlea and superior olivary complex. Those identified as coming from the inferior colliculus do not appear until P45–95 (but see Section 3.4.1). During this same time, Kane and Habib could first identify glomeruli and synaptic nests in the fusiform cell layer. In the octopus cell area of the posteroventral cochlear nucleus, the cochlear inputs are identifiable at P4–7 in the cat, but brain stem inputs are not identifiable until P20, well after hearing onset (Schwartz and Kane 1977). Authors who classify synaptic terminals into fewer classes (e.g., see Cant 1991) have reported the presence of all types during the weeks before hearing onset (Section 3.4). Because it is possible that some subtypes of synapses are not identifiable in young tissue, confirmation that particular sources do not provide synaptic input to a given nucleus before hearing onset should be confirmed with other methods.

One aspect of functional circuitry that undergoes important changes after hearing onset involves the auditory responses of neurons in the superior colliculus. The representation of auditory space by neurons of the superior colliculus (cf. Middlebrooks and Knudsen 1984) appears to develop entirely after hearing has begun. Neurons that respond to acoustic stimuli from limited regions of space are organized systematically to form "maps" of auditory space in the deep layers of the superior colliculus of guinea pigs (Palmer and King 1982; King and Palmer 1983) and ferrets (King and Hutchings 1987). These two species, of course, are at very different stages of development at the time of birth, but the representation of auditory space does not appear in either animal until over one month after the onset of function. From P0 to P15, neurons in the deep superior colliculus of the guinea pig respond vigorously to auditory stimuli in the contralateral field but otherwise evince no orientation preference (Withington-Wray, Binns, and Keating 1990a). From P16 to P30, an increasing proportion of these neurons displays spatial tuning, but there is no evidence of a systematic map. Such ordering first becomes evident at P31–32, the time at which the interaural distance first approaches adult size. The transition from no discernible map to a well-ordered map is rapid. After this time, there is a further maturation leading to a higher degree of spatial tuning. Thus it appears that the production of a map is delayed until an adult auditory periphery and head size are reached. Very little is known about structural correlates of these ontogenetic changes in physiological response properties. The circuitry that gives rise to the mapping is also unknown, although there may be a similar segregation of responsiveness to spatial location in the external nucleus of the inferior colliculus in guinea pigs (Binns et al. 1990), a part of the inferior colliculus that projects to the superior colliculus (e.g., Edwards et al. 1979). Development of normal response properties appears to depend on appropriate activity in both the auditory and visual inputs (Withington-Wray, Binns and Keating 1990b; Withington-Wray et al. 1990).

4.1.4 Myelination

The course of myelination in auditory pathways must have an important influence on the development of evoked potentials and latency of responses. In humans, myelination of the auditory pathways begins before birth and continues for some time after birth. At 22–23 weeks of gestational age, no myelinated axons are present in the brain stem (Moore, Perrazo, and Braun 1995). In the 26- to 29-week-old fetus, light myelination appears in the cochlear nerve, dorsal acoustic stria, trapezoid body, and lateral lemniscus. These are among some of the earliest fiber systems in the central nervous system to complete myelination (Langworthy 1933). The time of onset of myelination in the human coincides with the onset of acousticomotor reflexes and brain stem evoked responses (discussed by Moore, Perrazo, and Braun 1995). Two months after birth, the brachium of the inferior colliculus and the medial geniculate nucleus have become lightly myelinated. Few fibers ascending to the thalamus and none ascending to the cortex are myelinated at birth (Langworthy 1933). The acoustic radiations continue to acquire myelin from about birth through 3–4 years of age (Yakolev and LeCours 1967).

In the auditory nerve of kittens, myelination begins before birth (about E57), proceeds rapidly during the first 18 postnatal days, and then slows (Romand et al. 1976). Myelination of the nerve is not complete until sometime after the second year of life (Walsh and McGee 1986). In newborn cats and in opossums 46 days old, the trapezoid body is only lightly myelinated (Langworthy, quoted by Morest 1968). In cats, the fibers begin to myelinate during the first week of life (Held, quoted by Morest 1968). The ventral third, which contains the largest axons, begins myelination first. The intermediate third, which contains many fine axons, begins to myelinate later. Although the course of myelination of the calyciferous axons has not been studied directly, Morest (1968) estimates that they become heavily myelinated in a proximal to distal direction during the first two postnatal weeks because that is the period of time during which less and less of the preterminal axon is impregnated by the rapid Golgi method (which is effective only until fibers are myelinated). There is a temporal correlation between a rapid decline in the latency of the auditory evoked potential from P15 to P26 in the rat and an increase in the concentration of cerebrosides (myelin lipids) in the inferior colliculus (Shah, Bhargava, and McKean 1978).

4.1.5 Angiogenesis

Angiogenesis appears to contribute to the growth in auditory structures after function begins. Craigie (1924) documented significant changes in the vascularity (i.e., the total length of capillaries per square unit area) of the dorsal cochlear nucleus and the superior olivary complex of rats between P10 and P22. (These are two of the most highly vascular nuclei in the brain stem of the adult, the dorsal cochlear nucleus being the highest among many structures measured by Craigie.) Capillary length in both structures increases about threefold to a peak that is maintained for some period of time but then gradually falls off over the first year of life. Craigie notes, but does not document, his impression that the capillaries grow more rapidly during this time than the nervous elements. Differences in vascularity among nuclei are not evident at birth but become so by P22, leading Craigie to suggest that there is a direct relationship between growth of the vasculature and functional activity. Growth of the vasculature follows essentially the same pattern in the inferior colliculus of rats and is complete by P24, with much of the growth occurring after the animals begin to hear (Andrew and Paterson 1989).

4.2 Development After Hearing Onset: Discussion

After the auditory system becomes functional, structural maturation of neurons continues. During this time, the major changes taking place are in the distal dendrites and in the axonal arbors that presumably innervate them. It is also during this time that the development of the auditory system becomes sensitive to environmental influences. The period of sensitivity, often referred to as a "critical period," lasts for some finite period of time and may function to calibrate the system or to coordinate activity among different sensory systems. At present, the mechanisms operating during critical periods are largely unknown. Indeed, as Rubel (1984) has pointed out, the use of the term "critical period" serves merely to restate the results obtained in experiments in which environmental manipulations lead to changes only during restricted time periods. Knowledge of the timing of specific developmental events relative to the extent of any given critical period should help in the formulation of hypotheses concerning the underlying mechanisms. Given the available experimental results, the period during which the auditory system seems most vulnerable to environmental influences occurs just at the time of hearing onset. During this time, cell size as well as dendritic and axonal morphology is vulnerable to experimental manipulation as discussed briefly below.

Although many neurons have reached their adult size before hearing begins, activity introduced due to environmental stimuli at hearing onset appears to be necessary for the maintenance of cell size (mouse: Webster 1983a,b,c,d, 1984, 1988a,b; rat: Coleman and O'Connor 1979; Coleman 1981; Coleman, Blatchley, and Williams 1982; Blatchley, Williams, and Coleman 1983; gerbil: Pasic and Rubel 1989, 1991; Pasic, Moore, and Rubel 1994; chick: Born and Rubel 1985; Tucci and Rubel 1985). The dependence of cell size on environmental input seems to decline gradually with age.

The extent to which the effect of cochlear ablation on cell size is a result of the loss of afferent electrical activity in the auditory nerve has been assessed by Rubel and his colleagues in gerbils (Pasic and Rubel 1989, 1991; Sie and Rubel 1992). If the inner ear is continuously exposed to tetrodotoxin (TTX) to block all electrical activity in the nerve from 4 to 6 weeks of age, a 21% decrease in the cross-sectional area of ipsilateral spherical cells is observed within 48 hours (Pasic and Rubel 1989; a 25% decrease is seen with cochlear ablation). If one cochlea is exposed to TTX and the other is ablated, there is no difference in cell size on the two sides. If the TTX infusion is followed by one week of recovery, the effect of the infusion is reversed (Pasic and Rubel 1991). These results imply that the signal associated with activity in the auditory nerve fibers is a dynamic one and that the balance of input from the two ears is an important variable. Both unilateral exposure to TTX and cochlear ablation at P14 followed by a survival of one hour lead to a 30–40% decrease in protein synthesis in cells of the ventral cochlear nucleus of gerbils (Sie and Rubel 1992; see also Trune and Kiessling 1988, mouse). The afferent regulation of protein synthesis may be a precursor to effects on cell morphology. In the chick, cochlear ablation also leads to rapid changes in cytoskeletal proteins (Rubel and Parks 1988), metabolic enzymes (Durham and Rubel 1985; Durham, Matschinsky, and Rubel 1993), mitochondria (Hyde and Durham 1994), and glial morphology (Canady and Rubel 1992; Rubel and MacDonald 1992).

The growth or loss of dendrites after hearing onset is limited to elongation or retraction of terminal dendritic segments. There may also be changes in the number of higher order dendritic branches, which may lead to changes in the spatial orientation of the dendritic tree. Normal development of both the length and spatial orientation of dendrites is disrupted if input from one cochlea is decreased through cochlear ablation, induced conductive hearing loss or other experimental manipulations (e.g., Feng and Rogowski 1980; Trune 1982b; Schweitzer and Cant 1985a; McMullen and Glaser 1988; McMullen et al. 1988; Sanes and Chokshi 1992; Sanes et al. 1992). Although only a few studies are available, they are consistent in suggesting that symmetry of input from the two ears is especially important (Feng and Rogowski 1980; Parks, Gill, and Jackson 1987).

Very little is known about the normal course of maturation of axonal arbors and synapses after hearing begins, but there is some evidence that an intact cochlea may be important for the normal development of axons. Neonatal cochlear ablation affects axonal branching in the lateral superior olivary nucleus (Sanes and Takács 1993) and in the inferior colliculus (Moore and Kitzes 1985; Kitzes 1986; Moore and Kowalchuk 1988; Moore et al. 1989; Moore 1990b, 1994). The effect in the inferior colliculus seems to be due to unequal activity from the two ears because it is not seen with bilateral cochlear ablation (Moore 1990b). There is one report suggesting that an experimentally induced conductive hearing loss can lead to synaptic rearrangements on single neurons (Trune and Morgan 1988).

5. Summary

Structural development of the auditory pathways can be divided into two phases, the first occurring before the system is subject to acoustic stimulation and the second beginning at the time the cochlea nears maturity and hearing begins. In all animals studied, the bulk of structural development of the auditory system occurs during the first phase. Generation of the neurons of the auditory system begins as the neural tube is closing. Axonal outgrowth begins as soon as cells are generated and begin to migrate. Dendritic differentiation begins soon thereafter, and synaptogenesis is well underway before hearing onset. Cochleotopic order in the connections among nuclei is established as the connections are made; most connections appear to be precise when the system becomes operational. Development after hearing onset involves stabilization of cell size and continued maturation of axons and dendrites; acoustic input appears to be necessary for these elements to complete development normally.

The results of most experimental studies to date provide only limited insight into the mechanisms operating during normal development. Generally, the effects of manipulation of the acoustic environment in the neonate have not been assessed until the experimental animals have attained adulthood. Consequently, although these studies have served to establish that acoustic input plays an important role in normal development, the time course and period of susceptibility to manipulation are not known in any detail. To further confound interpretation of results, in many studies, manipulations have been performed well before hearing onset and the effects assayed well after. Because different mechanisms may operate during the time before acoustic stimulation activates the system, interpretation of the results of such studies cannot be definitive. Further progress will depend on systematic studies of normal ontogeny in combination with carefully timed experimental studies that address specific aspects of the developmental sequence.

Acknowledgments. I am grateful to Laura Poole and Debra Evenson, who provided essential assistance in gathering and copying references. I benefited greatly from conversations with numerous colleagues who know much more than I about development of the auditory system. Many of them were also kind enough to share their unpublished work. Among the latter, I am grateful to Drs. Karl Kandler, Leonard Kitzes, Motoi Kudo, Rudolf Rübsamen, Laura Schweitzer, and, especially, Frank Willard. During the time that this manuscript was in preparation, work in my laboratory was supported by National Institute on Deafness and Other Communication Disorders Grant RO1-DC-00135.

References

Aitkin LM (1986) The development of the mammalian inferior colliculus. In: Aitkin LM (ed) The Auditory Midbrain. Structure and Function in the Central Auditory Pathway. Clifton, NJ: Humana, pp. 129–144.

Aitkin LM, Moore DR (1975) Inferior colliculus. II. Development of tuning characteristics and tonotopic organization in central nucleus of the neonatal cat. J Neurophysiol 38:1208–1216.

Aitkin L, Nelson J, Farrington M, Swann S (1991) Neurogenesis in the brain auditory pathway of a marsupial, the northern native cat (*Dasyurus hallucatus*). J Comp Neurol 309:250–260.

Aitkin L, Nelson J, Farrington M, Swann S (1994) The morphological development of the inferior colliculus in a marsupial, the Northern quoll (*Dasyurus hallucatus*). J Comp Neurol 343:532–541.

Alford BR, Ruben RJ (1963) Physiological, behavioral and anatomical correlates of the development of hearing in the mouse. Ann Otol Rhinol Laryngol 72:237–247.

Altman J, Bayer SA (1979) Development of the diencephalon in the rat. IV. Quantitative study of the time of origin of neurons and the internuclear chronological gradients in the thalamus. J Comp Neurol 188:455–472.

Altman J, Bayer SA (1980a) Development of the brain stem in the rat. III. Thymidine-radiographic study of the time of origin of neurons of the vestibular and auditory nuclei of the upper medulla. J Comp Neurol 194:877–904.

Altman J, Bayer SA (1980b) Development of the brain stem in the rat. IV. Thymidine-radiographic study of the time of origin of neurons in the pontine region. J Comp Neurol 194:905–929.

Altman J, Bayer SA (1981) Time of origin of neurons of the rat inferior colliculus and the relations between cytogenesis and tonotopic order in the auditory pathway. Exp Brain Res 42:411–423.

Altman J, Bayer SA (1982) Development of the cranial nerve ganglia and related nuclei in the rat. Adv Anat Embryol Cell Biol 74:1–90.

Altman J, Bayer SA (1987) Development of the precerebellar nuclei in the rat. I. The precerebellar neuroepithelium of the rhombencephalon. J Comp Neurol 257:477–489.

Altman J, Bayer SA (1989) Development of the rat thalamus. V. The posterior lobule of the thalamic neuroepithelium and the time and site of origin and settling pattern of neurons of the medial geniculate body. J Comp Neurol 284:567–580.

Altman J, Das GD (1966) Autoradiographic and histological studies of postnatal neurogenesis. I. A longitudinal investigation of the kinetics, migration and transformation of cells incorporating tritiated thymidine in neonate rats, with special reference to postnatal neurogenesis in some brain regions. J Comp Neurol 126:337–390.

Andrew DLE, Paterson JA (1989) Postnatal development of vascularity in the inferior colliculus of the young rat. Am J Anat 186:389–396.

Angevine JB Jr (1970) Time of neuron origin in the diencephalon of the mouse. An autoradiographic study. J Comp Neurol 139:129–188.

Angevine JB, Sidman RL (1961) Autoradiographic study of cell migration during histogenesis of cerebral cortex in the mouse. Nature 192:766–768.

Angevine JB, Bodian D, Coulombre AJ, Edds MV, Hamburger V, Jacobson M, Lyser K, Prestige MC, Sidman RL, Varon S, Weiss PA (1970) Embryonic vertebrate central nervous system: Revised terminology. Anat Rec 166:257–262.

Angulo A, Merchán JA, Merchán MA (1990) Morphology of the rat cochlear primary afferents during prenatal development: a Cajal's reduced silver and rapid Golgi study. J Anat 168:241–255.

Asanuma C, Ohkawa R, Stanfield BB, Cowan WM (1986) Pre- and postnatal development of the medial lemniscus (ML), the brachium conjunctivum (BC) and the brachium of the inferior colliculus (BIC) in rats. Soc Neurosci Abstr 12:953.

Asanuma C, Ohkawa R, Stanfield BB, Cowan WM (1988) Observations on the development of certain ascending inputs to the thalamus in rats. I. Postnatal development. Dev Brain Res 41:159–170.

Austin CP, Cepko CK (1990) Cellular migration patterns in the developing mouse cerebral cortex. Development 110:713–732.

Bartelmez GW, Dekaban AS (1962) The early development of the human brain. Contrib Embryol 37:13–32.

Bayer SA (1980) The development of the hippocampal region in the rat. I. Neurogenesis examined with [³H]thymidine autoradiography. J Comp Neurol

190:87–114.

Bayer SA, Altman J (1991) Neocortical development. In: Development of the Auditory Areas. New York: Raven Press, chapt. 12, pp. 161–166.

Bayer SA, Altman J, Russo RJ, Dai X, Simmons JA (1991) Cell migration in the rat embryonic neocortex. J Comp Neurol 307:499–516.

Binns KE, Grant S, Keating MJ, Withington-Wray DJ (1990) In the anaesthetized guinea-pig, the external nucleus of the inferior colliculus contains a spatial map of the auditory azimuth (Abstract). J Physiol (Lond) 126:106P.

Blake JA (1900) The roof and lateral recesses of the fourth ventricle, considered morphologically and embryologically. J Comp Neurol 10:79–108.

Blatchley BJ, Brugge JF (1990) Sensitivity to binaural intensity and phase difference cues in kitten inferior colliculus. J Neurophysiol 64:582–597.

Blatchley BJ, Williams JE, Coleman JR (1983) Age-dependent effects of acoustic deprivation on spherical cells of the rat anteroventral cochlear nucleus. Exp Neurol 80:81–93.

Blatchley BJ, Cooper WA, Coleman JR (1987) Development of auditory brainstem response to tone pip stimuli in the rat. Dev Brain Res 32:75–84.

Book KJ, Morest DK (1990) Migration of neuroblasts by perikaryal translocation: role of cellular elongation and axonal outgrowth in the acoustic nuclei of the chick embryo medulla. J Comp Neurol 297:55–76.

Born DE, Rubel EW (1985) Afferent influences on brain stem auditory nuclei of the chicken: neuron number and size following cochlea removal. J Comp Neurol 231:435–445.

Bourrat F, Sotelo C (1988) Migratory pathways and neuritic differentiation of inferior olivary neurons in the rat embryo. Axonal tracing study using the in vitro slab technique. Dev Brain Res 39:19–37.

Bourrat F, Sotelo C (1990) Migratory pathways and selective aggregation of the lateral reticular neurons in the rat embryo: a horseradish peroxidase in vitro study, with special reference to migration patterns of the precerebellar nuclei. J Comp Neurol 294:1–13.

Brugge JF (1983) Development of the lower brainstem auditory nuclei. In: Romand R (ed) Development of Auditory and Vestibular Systems. New York: Academic Press, pp. 89–120.

Brugge JF (1988) Stimulus coding in the developing auditory system. In: Edelman GM, Gall WE, Cowan WM (eds) Auditory Function. Neurobiological Bases of Hearing. New York: John Wiley and Sons, pp. 113–136.

Brugge JF (1992) Development of the lower auditory brainstem of the cat. In: Romand R (ed) Development of Auditory and Vestibular Systems 2. Amsterdam: Elsevier Science Publishers, pp. 273–296.

Brugge JF, O'Connor TA (1984) Postnatal functional development of the dorsal and posteroventral cochlear nuclei of the cat. J Acoust Soc Am 75:1548–1562.

Brugge JF, Javel E, Kitzes LM (1978) Signs of functional maturation of peripheral auditory system in discharge patterns of neurons in anteroventral cochlear nucleus of kitten. J Neurophysiol 41:1557–1579.

Brugge JF, Reale RA, Wilson GF (1988) Sensitivity of auditory cortical neurons of kittens to monaural and binaural high frequency sound. Hear Res 34:127–140.

Brunso-Bechtold JK, Henkel CK, Linville C (1987) Ultrastructure of the developing medial superior olive in the ferret. Soc Neurosci Abstr 13:79.

Brunso-Bechtold JK, Henkel CK, Linville C (1992) Ultrastructural development of the medial superior olive (MSO) in the ferret. J Comp Neurol 324:539–556.

Canady KS, Rubel EW (1992) Rapid and reversible astrocytic reaction to afferent activity blockade in chick cochlear nucleus. J Neurosci 12:1001–1009.

Cant NB (1991) The cochlear nucleus: neuronal types and their synaptic organization. In: Webster WB, Popper AN, Fay RR (eds) The Mammalian Auditory Pathway: Neuroanatomy. New York: Springer-Verlag, pp. 66–116.

Carlier E, Abonnenc M, Pujol R (1975) Maturation des responses unitaires a la stimulation tonale dans le nerf cochleaire du chaton. J Physiol Paris 70:129–138.

Carlier E, Lenoir M, Pujol R (1979) Development of cochlear frequency selectivity tested by compound action potential tuning curves. Hear Res 1:197–201.

Carlile S (1991) The auditory periphery of the ferret: postnatal development of acoustic properties. Hear Res 51:265–278.

Carney PR, Silver J (1983) Studies on cell migration and axon guidance in the developing distal auditory system of the mouse. J Comp Neurol 215:359–369.

Chamak B, Fellous A, Glowinski J, Prochiantz A (1987) MAP2 expression and neuritic outgrowth and branching are coregulated through region-specific neuro-astroglial interactions. J Neurosci 7:3163–3170.

Chisaka O, Musci TS, Capecchi MR (1992) Developmental defects of the ear, cranial nerves and hindbrain resulting from targeted disruption of the mouse homeobox gene Hox-1.6. Nature 355:516–520.

Church MW, Williams HL, Holloway JA (1984) Postnatal development of the brainstem auditory evoked potential and far-field cochlear microphonic in non-sedated rat pups. Dev Brain Res 14:23–31.

Clarke S, Innocenti GM (1986) Organization of immature intrahemispheric connections. J Comp Neurol 251:1–22.

Clarke S, Innocenti GM (1990) Auditory neurons with transitory axons to visual areas form short permanent projections. Eur J Neurosci 2:227–242.

Clements M, Kelly JB (1978) Directional responses by kittens to an auditory stimulus. Dev Psychobiol 11:505–511.

Clerici WJ, Coleman JR (1986) Resting and high-frequency evoked 2-deoxyglucose uptake in the rat inferior colliculus: developmental changes and effects of short-term conduction blockade. Dev Brain Res 27:127–137.

Clerici WJ, Maxwell B, Coleman JR (1988) Cytoarchitecture of the medial geniculate body of adult and infant rats (Abstract). Anat Rec 218:23A.

Clopton BM, Snead CR (1990) Experiential factors in auditory development. In: Coleman JR (ed) Development of Sensory Systems in Mammals. New York: John Wiley and Sons, pp. 317–338.

Code RA, Burd GD, Rubel EW (1989) Development of GABA immunoreactivity in brainstem auditory nuclei of the chick: ontogeny of gradients in terminal staining. J Comp Neurol 284:504–518.

Coggeshall RE (1964) A study of diencephalic development in the albino rat. J Comp Neurol 122:241–269.

Coleman J (1981) Effects of acoustic deprivation on morphological parameters of development of auditory neurons in rat. In: Syka J, Aitkin L (eds) Neuronal Mechanisms in Hearing. New York: Plenum Press, pp. 359–362.

Coleman J (1990) Development of auditory system structures. In: Coleman JR (ed) Development of Sensory Systems in Mammals. New York: John Wiley and Sons, pp. 205–247.

Coleman JR, O'Connor P (1979) Effects of monaural and binaural sound deprivation on cell development in the anteroventral cochlear nucleus of rats. Exp Neurol 64:553–566.

Coleman JR, Blatchley BJ, Williams JE (1982) Development of the dorsal and ventral cochlear nuclei in rat and effects of acoustic deprivation. Dev Brain Res 4:119–123.

Coleman JR, Ding J-M, Wei A, Dorn H (1987) Postnatal development of cortical cytoarchitecture and lamina V pyramidal cells in area 41 of rat. Soc Neurosci Abstr 13:325.

Coleman JR, Clerici WJ, Maxwell B, Zrull MC (1989) Neural growth in the medial geniculate body of the postnatal rat. Soc Neurosci Abstr 15:747.

Constantine-Paton M, Cline HT, Debski E (1990) Patterned activity, synaptic convergence, and the NMDA receptor in developing visual pathways. Ann Rev Neurosci 13:129–154.

Cooper ERA (1948) The development of the human auditory pathway from the cochlear ganglion to the medial geniculate body. Acta Anat 5:99–122.

Cooper ML, Rakic P (1981) Neurogenetic gradients in the superior and inferior colliculi of the rhesus monkey. J Comp Neurol 202:309–334.

Cornwell P, Ravizza R, Payne B (1984) Extrinsic visual and auditory cortical connections in the 4-day-old kitten. J Comp Neurol 229:97–120.

Craigie EH (1924) Changes in vascularity in the brain stem and cerebellum of the albino rat between birth and maturity. J Comp Neurol 38:27–48.

Crowley DE, Hepp-Reymond M-C (1966) Development of cochlear function in the ear of the infant rat. J Comp Physiol Psychol 3:427–432.

Dardennes R, Jarreau PH, Meininger V (1984) A quantitative Golgi analysis of the postnatal maturation of dendrites in the central nucleus of the inferior colliculus of the rat. Dev Brain Res 16:159–169.

Das GD, Nornes HO (1972) Neurogenesis in the cerebellum of the rat: an autoradiographic study. Z Anat Entwicklungsgesch 138:155–165.

De Carlos JA, Lopez-Mascaraque L, Valverde F (1985) Development, morphology and topography of chandelier cells in the auditory cortex of the cat. Dev Brain Res 22:293–300.

Dehay C, Kennedy H, Bullier J (1988) Characterization of transient cortical projections from auditory, somatosensory, and motor cortices to visual areas 17, 18, and 19 in the kitten. J Comp Neurol 272:68–89.

Dekaban A (1954) Human thalamus. An anatomical, developmental and pathological study. II. Development of the human thalamic nuclei. J Comp Neurol 100:63–97.

Dekaban AS (1978) Changes in brain weights during the span of human life: relation of brain weights to body heights and body weights. Ann Neurol 4:345–356.

Dobbing J, Sands J (1971) Vulnerability of developing brain. IX. The effect of nutritional growth retardation on the timing of the brain growth-spurt. Biol Neonate 19:363–378.

Domesick VB, Morest DK (1977a) Migration and differentiation of ganglion cells in the optic tectum of the chick embryo. Neuroscience 2:459–476.

Domesick VB, Morest DK (1977b) Migration and differentiation of shepherd's crook cells in the optic tectum of the chick embryo. Neuroscience 2:477–491.

Dubin MW, Stark LA, Archer SM (1986) A role for action-potential activity in the development of neuronal connections in the kitten retinogeniculate pathway. J Neurosci 6:1021–1036.

Durham D, Rubel WE (1985) Afferent influences on brain stem auditory nuclei of the chicken: changes in succinate dehydrogenase activity following cochlea removal. J Comp Neurol 231:446–456.

Durham D, Matschinsky FM, Rubel EW (1993) Altered malate dehydrogenase activity in nucleus magnocellularis of the chicken following cochlea removal. Hear Res 70:151–159.

Edwards SB, Ginsburgh CL, Henkel CK, Stein BE (1979) Sources of subcortical projections to the superior colliculus in the cat. J Comp Neurol 184:309–330.

Ehret G (1988) Auditory development: Psychophysical and behavioral aspects. In: Meisami E, Timiras PS (eds) Handbook of Human Growth and Developmental Biology. Vol I. Neural, Sensory, Motor, and Integrative Development, Part B. Boca Raton, FL: CRC Press, pp. 141–154.

Ehret G (1990) Development of behavioral responses to sound. In: Coleman JR (ed) Development of Sensory Systems in Mammals. New York: John Wiley and Sons, pp. 289–315.

Ellenberger C, Hanaway J, Netsky MG (1969) Embryogenesis of the inferior olivary nucleus in the rat: a radioautographic study and a re-evaluation of the rhombic lip. J Comp Neurol 137:71–88.

Essick CR (1912) The development of the nuclei pontis and the nucleus arcuatus in man. Am J Anat 13:25–54.

Fawcett JW, O'Leary DDM (1985) The role of electrical activity in the formation of topographic maps in the nervous system. Trends Neurosci 8:201–206.

Faye-Lund H, Osen KK (1985) Anatomy of the inferior colliculus in rat. Anat Embryol 171:1–20.

Feng AS, Rogowski BA (1980) Effects of monaural and binaural occlusion on the morphology of neurons in the medial superior olivary nucleus of the rat. Brain Res 189:530–534.

Feng JZ, Brugge JF (1983) Postnatal development of auditory callosal connections in the kitten. J Comp Neurol 214:416–426.

Finck A, Schneck CD, Hartman AF (1972) Development of cochlear function in the neonate Mongolian gerbil (*Meriones unguiculatus*). J Comp Physiol Psychol 78:375–380.

Fishman R, Hatten ME (1993) Multiple receptor systems promote CNS neuronal migration. J Neurosci 13:3485–3495.

Fitzgerald M (1987) Spontaneous and evoked activity of fetal primary afferents in vivo. Nature 326:603–605.

Foss I, Flottorp G (1974) A comparative study of the development of hearing and vision in various species commonly used in experiments. Acta Otolaryngol 77:202–214.

Fox MW (1968) Neuronal development and ontogeny of evoked potentials in auditory and visual cortex of the dog. Electroencephalogr Clin Neurophysiol 24:213–226.

Frank E, Wenner P (1993) Environmental specification of neuronal connectivity. Neuron 10:779–785.

Friauf E (1992) Tonotopic order in the adult and developing auditory system of the rat as shown by c-*fos* immunocytochemistry. Eur J Neurosci 4:798–812.

Friauf E (1993) Transient appearance of calbindin-D_{28k}-positive neurons in the superior olivary complex of developing rats. J Comp Neurol 334:59–74.

Friauf E, Kandler K (1990) Auditory projections to the inferior colliculus of the rat are present by birth. Neurosci Lett 120:58–61.

Friauf E, Kandler K (1993) Cell birth, formation of efferent connections, and establishment of tonotopic order in the rat cochlear nucleus. In: Merchán MA, Juiz JM, Godfrey DA, Mugnaini E (eds.) The Mammalian Cochlear Nuclei:

Organization and Function. New York: Plenum Press, pp. 19–28.

Fritzsch B, Nichols DH (1993) DiI reveals a prenatal arrival of efferents at the differentiating otocyst of mice. Hear Res 65:51–60.

Galli L, Maffei L (1988) Spontaneous impulse activity of rat retinal ganglion cells in prenatal life. Science 242:90–91.

Gardette R, Courtois M, Bisconte J-C (1982) Prenatal development of mouse central nervous structures: time of neuron origin and gradients of neuronal production. A radioautographic study. J Hirnforsch 23:415–431.

Geal-Dor M, Freeman S, Li G, Sohmer H (1993) Development of hearing in neonatal rats: air and bone conducted ABR thresholds. Hear Res 69:236–242.

Gemmell RT, Nelson J (1992) Development of the vestibular and auditory system of the Northern native cat, *Dasyurus hallucatus*. Anat Rec 234:136–143.

Ghosh A, Antonini A, McConnell SK, Shatz CJ (1990) Requirement for subplate neurons in the formation of thalamocortical connections. Nature 347:179–181.

Gilbert MS (1935) The early development of the human diencephalon. J Comp Neurol 62:81–115.

González-Hernández T, Meyer G, Ferres-Torres R (1989) Development of neuronal types and laminar organization in the central nucleus of the inferior colliculus in the cat. Neuroscience 30:127–141.

Goodman CS, Shatz CJ (1993) Developmental mechanisms that generate precise patterns of neuronal connectivity. Cell 72/Neuron 10, Suppl:77–98.

Gottlieb G (1971) Ontogenesis of sensory function in birds and mammals. In: Tobach E, Aronson LR, Shaw E (eds) The Biopsychology of Development. New York: Academic Press, pp. 67–128.

Hahn ME, Walters JK, Lavooy J, DeLuca J (1983) Brain growth in young mice: evidence on the theory of phrenoblysis. Dev Psychobiol 16:377–383.

Harkmark W (1954a) Cell migrations from the rhombic lip to the inferior olive, the nucleus raphe and the pons. A morphological and experimental investigation on chick embryos. J Comp Neurol 100:115–209.

Harkmark W (1954b) The rhombic lip and its derivatives in relation to the theory of neurobiotaxis. In: Jansen J, Brodal A (eds) Aspects of Cerebellar Anatomy. Oslo: Johan Grundt Tanum Forlag, pp. 264–284.

Harris DM, Dallos P (1984) Ontogenetic changes in frequency mapping of a mammalian ear. Science 225:741–743.

Harvey AR, Robertson D, Cole KS (1990) Direct visualization of death of neurones projecting to specific targets in the developing rat brain. Exp Brain Res 80:213–217.

Hashisaki GT, Rubel EW (1989) Effects of unilateral cochlea removal on antero-ventral cochlear nucleus neurons in developing gerbils. J Comp Neurol 283:465–473.

Hassmannová J, Mysliveček J (1967) Maturation of the primary cortical response to stimulation of medial geniculate body. Electroencephalogr Clin Neurophysiol 22:547–555.

Hatten ME, Mason CA (1990) Mechanisms of glial-guided neuronal migration in vitro and in vivo. Experentia 46:907–916.

Hebel R, Stromberg MW (1986) Anatomy and Embryology of the Laboratory Rat. Wörthsee, Germany: BioMed Verlag, pp. 237.

Heffner CD, Lumsden AGS, O'Leary DDM (1990) Target control of collateral extension and directional growth in the mammalian brain. Science 247:217–220.

Henderson-Smart DJ, Pettigrew AG, Edwards DA, Jiang ZD (1990) Brain stem auditory evoked responses: physiological and clinical issues. In: Hanson MA (ed) The Fetal and Neonatal Brain Stem. Cambridge: Cambridge University Press, pp. 211–229.

Henkel CK, Brunso-Bechtold JK (1990) Dendritic morphology and development in the ferret medial superior olivary nucleus. J Comp Neurol 294:377–388.

Henkel CK, Brunso-Bechtold JK (1991) Dendritic morphology and development in the ferret lateral superior olivary nucleus. J Comp Neurol 313:259–272.

Henkel CK, Brunso-Bechtold JK (1993) Laterality of superior olive projections to the inferior colliculus in adult and developing ferret. J Comp Neurol 331:458–468.

Henkel CK, Brunso-Bechtold JK (1995) Development of glycinergic cells and puncta in nuclei of the superior olivary complex of the postnatal ferret. J Comp Neurol 354:470–480.

Hicks SP, D'Amato CJ (1968) Cell migrations to the isocortex in the rat. Anat Rec 160:619–634.

His W (1888) Zur Geschichte des Gehirns sowie der centralen und peripherischen Nervenbahnen. Abh Math-Phys Kl Saechs Akad Wiss 24:341–392.

Hugosson R (1957) Morphologic and experimental studies on the development and significance of the rhombencephalic longitudinal cell columns. Lund, Sweden: Håkan Ohlssons Boktryckeri.

Hyde GE, Durham D (1994) Increased deafferentation-induced cell death in chick brainstem auditory neurons following blockade of mitochondrial protein synthesis with chloramphenicol. J Neurosci 14:291–300.

Hyson RL, Rudy JW (1987) Ontogenetic change in the analysis of sound frequency in the infant rat. Dev Psychobiol 20:189–207.

Innocenti GM (1988) Loss of axonal projections in the development of the mammalian brain. In: Parnavelas JG, Stern CD, Stirling RV (eds) The Making of the Nervous System. Oxford: Oxford University Press, pp. 319–339.

Innocenti GM, Clarke S (1984) Bilateral transitory projection to visual areas from auditory cortex in kittens. Dev Brain Res 14:143–148.

Innocenti GM, Berbel P, Clarke S (1988) Development of projections from auditory to visual areas in the cat. J Comp Neurol 272:242–259.

Jackson CA, Peduzzi JD, Hickey TL (1989) Visual cortex development in the ferret. I. Genesis and migration of visual cortex neurons. J Neurosci 9:1242–1253.

Jacobson M (1991) Developmental Neurobiology (3rd Ed). New York: Plenum Press.

Jaeger CB, Joh TH (1983) Transient expression of tyrosine hydroxylase in some neurons of the developing inferior colliculus of the rat. Dev Brain Res 11:128–132.

Jewett DL, Romano MN (1972) Neonatal development of auditory system potentials averaged from the scalp of rat and cat. Brain Res 36:101–115.

Jhaveri S, Morest DK (1982) Sequential alterations of neuronal architecture in nucleus magnocellularis of the developing chicken: a Golgi study. Neuroscience 7:837–853.

Kandler K (1993) Die Entwicklung von erregenden und hemmenden Verbindungen im auditorischen Hirnstamm der Ratte. Dissertation. Fakultät für Biologie der Eberhard-Karls-Universität Tübingen, Germany.

Kandler K, Friauf E (1993) Pre- and postnatal development of efferent connections of the cochlear nucleus in the rat. J Comp Neurol 328:161–184.

Kandler K, Friauf E (1995) Development of glycinergic and glutamatergic synaptic transmission in the auditory brainstem of perinatal rats. J Neurosci 15:6890–6904.

Kane ES, Habib CP (1978) Development of the dorsal cochlear nucleus of the cat: an electron microscopic study. Am J Anat 153:321–344.

Kato T (1983) Transient retinal fibers to the inferior colliculus in the newborn albino rat. Neurosci Lett 37:7–9.

Katz LC (1993) Coordinate activity in retinal and cortical development. Curr Opin Neurobiol 3:93–99.

Kaufman MH (1992) The Atlas of Mouse Development. San Diego, CA: Academic Press.

Kelly JB (1992) Behavioral development of the auditory orientation response. In: Romand R (ed) Development of Auditory and Vestibular Systems 2. Amsterdam: Elsevier Science Publishers, pp. 391–417.

Kelly JB, Potash M (1986) Directional responses to sounds in young gerbils (*Meriones unguiculatus*). J Comp Psychol 100:37–45.

Kelly JB, Judge PW, Fraser IH (1987) Development of the auditory orientation response in the albino rat (*Rattus norvegicus*). J Comp Psychol 101:60–66.

Kettner RE, Feng J-Z, Brugge JF (1985) Postnatal development of the phase-locked response to low frequency tones of auditory nerve fibers in the cat. J Neurosci 5:275–283.

Keyser A (1972) The development of the diencephalon of the Chinese hamster. Acta Anat 83, Suppl. 59:1–178.

Kikuchi K, Hilding D (1965) The development of the organ of Corti in the mouse. Acta Otolaryngol 60:207–222.

Kil J, Kageyama GH, Semple MN, Kitzes LM (1995) Development of ventral cochlear nucleus projections to the superior olivary complex in gerbil. J Comp Neurol 353:317–340.

King AJ, Hutchings ME (1987) Spatial response properties of acoustically responsive neurons in the superior colliculus of the ferret: a map of auditory space. J Neurophysiol 57:596–624.

King AJ, Moore DR (1991) Plasticity of auditory maps in the brain. Trends Neurosci 14:31–37.

King AJ, Palmer AR (1983) Cells responsive to free-field auditory stimuli in guinea-pig superior colliculus: distribution and response properties. J Physiol (Lond) 342:361–381.

Kitzes LM (1986) The role of binaural innervation in the development of the auditory brainstem. In: Ruben RJ, van de Water TN, Rubel EW (eds) The Biology of Change in Otolaryngology. Amsterdam: Elsevier Science Publishers, pp. 185–199.

Kitzes LM (1990) Development of auditory system physiology. In: Coleman JR (ed) Development of Sensory Systems in Mammals. New York: John Wiley and Sons, pp. 249–288.

Kitzes LM, Kageyama GH, Semple MN, Kil J (1995) Development of ectopic projections from the ventral cochlear nucleus to the superior olivary complex induced by neonatal ablation of the contralateral cochlea. J Comp Neurol 353:341–363.

Kobayashi T (1963) Brain-to-body ratios and time of maturation of the mouse brain. Am J Physiol 204:343–346.

König N, Marty R (1974) On functions and structure of deep layers of immature auditory cortex. J Physiol Paris 68:145–155.

König N, Roch G, Marty R (1975) The onset of synaptogenesis in rat temporal cortex. Anat Embryol 148:73–87.

Konigsmark BW, Murphy EA (1972) Volume of the ventral cochlear nucleus in man: its relationship to neuronal population and age. J Neuropathol Exp Neurol 31:304–316.

Kraus H-J, Aulbach-Kraus K (1981) Morphological changes in the cochlea of the mouse after the onset of hearing. Hear Res 4:89–102.

Krmpotić-Nemanić J, Kostović I, Nemanić D, Kelović Z (1979) The laminar organization of the prospective auditory cortex in the human fetus. Acta Otolaryngol 87:241–246.

Krmpotić-Nemanić J, Kostović I, Kelović Z, Nemanić D (1980) Development of acetylcholinesterase (AChE) staining in human fetal auditory cortex. Acta Otolaryngol 89:388–392.

Kudo M, Kitao Y, Okoyama S, Moriizumi T (1992) Neurogenesis of the auditory brainstem in the rat: a double labeling study using BRDU and retrograde fluorescent tracers (Abstract). Neurosci Res Suppl 17:S249.

Kungel M, Friauf E (1995) Somatostatin and leu-enkephalin in the rat auditory brainstem during fetal and postnatal development. Anat Embryol 191:425–443.

Kuwabara N, DiCaprio RA, Zook JM (1991) Afferents to the medial nucleus of the trapezoid body and their collateral projections. J Comp Neurol 314:684–706.

Landmesser L, Pilar G (1972) The onset and development of transmission in the chick ciliary ganglion. J Physiol (Lond) 222:691–713.

Langworthy OR (1933) Development of behavior patterns and myelinization of the nervous system in the human fetus and infant. Contrib Embryol 139:3–57.

Larsen SA (1984) Postnatal maturation of the cat cochlear nuclear complex. Acta Otolaryngol Suppl 417:1–43.

Larsen WJ (1993) Human Embryology. New York: Churchill Livingstone.

Lauder JM (1988) Roles for neurotransmitters in neurogenesis and development. In: Meisami E, Timiras PS (eds) Handbook of Human Growth and Developmental Biology. Vol I. Neural, Sensory, Motor, and Integrative Development, Part B. Boca Raton, FL: CRC Press, pp. 53–66.

Lauder JM (1993) Neurotransmitters as growth regulatory signals: role of receptors and second messengers. Trends Neurosci 16:233–240.

Leber SM, Sanes JR (1995) Migratory paths of neurons and glia in the embryonic chick spinal cord. J Neurosci 15:1236–1248.

Lenoir M, Shnerson A, Pujol R (1980) Cochlear receptor development in the rat with emphasis on synaptogenesis. Anat Embryol 160:253–262.

Levi-Montalcini R (1949) The development of the acoustico-vestibular centers in the chick embryo in the absence of the afferent root fibers and of descending fiber tracts. J Comp Neurol 91:209–242.

Lippe WR (1994) Rhythmic spontaneous activity in the developing avian auditory system. J Neurosci 14:1486–1495.

Luo L, Moore JK, Baird A, Ryan AF (1995) Expression of acidic FGF mRNA in rat auditory brainstem during postnatal maturation. Dev Brain Res 86:24–34.

Luskin MB, Shatz CJ (1985) Studies of the earliest generated cells of the cat's visual cortex: cogeneration of subplate and marginal zones. J Neurosci 5:1062–1075.

Mair IWS, Elverland HH, Laukli E (1978) Development of early auditory-evoked responses in the cat. Audiology 17:469–488.

Marin-Padilla M (1970a) Prenatal and early postnatal ontogenesis of the human motor cortex: a Golgi study. I. The sequential development of the cortical layers.

Brain Res 23:167–183.

Marin-Padilla M (1970b) Prenatal and early postnatal ontogenesis of the human motor cortex: a Golgi study. II. The basket-pyramidal system. Brain Res 23:185–191.

Marin-Padilla M (1971) Early prenatal ontogenesis of the cerebral cortex (neocortex) of the cat (*Felis domestica*): a Golgi study. I. The primordial neocortical organization. Z Anat Entwicklungsgesch 134:117–145.

Marin-Padilla M (1972) Prenatal ontogenetic history of the principal neurons of the neocortex of the cat (*Felis domestica*): a Golgi study. II. Developmental differences and their significances. Z Anat Entwicklungsgesch 136:125–142.

Marin-Padilla M (1988) Early ontogenesis of the human cerebral cortex. In: Peters A, Jones EG (eds) Cerebral Cortex. Vol. 7. Development and Maturation of Cerebral Cortex. New York: Plenum Press, pp. 1–34.

Martin MR, Rickets C (1981) Histogenesis of the cochlear nucleus of the mouse. J Comp Neurol 197:169–184.

Marty R, Thomas J (1973) Réponse électro-corticale à la stimulation du nerf cochléaire chez le chat nouveau-né. J Physiol (Paris) 55:165–166.

Mastronarde DN (1983) Correlated firing of cat retinal ganglion cells. I. Spontaneously active inputs to X- and Y-cells. J Neurophysiol 49:303–324.

Mattox DE, Neises GR, Gulley RL (1982) A freeze-fracture study of the maturation of synapses in the anteroventral cochlear nucleus of the developing rat. Anat Rec 204:281–287.

Mattson MP (1988) Neurotransmitters in the regulation of neuronal cytoarchitecture. Brain Res Rev 13:179–212.

Maxwell B, Coleman JR (1989) Differential timetable of projections into the developing inferior colliculus in rat. Soc Neurosci Abstr 15:747.

Maxwell B, Clerici WJ, Brady J, McDonald AJ, Coleman JR (1988) Sources of connections to the inferior colliculus in the immature rat (Abstract). Anat Rec 220:62A.

McAllister JP II, Das GD (1977) Neurogenesis in the epithalamus, dorsal thalamus and ventral thalamus of the rat: an autoradiographic and cytological study. J Comp Neurol 172:647–686.

McConnell SK, Ghosh A, Shatz CJ (1994) Subplate pioneers and the formation of descending connections from cerebral cortex. J Neurosci 14:1892–1907.

McCrady E Jr (1938) The Embryology of the Opossum. Vol. 16. The American Anatomical Memoirs. Philadelphia, PA: The Wistar Institute of Anatomy and Biology.

McCrady E Jr, Wever EG, Bray CW (1937) The development of hearing in the opossum. J Exp Zool 75:503–517.

McCrady E Jr, Wever EG, Bray CW (1940) A further investigation of the development of hearing in the opossum. J Comp Psychol 30:17–21.

McMahon AP, Bradley A (1990) The *Wnt*-1 (*int*-1) proto-oncogene is required for development of a large region of the mouse brain. Cell 62:1073–1085.

McMullen NT, Glaser EM (1988) Auditory cortical responses to neonatal deafening: pyramidal neuron spine loss without changes in growth or orientation. Exp Brain Res 72:195–200.

McMullen NT, Goldberger B, Glaser EM (1988) Postnatal development of lamina III/IV nonpyramidal neurons in rabbit auditory cortex: quantitative and spatial analyses of Golgi-impregnated material. J Comp Neurol 278:139–155.

McMullen NT, Goldberger B, Suter CM, Glaser EM (1988) Neonatal deafening alters nonpyramidal dendrite orientation in rabbit auditory cortex: a computer microscope study in the rabbit. J Comp Neurol 267:92–106.

Meininger V, Baudrimont M (1981) Postnatal modifications of the dendritic tree of cells in the inferior coliculus of the cat. A quantitative Golgi analysis. J Comp Neurol 200:339–355.

Meister M, Wong ROL, Baylor DA, Shatz CJ (1991) Synchronous bursts of action potentials in ganglion cells of the developing mammalian retina. Science 252:939–943.

Merzenich MM, Roth GL, Andersen RA, Knight PL, Colwell SA (1977) Some basic features of organization of the central auditory nervous system. In: Evans EF, Wilson JP (eds) Psychophysics and Physiology of Hearing. New York: Academic Press, pp. 485–497.

Middlebrooks JC, Knudsen EI (1984) A neural code for auditory space in the cat's superior colliculus. J Neurosci 4:2621–2634.

Mikaelian D, Ruben RJ (1965) Development of hearing in the normal CBA-J mouse. Acta Otolaryngol 59:451–461.

Mlonyeni M (1967) The late stages of the development of the primary cochlear nuclei in mice. Brain Res 4:334–344.

Molliver ME, Kostović I, van der Loos H (1973) The development of synapses in cerebral cortex of the human fetus. Brain Res 50:403–407.

Moore DR (1982) Late onset of hearing in the ferret. Brain Res 253:309–311.

Moore DR (1985) Postnatal development of mammalian central auditory system and the neural consequences of auditory deprivation. Acta Otolaryngol Suppl 421:19–30.

Moore DR (1988) Auditory development: central nervous aspects. In: Meisami E, Timiras PS (eds) Handbook of Human Growth and Developmental Biology. Vol I. Neural, Sensory, Motor, and Integrative Development, Part B, Boca Raton, FL: CRC Press, pp. 131–140.

Moore DR (1990a) Auditory brainstem of the ferret: early cessation of developmental sensitivity of neurons in the cochlear nucleus to removal of the cochlea. J Comp Neurol 302:810–823.

Moore DR (1990b) Auditory brainstem of the ferret: bilateral cochlear lesions in infancy do not affect the number of neurons projecting from the cochlear nucleus to the inferior colliculus. Dev Brain Res 54:125–130.

Moore DR (1990c) Hearing loss and auditory brain stem development. In: Hanson, MA (ed) The Fetal and Neonatal Brainstem. Cambridge: Cambridge University Press, pp. 161–184.

Moore DR (1991) Development and plasticity of the ferret auditory system. In: Altschuler RA, Bobbin RP, Clopton BM, Hoffman DW (eds) Neurobiology of Hearing: The Central Auditory System. New York: Raven Press, pp. 461–475.

Moore DR (1992a) Trophic influences of excitatory and inhibitory synapses on neurones in the auditory brain stem. NeuroReport 3:269–272.

Moore DR (1992b) Developmental plasticity of the brainstem and midbrain auditory nuclei. In: Romand R (ed) Development of Auditory and Vestibular Systems 2. Amsterdam: Elsevier Science Publishers, pp. 297–320.

Moore DR (1994) Auditory brainstem of the ferret: long survival following cochlear removal progressively changes projections from the cochlear nucleus to the inferior colliculus. J Comp Neurol 339:301–310.

Moore DR, Hine JE (1992) Rapid development of the auditory brainstem response threshold in individual ferrets. Dev Brain Res 66:229–235.

Moore DR, Kitzes LM (1985) Projections from the cochlear nucleus to the inferior colliculus in normal and neonatally cochlea-ablated gerbils. J Comp Neurol 240:180–195.

Moore DR, Kowalchuk NE (1988) Auditory brainstem of the ferret: effects of unilateral cochlear lesions on cochlear nucleus volume and projections to the inferior colliculus. J Comp Neurol 272:503–515.

Moore DR, Hutchings ME, King AJ, Kowalchuk NE (1989) Auditory brain stem of the ferret: some effects of rearing with a unilateral ear plug on the cochlea, cochlear nucleus, and projections to the inferior colliculus. J Neurosci 9:1213–1222.

Moore JK, Perazzo LM, Braun A (1995) Time course of axonal myelination in the human brainstem auditory pathway. Hear Res 87:21–31.

Morest DK (1968) The growth of synaptic endings in the mammalian brain: a study of the calyces of the trapezoid body. Z Anat Entwicklungsgesch 127:201–220.

Morest DK (1969a) The differentiation of cerebral dendrites: a study of the post-migratory neuroblast in the medial nucleus of the trapezoid body. Z Anat Entwicklungsgesch 128:271–289.

Morest DK (1969b) The growth of dendrites in the mammalian brain. Z Anat Entwicklungsgesch 128:290–317.

Morest DK (1970) A study of neurogenesis in the forebrain of opossum pouch young. Z Anat Entwicklungsgesch 130:265–305.

Morest DK (1973) Auditory neurons of the brain stem. Adv OtorhinoLaryngol 20:337–356.

Morest DK, Zhou X, Brennan A, Baier C (1993) Basic FGF affects development of chick embryo acoustico-vestibular neurons in vitro. Soc Neurosci Abstr 19:1101.

Morey AL, Carlile S (1990) Auditory brainstem of the ferret: maturation of the brainstem auditory evoked response. Dev Brain Res 52:279–288.

Mysliveček J (1983) Development of the auditory evoked responses in the auditory cortex in mammals. In: Romand R (ed) Development of Auditory and Vestibular Systems. New York: Academic Press, pp. 167–209.

Neises GR, Mattox DE, Gulley RL (1982) The maturation of the end bulb of Held in the rat anteroventral cochlear nucleus. Anat Rec 204:271–279.

Nelson JE (1992) Developmental staging in a marsupial *Dasyurus hallucatus*. Anat Embryol 185:335–354.

Nordeen KW, Killackey HP, Kitzes LM (1983) Ascending projections to the inferior colliculus following unilateral cochlear ablation in the neonatal gerbil, *Meriones unguiculatus*. J Comp Neurol 214:144–153.

Nornes HO, Morita M (1979) Time of origin of the neurons in the caudal brain stem of rat. Dev Neurosci 2:101–114.

Norton TT (1974) Receptive-field properties of superior colliculus cells and development of visual behavior in kittens. J Neurophysiol 37:674–690.

Oblinger MM, Das GD (1981) Neurogenesis in the brain stem of the rabbit: an autoradiographic study. J Comp Neurol 197:45–62.

Ochs MT, Brunso-Bechtold JK (1982)The postnatal development of the dorsal cochlear nucleus in the tree shrew. Soc Neurosci Abstr 8:752.

Ollo C, Schwartz IR (1979) The superior olivary complex in C57BL/6 mice. Am J Anat 155:349–374.

Olmstead CE, Villablanca JR (1980) Development of behavioral audition in the kitten. Physiol Behav 24:705–712.

Ono K, Kawamura K (1989) Migration of immature neurons along tangentially oriented fibers in the subpial part of the fetal mouse medulla oblongata. Exp Brain Res 78:290–300.

Ono K, Kawamura K (1990) Mode of neuronal migration of the pontine stream in fetal mice. Anat Embryol 182:11–19.

O'Rahilly R, Gardner E (1971) The timing and sequence of events in the development of the human nervous system during the embryonic period proper. Z Anat Entwicklungsgesch 134:1–12.

Otis EM, Brent R (1954) Equivalent ages in mouse and human embryos. Anat Rec 120:33–63.

Palmer AR, King AJ (1982) The representation of auditory space in the mammalian superior colliculus. Nature 299:248–249.

Parks TN (1979) Afferent influences on the development of the brain stem auditory nuclei of the chicken: otocyst ablation. J Comp Neurol 183:665–578.

Parks TN, Jackson H (1984) A developmental gradient of dendritic loss in the avian cochlear nucleus occurring independently of primary afferents. J Comp Neurol 227:459–466.

Parks TN, Gill SS, Jackson H (1987) Experience-independent development of dendritic organization in the avian nucleus laminaris. J Comp Neurol 260:312–319.

Parks TN, Jackson H, Conlee JW (1987) Axon-target cell interactions in the developing auditory system. In: Moscona AA, Monroy A (eds) Current Topics in Developmental Biology. Vol 21. Hunt RK (ed) Neural Development. Part IV. Cellular and Molecular Differentiation, New York: Academic Press, pp. 309–340.

Pasic TR, Rubel EW (1989) Rapid changes in cochlear nucleus cell size following blockade of auditory nerve electrical activity in gerbils. J Comp Neurol 283:474–480.

Pasic TR, Rubel EW (1991) Cochlear nucleus cell size is regulated by auditory nerve electrical activity. Otolaryngol Head Neck Surg 104:6–13.

Pasic TR, Moore DR, Rubel EW (1994) Effect of altered neuronal activity on cell size in the medial nucleus of the trapezoid body and ventral cochlear nucleus of the gerbil. J Comp Neurol 348:111–120.

Patterson PH, Nawa H (1993) Neuronal differentiation factors/cytokines and synaptic plasticity. Cell 72/Neuron 10 Suppl:123–137.

Payne BR (1992) Development of the auditory cortex. In: Romand R (ed) Development of Auditory and Vestibular Systems 2. Amsterdam: Elsevier Science Publishers, pp. 357–389.

Payne B, Pearson H, Cornwell P (1988a) Development of visual and auditory cortical connections in the cat. In: Peters A, Jones EG (eds) Cerebral Cortex. Vol. 7. Development and Maturation of Cerebral Cortex. New York: Plenum Press, pp. 309–389.

Payne BR, Pearson HE, Cornwell P (1988b) Neocortical connections in fetal cats. Neurosci Res 5:513–543.

Penn AA, Wong ROL, Shatz CJ (1994) Neuronal coupling in the developing mammalian retina. J Neurosci 14:3805–3815.

Perazzo LM, Moore JK (1991) Ontogeny of the human brainstem auditory nuclei. Assoc Res Otolaryngol Abstr 14:21.

Poston MR, Fredieu J, Carney PR, Silver J (1988) Roles of glia and neural crest cells in creating axon pathways and boundaries in the vertebrate central and peripheral nervous systems. In Parnavelas JG, Stern CD, Stirling RV (eds) The Making of the Nervous System. Oxford: Oxford University Press, pp 282–313.

Puel J-L, Uziel A (1987) Correlative development of cochlear action potential sensitivity, latency, and frequency selectivity. Dev Brain Res 37:179–188.

Pujol, R (1969) Développement des réponses à la stimulation sonore dans le colliculus inférieur chez le chat. J Physiol Paris 61:411–421.

Pujol R (1972) Development of tone-burst responses along the auditory pathway in the cat. Acta Otolaryngol 74:383–391.

Pujol R (1985) Morphology, synaptology and electrophysiology of the developing cochlea. Acta Otolaryngol Suppl 421:5–9.

Pujol R, Hilding D (1973) Anatomy and physiology of the onset of auditory function. Acta Otolaryngol 76:1–10.

Pujol R, Marty R (1970) Postnatal maturation in the cochlea of the cat. J Comp Neurol 139:115–126.

Pujol R, Uziel A (1988) Auditory development: peripheral aspects. In: Meisami E, Timiras PS (eds) Handbook of Human Growth and Developmental Biology. Vol I. Neural, Sensory, Motor, and Integrative Development. Part B. Boca Raton, FL: CRC Press, pp. 109–130.

Pujol R, Carlier E, Devigne C (1978) Different patterns of cochlear innervation during the development of the kitten. J Comp Neurol 177:529–536.

Pujol R, Carlier E, Lenoir M (1980) Ontogenetic approach to inner and outer hair cell function. Hear Res 2:423–430.

Purves D, Lichtman JW (1985) Principles of Neural Development. Sunderland, MA: Sinauer Associates.

Purves D, Snider WD, Voyvodic JT (1988) Trophic regulation of nerve cell morphology and innervation in the autonomic nervous system. Nature 336:123–128.

Pysh JJ (1969) The development of the extracellular space in neonatal rat inferior colliculus: an electron microscopic study. Am J Anat 124:411–430.

Rakic P (1972) Mode of cell migration to the superficial layers of fetal monkey neocortex. J Comp Neurol 145:61–83.

Rakic P (1974) Neurons in Rhesus monkey visual cortex: systematic relation between time of origin and eventual disposition. Science 183:425–427.

Rakic P, Bourgeois J-P, Eckenhoff MF, Zecevic N, Goldman-Rakic PS (1986) Concurrent overproduction of synapses in diverse regions of the primate cerebral cortex. Science 232:232–235.

Ramón y Cajal S (1909) Histologie du systéme nerveaux de l'homme et des vertèbrès (1952 reprint). Madrid: Instituto Ramón y Cajal.

Ravizza R, Garlitz B, Cornwell P (1978) Extrinsic connections of auditory and visual cortex in infant kittens: an HRP analysis. Brain Res 149:508–510.

Repetto-Antoine M, Meininger V (1982) Histogenesis of the inferior colliculus in rat. Anat Embryol 165:19–37.

Riedel B, Friauf E, Grothe C, Unsicker K (1995) Fibroblast growth factor-2-like immunoreactivity in auditory brainstem nuclei of the developing and adult rat: correlation with onset and loss of hearing. J Comp Neurol 354:353–360.

Rietzel H-J, Friauf E (1994) Development of morphological properties of intracellularly labeled neurons in the rat lateral superior olive (Abstract). In: Elsner N, Breer H (eds) Proceedings of the 22nd Göttingen Neurobiology Conference. Vol. II. Stuttgart, Germany: Thieme-Verlag, p. 172.

Rogowski BA, Feng AS (1981) Normal postnatal development of medial superior olivary neurons in the albino rat: a Golgi and Nissl study. J Comp Neurol

196:85–97.

Romand R (1984) Functional properties of auditory-nerve fibers during postnatal development in the kitten. Exp Brain Res 56:395–402.

Romand R, Ehret G (1990) Development of tonotopy in the inferior colliculus. I. Electrophysiological mapping in house mice. Dev Brain Res 54:221–234.

Romand R, Marty R (1975) Postnatal maturation of the cochlear nuclei in the cat: a neurophysiological study. Brain Res 83:225–233.

Romand R, Granier M, Marty R (1973) Développement postnatal de l'activité provoquée dans l'olive supérieure latérale chez le chat par la stimulation sonore. J Physiol Paris 66:303–315.

Romand R, Sans A, Romand MR, Marty R (1976) The structural maturation of the stato-acoustic nerve in the cat. J Comp Neurol 170:1–16.

Rose JE, Adrian H, Santibañez (1957) Electrical signs of maturation in the auditory system of the kitten Acta Neurol Latinoam 3:133–143.

Rubel EW (1978) Ontogeny of structure and function in the vertebrate auditory system. In: Jacobson M (ed) Handbook of Sensory Physiology IX. Development of Sensory Systems. Berlin: Springer-Verlag, pp. 135–237.

Rubel EW (1984) Ontogeny of auditory system function. Ann Rev Physiol 46:213–229.

Rubel EW (1985) Auditory system development. In: Gottlieb G, Krasnegor NA (eds) Measurement of Audition and Vision During the First Year of Life: A Methodological Overview. Norwood, NJ: Ablex Publishing, pp. 53–90.

Rubel EW, MacDonald GH (1992) Rapid growth of astrocytic processes in n. magnocellularis following cochlea removal. J Comp Neurol 318:415–425.

Rubel EW, Parks TN (1975) Organization and development of brain stem auditory nuclei of the chicken: tonotopic organization of n. magnocellularis and n. laminaris. J Comp Neurol 164:411–434.

Rubel EW, Parks TN (1988) Organization and development of the avian brain-stem auditory system. In: Edelman GM, Gall WE, Cowan WM (eds) Auditory Function: Neurological Bases of Hearing. New York: John Wiley and Sons, pp. 3–92.

Rubel EW, Hyson RL, Durham D (1990) Afferent regulation of neurons in the brain stem auditory system. J Neurobiol 21:169–196.

Ruben RJ (1967) Development of the inner ear of the mouse: a radioautographic study of terminal mitoses. Acta Otolaryngol Suppl 220:1–44.

Rübsamen R (1987) Ontogenesis of the echolocation system in the rufous horseshoe bat, *Rhinolophus rouxi* (Audition and vocalization in early postnatal development). J Comp Physiol A Sens Neural Behav Physiol 161:899–913.

Rübsamen R (1992) Postnatal development of central auditory frequency maps. J Comp Physiol A Sens Neural Behav Physiol 170:129–143.

Rübsamen R, Schäfer M (1990) Ontogenesis of auditory fovea representation in the inferior colliculus of the Sri Lankan rufous horseshoe bat, *Rhinolophus rouxi*. J Comp Physiol A Sens Neural Behav Physiol 167:757–769.

Rübsamen R, Gutowski M, Langkau J, Dörrscheidt (1994) Growth of central nervous system auditory and visual nuclei in the postnatal gerbil (*Meriones unguiculatus*). J Comp Neurol 346:289–305.

Russell FA, Moore DR (1995) Afferent reorganization within the superior olivary complex of the gerbil: development and induction by neonatal, unilateral cochlear removal. J Comp Neurol 352:607–625.

Ryan AF, Woolf NK (1992) Development of the lower auditory system in the gerbil.

In: Romand R (ed) Development of Auditory and Vestibular Systems 2. Amsterdam: Elsevier Science Publishers, pp. 243–271.

Ryan AF, Woolf NK, Sharp FR (1982) Functional ontogeny in the central auditory pathway of the Mongolian gerbil. A 2-deoxyglucose study. Exp Brain Res 47:428–436.

Ryan AF, Woolf NK, Catanzaro A, Braverman S, Sharp FR (1985) Deoxyglucose uptake patterns in the auditory system: metabolic response to sound stimulation in the adult and neonate. In: Drescher DG (ed) Auditory Biochemistry. Springfield, IL: Charles C Thomas, pp. 401–421.

Ryugo DK, Fekete DM (1982) Morphology of primary axosomatic endings in the anteroventral cochlear nucleus of the cat: a study of the endbulbs of Held. J Comp Neurol 210:239–257.

Sanderson KJ, Aitkin LM (1990) Neurogenesis in a marsupial: the brush-tailed possum (*Trichosurus vulpecula*). I. Visual and auditory pathways. Brain Behav Evol 35:325–338.

Sanderson KJ, Weller WL (1990) Gradients of neurogenesis in possum neocortex. Dev Brain Res 55:269–274.

Sanes DH (1992) The refinement of central auditory form and function during development. In: Werner LA, Rubel EW (eds) Developmental Psychoacoustics. Washington, DC: American Psychological Association, pp. 257–279.

Sanes DH (1993) The development of synaptic function and integration in the central auditory system. J Neurosci 13:2627–2637.

Sanes DH, Chokshi P (1992) Glycinergic transmission influences the development of dendrite shape. NeuroReport 3:323–326.

Sanes DH, Rubel EW (1988) The ontogeny of inhibition and excitation in the gerbil lateral superior olive. J Neurosci 8:682–700.

Sanes DH, Siverls V (1991) Development and specificity of inhibitory terminal arborizations in the central nervous system. J Neurobiol 22:837–854.

Sanes DH, Takács C (1993) Activity-dependent refinement of inhibitory connections. Eur J Neurosci 5:570–574.

Sanes DH, Wooten GF (1987) Development of glycine receptor distribution in the lateral superior olive of the gerbil. J Neurosci 7:3803–3811.

Sanes DH, Merickel M, Rubel EW (1989) Evidence for an alteration of the tonotopic map in the gerbil cochlea during development. J Comp Neurol 279:436–444.

Sanes DH, Markowitz S, Bernstein J, Wardlow J (1992) The influence of inhibitory afferents on the development of postsynaptic dendritic arbors. J Comp Neurol 321:637–644.

Sanes DH, Song J, Tyson J (1992) Refinement of dendritic arbors along the tonotopic axis of the gerbil lateral superior olive. Brain Res 67:47–55.

Sanes JR (1993) Topographic maps and molecular gradients. Curr Opin Neurobiol 3:67–74.

Scheibel ME, Scheibel AB (1974) Neuropil organization in the superior olive of the cat. Exp Neurol 43:339–348.

Schwartz AM, Kane ES (1977) Development of the octopus cell area in the cat ventral cochlear nucleus. Am J Anat 148:1–18.

Schwartz IR (1972) The development of terminals in the cat medial superior olive (Abstract). Anat Rec 172:401.

Schwartz IR (1977) Dendritic arrangements in the cat medial superior olive. Neuroscience 2:81–101.

Schwartz IR, Ryan AF (1985) Development of synaptic terminals in the cochlear nucleus of the mongolian gerbil. Assoc Res Otolaryngol Abstr 8:134.

Schweitzer L (1987) Development of brainstem auditory evoked responses in the hamster. Hear Res 25:249–255.

Schweitzer L (1990) Differentiation of apical, basal and mixed dendrites of fusiform cells in the cochlear nucleus. Dev Brain Res 56:19–27.

Schweitzer L (1991) Morphometric analysis of developing neuronal geometry in the dorsal cochlear nucleus of the hamster. Dev Brain Res 59:39–47.

Schweitzer L, Cant NB (1984) Development of the cochlear innervation of the dorsal cochlear nucleus of the hamster. J Comp Neurol 225:228–243.

Schweitzer L, Cant NB (1985a) Development of oriented dendritic fields in the dorsal cochlear nucleus of the hamster. Neuroscience 16:969–978.

Schweitzer L, Cant NB (1985b) Differentiation of the giant and fusiform cells in the dorsal cochlear nucleus of the hamster. Dev Brain Res 20:69–82.

Schweitzer L, Cecil T (1992) Morphology of HRP-labelled cochlear nerve axons in the dorsal cochlear nucleus of the developing hamster. Hear Res 60:34–44.

Schweitzer L, Savells KL (1991) Development of projections between the inferior colliculus and the dorsal cochlear nucleus of the hamster (Abstract). Anat Rec 229:79A.

Schweitzer L, Bell JM, Slotkin TA (1987) Impaired morphological development of the dorsal cochlear nucleus in hamsters treated postnatally with α-difluoromethylornithine. Neuroscience 23:1123–1132.

Schweitzer L, Cecil T, Walsh EJ (1993) Development of GAD-immunoreactivity in the dorsal cochlear nucleus of the hamster and cat: light and electron microscopic observations. Hear Res 65:240–251.

Seto-Ohshima A, Aoki E, Semba R, Emson PC, Heizmann CW (1990) Parvalbumin immunoreactivity in the central auditory system of the gerbil: a developmental study. Neurosci Lett 119:60–63.

Shah SN, Bhargava VK, McKean CM (1978) Maturational changes in early auditory evoked potentials and myelination of the inferior colliculus in rats. Neuroscience 3:561–563.

Shaner RF (1934) The development of the nuclei and tracts related to the acoustic nerve in the pig. J Comp Neurol 60:5–19.

Shatz CJ (1990a) Competitive interactions between retinal ganglion cells during prenatal development. J Neurobiol 21:197–211.

Shatz CJ (1990b) Impulse activity and the patterning of connections during CNS development. Neuron 5:745–756.

Shigemoto R, Nakanishi S, Mizuno N (1992) Distribution of the mRNA for a metabotropic glutamate receptor (mGluR1) in the central nervous system: an in situ hybridization study in adult and developing rat. J Comp Neurol 322:121–135.

Shipley C, Buchwald JS, Norman R, Guthrie D (1980) Brain stem auditory evoked response development in the kitten. Brain Res 182:313–326.

Shnerson A, Pujol R (1983) Development: anatomy, electrophysiology and behavior. In: Willott JF (ed) The Auditory Psychobiology of the Mouse. Springfield, IL: Charles C Thomas, pp. 395–425.

Shnerson A, Willott JF (1979) Development of inferior colliculus response properties in C57BL/6J mouse pups. Exp Brain Res 37:373–385.

Sie KCY, Rubel EW (1992) Rapid changes in protein synthesis and cell size in the cochlear nucleus following eighth nerve activity blockade or cochlea ablation. J Comp Neurol 320:501–508.

Silver J, Lorenz SE, Wahlsten D, Coughlin J (1982) Axonal guidance during development of the great cerebral commissures: descriptive and experimental studies, in vivo, on the role of preformed glial pathways. J Comp Neurol 210:10–29.

Simmons DD, Raji-Kubba J, Strauss RS (1992) Postnatal immunoreactivity to calcitonin gene-related peptide in the lateral superior olive. Assoc Res Otolaryngol Abstr 15:58.

Singer M, Nordlander RH, Egar M (1979) Axonal guidance during embryogenesis and regeneration in the spinal cord of the newt. The blueprint hypothesis of neuronal pathway patterning. J Comp Neurol 185:1–22.

Smart IHM (1983) Three dimensional growth of the mouse isocortex. J Anat 137:683–694.

Smith DI, Kraus N (1987) Postnatal development of the auditory brainstem response (ABR) in the unanesthetized gerbil. Hear Res 27:157–164.

Smith ZDJ (1981) Organization and development of brain stem auditory nuclei of the chicken: dendritic development in n. laminaris. J Comp Neurol 203:309–333.

Sretavan DW, Shatz CJ, Stryker MP (1988) Modification of retinal ganglion cell axon morphology by prenatal infusion of tetrodotoxin. Nature 336:468–471.

Streeter GL (1906) On the development of the membranous labyrinth and the acoustic and facial nerves in the human embryo. Am J Anat 6:139–165.

Streeter GL (1912) The development of the nervous system. In: Keibel D, Mall FP (eds) Manual of Human Embryology. Vol. II. Philadelphia, PA: JB Lippincott, pp. 1–156.

Ströer WFH (1956) Studies on the diencephalon. I. The embryology of the diencephalon of the rat. J Comp Neurol 105:1–24.

Taber Pierce E (1966) Histogenesis of the nuclei griseum pontis, corporis pontobulbaris and reticularis tegmenti pontis (Bechterew) in the mouse. J Comp Neurol 126:219–240.

Taber Pierce E (1967) Histogenesis of the dorsal and ventral cochlear nuclei in the mouse. An autoradiographic study. J Comp Neurol 131:27–54.

Taber Pierce E (1973) Time of origin of neurons in the brain stem of the mouse. In: Ford DH (ed) Neurological Aspects of Maturation and Aging, Progress in Brain Research. Vol 40. Amsterdam: Elsevier Science Publishers, pp. 53–65.

Tessier-Lavigne M, Placzek M, Lumsden AGS, Dodd J, Jessell TM (1988) Chemotropic guidance of developing axons in the mammalian central nervous system. Nature 336:775–778.

Theiler K (1989) The House Mouse. Atlas of Embryonic Development. New York: Springer-Verlag.

Thomas KR, Capecchi MR (1990) Targeted disruption of the murine int-1 proto-oncogene resulting in severe abnormalities in midbrain and cerebellar development. Nature 346:847–850.

Tokimoto T, Osako S, Matsuura S (1977) Development of auditory evoked cortical and brain stem responses during the early postnatal period in the rat. Osaka City Med J 23:141–153.

Trune DR (1982a) Influence of neonatal cochlear removal on the development of mouse cochlear nucleus. I. Number, size, and density of its neurons. J Comp Neurol 209:409–424.

Trune DR (1982b) Influence of neonatal cochlear removal on the development of mouse cochlear nucleus. II. Dendritic morphometry of its neurons. J Comp Neurol 209:425–434.

Trune DR, Kiessling AA (1988) Decreased protein synthesis in cochlear nucleus following developmental auditory deprivation. Hear Res 35:259–264.

Trune DR. Morgan CR (1988) Influence of developmental auditory deprivation on neuronal ultrastructure in the mouse anteroventral cochlear nucleus. Dev Brain Res 42:304–308.

Tucci DL, Rubel EW (1985) Afferent influences on brain stem auditory nuclei of the chicken: effects of conductive and sensorineural hearing loss on n. magnocellularis. J Comp Neurol 238:371–381.

Udin SB, Fawcett JW (1988) Formation of topographic maps. Ann Rev Neurosci 11:289–327.

Uziel A, Romand R, Marot M (1981) Development of cochlear potentials in rats. Audiology 20:89–100.

Valverde F, Facal-Valverde MV (1988) Postnatal development of interstitial (subplate) cells in the white matter of the temporal cortex of kittens: a correlated Golgi and electron microscopic study. J Comp Neurol 269:168–192.

Varecka L, Wu C-H, Rotter A, Frostholm A (1994) GABA$_A$/benzodiazepine receptor α6 subunit mRNA in granule cells of the cerebellar cortex and cochlear nuclei: expression in developing and mutant mice. J Comp Neurol 339:341–352.

Vaughn JE (1989) Review: fine structure of synaptogenesis in the vertebrate central nervous system. Synapse 3:255–285.

Villablanca JR, Olmstead CE (1979) Neurological development of kittens. Dev Psychobiol 12:101–127.

Wada T (1923) Anatomical and physiological studies on the growth of the inner ear of the albino rat. Amer Anat Mem 10:1–74.

Walsh EJ, McGee J (1986) The development of function in the auditory periphery. In: Altschuler RA, Bobbin RP, Hoffman DW (eds) Neurobiology of Hearing: The Cochlea. New York: Raven Press, pp. 247–269.

Walsh EJ, McGee J (1987) Postnatal development of auditory nerve and cochlear nucleus neuronal responses in kittens. Hear Res 28:97–116.

Walsh EJ, McGee J (1988) Rhythmic discharge properties of caudal cochlear nucleus neurons during postnatal development in cats. Hear Res 36:233–248.

Walsh EJ, McGee J, Javel E (1986a) Development of auditory-evoked potentials in the cat. I. Onset of response and development of sensitivity. J Acoust Soc Am 79:712–724.

Walsh EJ, McGee J, Javel E (1986b) Development of auditory-evoked potentials in the cat. II. Wave latencies. J Acoust Soc Am 79:725–744.

Walsh EJ, McGee J, Javel E (1986c) Development of auditory-evoked potentials in the cat. III. Wave amplitudes. J Acoust Soc Am 79:745–754.

Walsh EJ, McGee J, Fitzakerley JL (1990) GABA actions within the caudal cochlear nucleus of developing kittens. J Neurophysiol 64:961–977.

Walsh EJ, McGee J, Fitzakerley JL (1993) Development of glutamate and NMDA sensitivity of neurons within the cochlear nuclear complex of kittens. J Neurophysiol 69:201–218.

Warr WB, White JS, Nyffeler MJ (1982) Olivocochlear neurons: quantitative comparisons of the lateral and medial efferent systems in adult and newborn cats. Soc Neurosci Abstr 8:346.

Webster DB (1983a) A critical period during postnatal auditory development of mice. Int J Pediatr Otorhinolaryngol 6:107–118.

Webster DB (1983b) Auditory neuronal sizes after a unilateral conductive hearing loss. Exp Neurol 79:130–140.

Webster DB (1983c) Effects of peripheral hearing losses on the auditory brainstem. In: Lasky EZ, Katz J (eds) Central Auditory Processing Disorders: Problems of Speech, Language, and Hearing. Baltimore, MD: University Park Press, pp. 185–200.

Webster DB (1983d) Late onset of auditory deprivation does not affect brainstem auditory neuron soma size. Hear Res 12:145–147.

Webster DB (1984) Conductive loss affects auditory neuronal soma size only during a sensitive postnatal period. In: Lim DJ, Bluestone CD, Klein JO, Nelson JD (eds) Recent Advances in Otitis Media with Effusion. Toronto: BC Decker, pp. 344–346.

Webster DB (1988a) Conductive hearing loss affects the growth of the cochlear nuclei over an extended period of time. Hear Res 32:185–192.

Webster DB (1988b) Sound amplification negates central effects of a neonatal conductive hearing loss. Hear Res 32:193–195.

Webster DB, Webster M (1980) Mouse brainstem auditory nuclei development. Ann Oto-Rhinol-Laryngol 89 (Suppl 68):254–256.

Webster DB, Sobin A, Anniko M (1986) Incomplete maturation of brainstem auditory nuclei in genetically induced early postnatal cochlear degeneration. Acta Otolaryngol 101:429–438.

Webster DB, Popper AN, Fay RR (Editors) (1992) The Mammalian Auditory Pathway: Neuroanatomy. New York: Springer-Verlag.

Webster WR, Martin RL (1991) The development of frequency representation in the inferior colliculus of the kitten. Hear Res 55:70–80.

Whitehead MC, Morest DK (1985) The growth of cochlear fibers and the formation of their synaptic endings in the avian inner ear: a study with the electron microscope. Neuroscience 14:277–300.

Wilkinson DG, Bailes JA, McMahon AP (1987) Expression of the proto-oncogene int-1 is restricted to specific neural cells in the developing mouse embryo. Cell 50:79–88.

Wilkinson F (1986) Eye and brain growth in the Mongolian gerbil (Meriones unguiculatus). Behav Brain Res 19:59–69.

Willard FH (1993) Postnatal development of auditory nerve projections to the cochlear nucleus in Monodelphis domestica. In: Merchán MA, Juiz JM, Godfrey DA, Mugnaini E (eds) The Mammalian Cochlear Nuclei: Organization and Function. New York: Plenum Press, pp. 29–42.

Willard FH (1995) Development of the mammalian auditory hindbrain. In: Malhotra S (ed) Advances in Neural Science. Vol. 2. Greenwich CT: JAI Press Inc., pp. 205–234.

Willard FH, Martin GF (1983) The auditory brainstem nuclei and some of their projections to the inferior colliculus in the North American opossum. Neuroscience 10:1203–1232.

Willard FH, Martin GF (1986) The development and migration of large multipolar neurons into the cochlear nucleus of the North American opossum. J Comp Neurol 248:119–132.

Willard FH, Martin GF (1987) Development of projections from the hindbrain to the inferior colliculus. Soc Neurosci Abstr 13:547.

Willott JF, Shnerson A (1978) Rapid development of tuning characteristics of inferior colliculus neurons of mouse pups. Brain Res 148:230–233.

Windle WF (1933) Neurofibrillar development in the central nervous system of cat embryos between 8 and 12 mm long. J Comp Neurol 58:643–723.

Withington-Wray DJ, Binns KE, Keating MJ (1990a) The developmental emergence of a map of auditory space in the superior colliculus of the guinea pig. Dev Brain Res 51:225–236.

Withington-Wray DJ, Binns KE, Keating MJ (1990b) The maturation of the superior collicular map of auditory space in the guinea pig is disrupted by developmental visual deprivation. Eur J Neurosci 2:682–692.

Withington-Wray DJ, Binns KE, Dhanjal SS, Brickley SG, Keating MJ (1990) The maturation of the superior collicular map of auditory space in the guinea pig is disrupted by developmental auditory deprivation. Eur J Neurosci 2:693–703.

Wong ROL, Meister M, Shatz CJ (1993) Transient period of correlated bursting activity during development of the mammalian retina. Neuron 11:923–938.

Woolf NK, Ryan AF (1984) The development of auditory function in the cochlea of the mongolian gerbil. Hear Res 13:277–283.

Woolf N, Ryan AF (1985) Ontogeny of neural discharge patterns in the ventral cochlear nucleus of the Mongolian gerbil. Dev Brain Res 17:131–147.

Wu SH, Oertel D (1987) Maturation of synapses and electrical properties of cells in the cochlear nuclei. Hear Res 30:99–110.

Yakolev PI, Lecours AR (1967) The myelogenetic cycles of regional maturation of the brain. In: Minkowski A (ed) Regional Development of the Brain in Early Life. Oxford UK: Blackwell, pp. 3–70.

Young SR, Rubel EW (1983) Frequency-specific projections of individual neurons in chick brainstem auditory nuclei. J Neurosci 3:1373–1378.

Young SR, Rubel EW (1986) Embryogenesis of arborization pattern and topography of individual axons in n. laminaris of the chicken brain stem. J Comp Neurol 254:425–459.

Zafra F, Castrén E, Thoenen H, Lindholm D (1991) Interplay between glutamate and γ-aminobutyric acid transmitter systems in the physiological regulation of brain-derived neurotrophic factor and nerve growth factor synthesis in hippocampal neurons. Proc Natl Acad Sci USA 88:10037–10041.

Index

Since the concept of development pervades this volume and all of the topics therein, we have chosen not to cross-index every term under development. Therefore, for topics of development (e.g., Cochlea, Mouse) the reader should not only look under the word "Development" but also under the specific aspect of development.

Readers should also note that this volume contains many useful tables that provide considerable insight into the literature on development. In addition to being indexed under individual topics, we have also indexed all of these tables under the heading "Tables."